The Galactic
Supermassive Black Hole

The Galactic
Supermassive Black Hole

Fulvio Melia

PRINCETON UNIVERSITY PRESS

PRINCETON AND OXFORD

ISBN-13: 978-0-691-09535-6 (cloth)
ISBN-10: 0-691-09535-3 (cloth)

ISBN-13: 978-0-691-13129-0 (pbk.)
ISBN-10: 0-691-13129-5 (pbk.)

Library of Congress Control Number: 2006937806

British Library Cataloging-in-Publication Data is available

This book has been composed in Sabon

pup.princeton.edu

10 9 8 7 6 5 4 3 2 1

A color version of the plate section may be downloaded as a PDF
from the following link: https://press.princeton.edu/books/
paperback/9780691131290/the-galactic-supermassive-black-hole

Contents

Foreword   vii

Preface   xiii

CHAPTER 1
The Galactic Center   1

    1.1   Discovery of Sagittarius A*   2
    1.2   Radio Morphology of the Central Region   5
    1.3   X-ray Morphology of the Central Region   10
    1.4   Sagittarius A East   17

CHAPTER 2
The Radio Source Sagittarius A*   25

    2.1   Position of Sagittarius A*   27
    2.2   Proper Motion   32
    2.3   Structure as a Function of Frequency   37

CHAPTER 3
Sagittarius A*'s Spectrum   42

    3.1   The Radio Spectrum   42
    3.2   Linear and Circular Polarization   46
    3.3   Infrared Observations   55
    3.4   X-ray Observations   58
    3.5   Observed High-Energy Characteristics   65

CHAPTER 4
Variability   71

    4.1   Short-Term Variability in the IR and X-rays   71
    4.2   Long-Term Variability in the Radio   81

CHAPTER 5

The Central Star Cluster 86

5.1 The Stellar Cusp Surrounding the Black Hole 87
5.2 Stellar Constituents and Dynamics 101
5.3 Stellar Orbits and the Enclosed Mass 109

CHAPTER 6

The Four-Dimensional Spacetime 114

6.1 The Flat Spacetime Metric 114
6.2 Relativistic Transformation of Physical Laws 126
6.3 Accelerated Frames 132
6.4 General Relativity 137
6.5 Particle Orbits and Trajectories 140
6.6 The Kerr Metric 149

CHAPTER 7

Mass Accretion and Expulsion 156

7.1 Sagittarius A*'s Gaseous Environment 158
7.2 Bondi-Hoyle Capture from Distributed Sources 162
7.3 Accretion Close to Sagittarius A* 165
7.4 Magnetic Field Dissipation 167
7.5 A Compact Magnetized Disk 176
7.6 Expulsion of Matter 204

CHAPTER 8

Flares 226

8.1 Flare Physics 226
8.2 Periodicity 234
8.3 General Relativistic Flux Modulations 240

CHAPTER 9

Strong Field Physics 245

9.1 Spin-Induced Disk Precession 246
9.2 Microlensing 251
9.3 Imaging the Shadow of the Black Hole 259

References 265

Index 287

Foreword

On **March 26, 2004**, at the end of an international meeting in Green Bank, West Virginia, a plaque commemorating the discovery of Sagittarius A* was fastened to the leg of a 14-meter radio telescope. One hundred scientists from around the world, and employees of the National Radio Astronomy Observatory (NRAO), stood in the unusually warm March sunshine to celebrate the occasion.

The 14-meter telescope had been the key instrument enabling the detection of what we now believe to be the Galaxy's central black hole. In the winter of 1972, the disk had been installed on top of a mountain near Huntersville, 35 kilometers south of Green Bank, where a line-of-sight radio link carried signals and operating commands between the telescope and the campus of the NRAO. The mountain is not high. Once spring arrived, the signals from the radio transmitter near the 14-meter antenna would be so badly attenuated by the intervening tree leaves that NRAO wouldn't operate it during the warm months. But in its short 1.5-year lifetime, the 14-meter telescope completed its task of unlocking the door to a treasure trove of exciting new science associated with the central source of strong gravity.

The 14-meter antenna became operational towards the end of 1972. Its signal was combined with those from three older 26-meter antennas at NRAO's campus to form a huge but very sparsely filled "aperture" 35 kilometers in diameter fixed to the Earth. As the Earth rotated, the four antennas strung out in a line turned with it, enabling the array to sweep out a large, odd wedge-shaped aperture and to form the sharpest—though certainly not the cleanest—images in radio astronomy at that time.

This strangely shaped telescope configuration did not produce pretty images. Even a single point source, such as a distant star, was smeared, elongated, and surrounded by a series of ripples like those produced by a rock falling into a still pond. This Huntersville–Green Bank interferometer would distort the image of a haystack beyond recognition.

But for the first time, it was uniquely capable of finding tiny bright needles lurking within that haystack.

Though the hot gas swirling around the central black hole emits light across the entire spectrum, very little of it reaches Earth. Estimates indicate that only one in a trillion visible photons radiated in our direction survives to reach us through the dense, dark clouds within the disk of the Galaxy. The discovery of the black hole had to await detection either in the radio spectrum, where small dust particles are highly ineffective absorbers, or in X-ray light. Radio astronomers were ready first.

Hiding within the messy haystack of radio emission—known as Sagittarius A—at the galactic center, was a bright golden needle that had never been recognized before. The hot gas orbiting the black hole, now known as Sagittarius A*, is the cause of intense radio emission emitted from a region smaller than our solar system, something like the light from a hot, intense welder's arc at night in a city full of streetlights. Cameras built prior to 1973 produced poorly focused images, so Sagittarius A* appeared as just one of many luminescent smears on an image that was lost in the glare of the other surrounding lights. The 35-kilometer baseline of the Huntersville–Green Bank interferometer, however, was ideally suited to produce images of small, intense points of light; the others were resolved out and did not register in the image that the interferometer produced. As we now know, Sagittarius A* would be easily detected the first time that the 35-km array interferometer looked in its direction.

By 1973, two proposals had been submitted to use the 35-kilometer aperture to look for bright structures buried deep within Sagittarius A. In December 1970, before anyone imagined that a huge black hole might be lurking at the center of the Milky Way, Bruce Balick, then a graduate student from Cornell working on his PhD thesis at Green Bank, proposed to NRAO to use the Huntersville–Green Bank interferometer to probe HII Regions for particularly bright sources that might reveal very hot young stars. Thirteen months later, he amended the proposal to add his close colleague Robert Brown of NRAO and asked to include in the proposal additional HII Regions seen in the southern sky. He didn't mention Sagittarius A by name, but it was high on his list of propitious targets. The amended request was approved.

A second proposal arrived at NRAO in June 1972 from Dennis Downes and Miller Goss, both Americans working at the Max-Planck-Institut in Germany at the time. The idea that a black hole ought to reside in the center of most large galaxies, such as the Milky Way, had been receiving a great deal of attention since late 1970, especially in Europe where Downes and Goss were working. They specifically mentioned the search for a huge black hole as their motivation and Sagittarius A as the target in their NRAO proposal. As it turns out, their proposal was far more prescient than that of Balick and Brown.

Two proposals to observe the same object at the same time constitute a proposal conflict—a situation that public observatories find very uncomfortable. Unless there are special circumstances, a target common to several proposals is observed just once. An adjudication process determines which proposing team is awarded time on the telescope: the team that proposes first, the team with the best scientific case, or a merger of individuals from both teams. However, these strategies are invoked only when the target conflict is recognized. Balick and Brown had not mentioned Sagittarius A by name, whereas Downes and Goss did. So to no one's surprise, NRAO missed the conflict entirely.

In the end, it was the team that was fortunate enough to get the observing time first that would discover the black hole in Sagittarius A. The other team was likely to be disappointed, frustrated, and perhaps angry. This is a telescope scheduler's worst nightmare.

Work on the interferometer's electronics ran late, so Downes and Goss were scheduled to go first, late in 1973. But as it turned out, they could not in that year abandon the responsibility of brand-new jobs and urgent projects. NRAO agreed to postpone their observation. Meanwhile, Balick and Brown were scheduled for their observation on February 13 and 15, 1974. Balick, by then a postdoctoral fellow at the University of California at Santa Cruz, flew to Green Bank, while Brown, a scientist at NRAO in Charlottesville, Virginia, remained at home to assist with the processing of the data on the large computers available there.

The weather was ideal for the observing run: the sky was sunny and the air stable, which meant that the radio signals coming through the atmosphere were only minimally distorted. (In stormy weather, radio sources can be badly distorted by atmospheric effects, making the detection of small objects very difficult.) The observation was flawless, and the detection of a small, bright object in Sagittarius A was obvious from the moment the telescopes were aimed in the right direction. Balick continued to observe it for almost six hours, letting the Earth rotate the aperture so that a better image of the region around Sagittarius A could be built up over time.

And thus began decades of intriguing research on the nature and characteristics of the massive black hole right here at the heart of the Milky Way. The plaque dedicated in 2004 enshrined Balick and Brown's radio observations in February 1974, the people who built the hardware, and the years of fascinating subsequent discoveries. Today Sagittarius A is being observed across the spectrum, except, of course, in the optical and ultraviolet, where the attenuation by the intervening dust forever blocks our view of the galactic center.

As you can imagine, however, the discovery of Sagittarius A* ignited the issue of the conflicting proposals. The NRAO scientist Dave Hogg, who was responsible for scheduling the interferometer, was informed by Balick of the stunning new results. Hogg was simultaneously elated and aghast. Downes and Goss were informed immediately. They were obviously disappointed that another group had been given time to perform the very same experiment that they had expected to conduct a few months later. Meanwhile, Balick and Brown exerted pressure on Hogg to permit them to rush their exciting results into print.

Hogg knew that the situation was delicate. With the passage of time, he no longer clearly remembers the sequence of events that ensued. Brown, whose office was near Hogg's, recalls that Hogg contacted Downes and Goss right away and, for lack of a more satisfactory resolution, offered them their time on the 35-kilometer Huntersville–Green Bank interferometer. Goss later recalled: "With [the pressure of our new jobs], the urgency to complete the Downes-Goss proposal decreased. Dave Hogg became aware of the proposal conflict in early 1974 and wrote a letter on February 15, 1974 (note the precise discovery date), proposing several ways to resolve the conflict. However, Goss and Downes seemed to have lost interest at this point."[1] In any event, Goss very kindly withdrew the Downes-Goss proposal.

From March through May, Balick and Brown pondered over their results. It was clear that the substantial intensity of the source and its small size showed conclusively that the radiating object was hotter and more intense than any other radio source then known in the Milky Way. The flux of radio emission in the beam of the synthesized telescope divided by the area of the beam, or the measured "surface brightness", was at least 1,000 times that of the Sun's. That pointed immediately to the significance of the discovery since no star and no hot gas then known had a surface brightness greater than ten times that of the Sun's.

[1] See Goss, Brown, and Lo (2003).

Sagittarius A* was recognized to be clearly something special at the time of its discovery. Subsequent radio measurements with far more sophisticated instruments have shown that the actual surface brightness of this source is even higher than the initial determination. This fact, its unique location at the center of the Milky Way, and its copious emission of radio waves underscore its uniqueness and importance. Sagittarius A* remains unchallenged as the most intriguing radio source in the Galaxy.

By June 3, 1974, Balick and Brown had completed the analysis of their data, reached a cautious conclusion about their significance, and written and submitted a short announcement for publication. The paper received expedited processing by the editor of *The Astrophysical Journal Letters*. It appeared in the December 1, 1974, issue.

At least a hundred papers reporting new data and interpretive ideas about Sagittarius A* have been written since 1974. The meeting in 2004 was a grand occasion for scientists to meet at Green Bank, to share their latest results and ideas, and to develop a strategy for coordinating their observing campaigns in the years ahead. It was also a time to look back circumspectly to see how far and how fast the research on Sagittarius A* had come. One thing is clear: Sagittarius A* is both an object and, as the 2004 Green Bank meeting demonstrated, a subject as well—a scientific subject in its own right.

Bruce Balick
Green Bank, West Virginia

Preface

A **sustained period of discovery** over the past two decades has fostered a growing interest in the galactic center. This region of the sky is now the focus of many observational campaigns and an ever-growing theoretical investigation, the former because it is by far the closest (active) galactic nucleus, the latter because Sagittarius A*— the supermassive black hole lurking there—offers us the most viable opportunity of studying the physics of strong fields.

One ought to approach the task of writing a book on this subject with some trepidation, knowing that what drives the excitement of new findings at the same time guarantees a rapid evolution in content. We have come far in understanding the behavior of Sagittarius A*, yet we all know that there is still much to be learned.

Unfortunately, the primary literature on this subject is now at such a mature level that young astronomers and physicists wishing to pursue its study and scientists in other disciplines find it daunting to bring themselves up to speed with current developments. My hope is that this book will assist them in their exploration.

With the many entry points created by investigators over the years, research on Sagittarius A* may at first appear to be a complex pattern of interwoven threads. I have tried to synthesize this extensive work into a crucible of essential ideas, while providing a coherent story overall. But for completeness, I have also compiled an extensive set of references to the original literature for the benefit of those wishing to study the various topics at greater depth.

I have had the good fortune over the years of being directly involved in galactic-center research, and in these endeavors I am very grateful to my students and collaborators for the pleasure of our joint efforts. They include Peter Tamblyn, Jack Hollywood, Laird Close, Sera Markoff, Alexei Khokhlov, Marco Fatuzzo, Robert Coker, Mike Fromerth, Siming Liu, Gabe Rockefeller, Brandon Wolfe, Chris Fryer, Susan Stolovy, Heino Falcke, Victor Kowalenko, Ray Volkas, Roland Crocker, Pasquale

Blasi, Don MacCarthy, George Rieke, Benjamin Bromley, Randy Jokipii, Vahé Petrosian, Martin Pessah, Martin Prescher, Andrea Goldwurm, Guillaume Bélanger, Eric Agol, Max Ruffert, Fred Baganoff, Joe Haller, and Daniel Wang.

I am particularly grateful to my close friend and longtime collaborator Farhad Yusef-Zadeh, whose early radio images of the galactic center inspired my interest in this field and whose ongoing drive and ground-breaking observations continue to be a fountain of enthusiasm and new ideas. And for generously supporting my research in this area for more than a decade and a half, I gratefully acknowledge the National Science Foundation, the National Aeronautics and Space Administration, and the Alfred P. Sloan Foundation.

Finally, to Patricia, Marcus, Eliana, and Adrian and to my parents, whose guidance has been priceless, I extend my enduring love and gratitude.

Fulvio Melia
Tucson, Arizona

The Galactic
Supermassive Black Hole

CHAPTER 1

The Galactic Center

Stellar radial velocity measurements, sensing the gravitational potential at the nucleus of our Galaxy,[1] and remarkable proper motion data acquired over eight years of observation have now allowed us to probe the central distribution of mass down to a field as small as 5 light-days. The heart of the Milky Way is evidently ensconced within two clusters of massive and evolved stellar systems orbiting with increasing velocity dispersion toward the middle, where 2.6–$3.6 \times 10^6 \, M_\odot$ of nonluminous matter is concentrated within a region no bigger than 0.015 pc—a mere 800 AU.[2]

The stellar kinematics in the central region is consistent with Keplerian motion—pointing to a supermassive black hole as the likely manifestation of this dark matter. Its inferred mass is arguably the most accurately known for such an object, with the possible exception of NGC 4258.[3]

But this condensation of matter is not alone at the galactic center; within a distance of only 20 light-years or so, several other principal components function in a mutually interactive coexistence, creating a rich tapestry of complexity in this unique portion of the sky. This assortment of players includes an enshrouding cluster of evolved stars, an assembly of young stars, molecular and gas clouds, and a powerful

[1] See McGinn et al. (1989), Rieke and Rieke (1989), Sellgren et al. (1990), and Haller et al. (1996).

[2] These results are based on measurements of the stellar velocity dispersion within the inner 0.1 pc of the Galaxy (reported by Genzel et al. 1996; Eckart and Genzel 1996, 1997; and Ghez et al. 1998) and, more recently, on the determination of specific stellar orbits, discussed extensively in chapter 5. See Schödel et al. (2002) and Ghez et al. (2003b).

[3] The spiral galaxy NGC 4258, in the constellation Canes Venatici, sits not too far from the Big Dipper, some 23 million lt-yr from Earth. Using Very Long Baseline Interferometry (VLBI), Miyoshi et al. (1995) identified microwave water maser emission from molecular material orbiting within the galaxy's nucleus at velocities of up to 650 miles per second. The disk within which these water molecules are trapped is tiny compared to the galaxy itself, but it is oriented fortuitously so that Doppler shifts can provide an unambiguous measure of the orbit's velocity and hence the enclosed mass. The black hole at the nucleus of NGC 4258 is thereby known to have a mass of $3.6 \times 10^7 \, M_\odot$.

supernova-like remnant, known as Sagittarius A East.[4] Some view
this assortment of objects as an indication that the galactic center
may be linked to the broader class of active galactic nuclei (AGN),
in which a supermassive black hole is thought to be a key participant
in the dynamics and energetics of the Galaxy's core. Thus, developing
a consistent picture of the primary interactions between the various
constituents at the galactic center not only enhances our appreciation
for the majesty of our nearby environment but also improves our
understanding of AGN machinery in a broader context.

In this chapter, we shall describe the principal components residing
within the Galaxy's inner core and account for the overall morphology
of this region, revealed primarily through the power of modern X-ray
and radio telescopes. The dark matter, it turns out, may not be so dark
after all, particularly if its inferred association with a point emitter of
radio waves proves to be correct.

1.1 DISCOVERY OF SAGITTARIUS A*

The radio source that would later be viewed as the most unusual object
in the Galaxy was discovered on February 13 and 15, 1974, under
excellent weather conditions and with virtually problem-free instru-
mental performance. Balick and Brown (1974) reported this "detection
of strong radio emission in the direction of the inner 1 pc core of
the galactic nucleus" later that year, adding that the structure had
a brightness temperature in excess of 10^7 K, that it was unresolved
at the level of $\sim 0.''1$, and that it was clearly distributed within just
a few arcseconds of the brightest radio and infrared emission seen
previously from this region. (At the 8 kpc distance to the galactic
center, $1'' \approx 0.04$ pc.) The novelty that permitted them to distinguish
pointlike objects from the overall radio emission in the inner $20''$ was
the newly commissioned 35-kilometer baseline interferometer of the
National Radio Astronomy Observatory (NRAO), consisting of three

---

[4] The heart of the Milky Way lies in the direction of the constellation Sagittarius, close to
the border with the neighboring constellation Scorpius. We tend to name celestial objects
and features after the constellation in which they are found, so the galactic center is said
to lie in the Sagittarius A complex, and gaseous structure within it is called, for example,
Sagittarius A East (or Sgr A East for short) and Sagittarius A West (Sgr A West). As we shall
see, the most unusual object in this region, discovered in 1974, stands out on a radio map
as a bright dot. Its name is Sagittarius A* (Sgr A*).

26-meter telescopes separable by up to 2.7 kilometers and a new 14-meter telescope located on a mountaintop about 35 kilometers southwest of the other dishes.

Balick and Brown had included the central infrared (IR)/radio complex as part of a program to identify "super-bright radio knots" in HII Regions, though in principle the motivation for establishing that the galactic center is active in ways similar to more powerful galactic nuclei had been discussed and developed over the previous three or four years. For example, Sanders and Prendergast (1974) had hypothesized earlier that year that, although now quiescent, the galactic center may once have housed energetic processes like those seen in BL Lac. And in 1971, Lynden-Bell and Rees used a prescient application of the then very speculative black hole model for quasars to point out that the galactic center also should contain a supermassive black hole, perhaps detectable with radio interferometry. Proposing that a central black hole may be currently emitting $\sim 1.5 \times 10^8 \, L_\odot$ of ultraviolet light and that it is blowing away a hot nuclear wind, they invoked a process first suggested by Salpeter (1964)— that gas circulating about the central object eventually flows viscously through the event horizon—to postulate a source for the required energy.

The argument made by Lynden-Bell and Rees was based on the implausibility of starlight alone ionizing the extended thermal source surrounding the central region, not to mention the difficulty of producing a "nuclear wind" with both ionized and neutral material moving at speeds exceeding $200 \, \mathrm{km \, s^{-1}}$. They proposed instead an ultraviolet nonstellar continuum produced by the hypothesized black hole, which presumably also created the observed efflux of mass. We shall see below that the actual picture is not quite as straightforward as this, but Lynden-Bell and Rees's proposal functioned as an influential catalyst in the early attempts to characterize the new radio source as a black hole phenomenon.

From the time of its discovery, the unusual nature of the sub-arcsecond structure and its positional coincidence with the inner 0.04 pc core of the Galaxy provided compelling evidence that it should be physically associated with the galactic center—perhaps even defining its location. Its high brightness temperature, small angular size, and nearby association with strong IR and radio continuum sources made it unique in the Galaxy. Other sources, such as pulsars, resemble the central compact radio structure in a few of its characteristics but not all. Its unusual properties were later confirmed by Westerbork[5] and Very Long Baseline Interferometry (VLBI) observations.[6]

Several years later, maps of 12.8 μm NeII fine-structure line emission from the galactic center[7] revealed that the ionized gas within the central parsec of the Galaxy is not only moving supersonically, but that it is also highly ordered. Regions of blueshifted NeII emission could be separated cleanly from preferentially redshifted streamers, and more precise high-resolution Very Large Array (VLA) observations by Brown, Johnston, and Lo (1981) placed the unresolved radio emitter very near the dynamical center of this implied circular motion. It was around this time that Brown (1982) named the unusual radio source Sagittarius A* (or Sgr A* for short) to distinguish it from the extended emission of the Sagittarius A complex and to emphasize its uniqueness and importance.[8]

Studies of the infrared fine-structure line emission of NeII were followed soon afterwards with mapping observations of the $^3P_1$–$^3P_2$ fine-structure line emission from neutral atomic oxygen at 63 μm.[9] It soon became apparent that the clouds producing the NeII emission and the gas containing the neutral oxygen were rotating about the galactic center with velocities corresponding to a Keplerian mass of $\sim 3 \times 10^6 \, M_\odot$ within the central parsec. Though not accepted immediately, these were early indications that Sagittarius A* might be a concentrated source of gravity, with an estimated mass remarkably close to the value inferred much later from the motion of stars in this region.

Sagittarius A*'s unusual character and possible association with quasar activity—albeit on a significantly smaller scale—were cemented soon thereafter with dual-frequency radio observations made on 25 epochs over a period of three years.[10] Sagittarius A*'s lightcurve clearly demonstrated a variability of 20%–40% in its centimeter-flux density on all timescales, from days to years. As we shall see in the remainder of this

[5] See "A Full Synthesis Map of Sgr A at 5GHz" by Ekers et al. (1975).

[6] These were presented by Lo et al. (1975) in a paper entitled "VLBI Observations of the Compact Radio Source in the Center of the Galaxy."

[7] These observations were first reported by Lacy et al. (1979) and Lacy et al. (1980).

[8] Ten years later, this nomenclature was also used to denote the central radio point source, now known as M31*, in the nucleus of M31 (Melia 1992b) and has since been generalized to identify all such sources in the nuclei of nearby galaxies.

[9] Both the NeII and OI observations and the early evidence they provided for a central massive object were reported in a series of papers by Townes and his collaborators, including Lacy, Townes, and Hollenbach (1982), Townes et al. (1983), and Genzel et al. (1984).

[10] See Brown and Lo (1982).

book, these temporal fluctuations and a wealth of evidence accumulated since the early 1980s have rendered Sagittarius A* the prime suspect in the radiative uncloaking of the putative supermassive black hole at the center of our Galaxy.

1.2 Radio Morphology of the Central Region

Before we focus our attention exclusively on Sagittarius A*, however, let us first widen the field of view and examine its position among the other key components of the galactic nucleus. Color plate 1 shows a wide-field, high-resolution 90 cm image centered on Sagittarius A, covering an area of $4° \times 5°$ with an angular resolution of 43". This map of the galactic center is based on archival data originally acquired and presented by Pedlar et al. (1989) and Anantharamaiah et al. (1991), who observed the galactic center with the VLA 333 MHz system in all four array configurations between 1986 and 1989. But it was only the use of a wide-field algorithm that properly compensates for the nonplanar baseline effects seen at long wavelengths that permitted LaRosa, Kassim, and Lazio (2000) to properly image such a large field of view and obtain increased image fidelity and sensitivity. A schematic diagram in galactic coordinates of the extended sources seen in the 90 cm image is shown in figure 1.1.

With the exception of the Sagittarius A complex centered on Sagittarius A*, nearly all of the sources in color plate 1 are detected in emission, providing for the first time a view of the large-scale radio structure in the galactic center. However, of the seventy-eight small-diameter (<1') sources concentrated toward the galactic plane, about half have steep spectra ($\alpha \approx -0.8$) and are therefore probably extragalactic, though a small population of radio pulsars and young supernova remnants cannot be excluded. The other half are concentrated even more toward the galactic plane and are thus probably HII Regions.

Within the central 15' (or roughly 37 pc for an assumed galactic-center distance of 8 kpc), the most notable structure is the Sagittarius A complex, consisting of the compact nonthermal source Sagittarius A*, surrounded by an orbiting spiral of thermal gas known as Sagittarius A West.[11] Along the same line of sight lies the nonthermal shell source known as Sagittarius A East, which appears to be the remnant of an energetic explosion. In their initial analysis, Pedlar et al. (1989) found

---

[11] A description of this structure may be found in Ekers et al. (1983) and Lo and Claussen (1983).

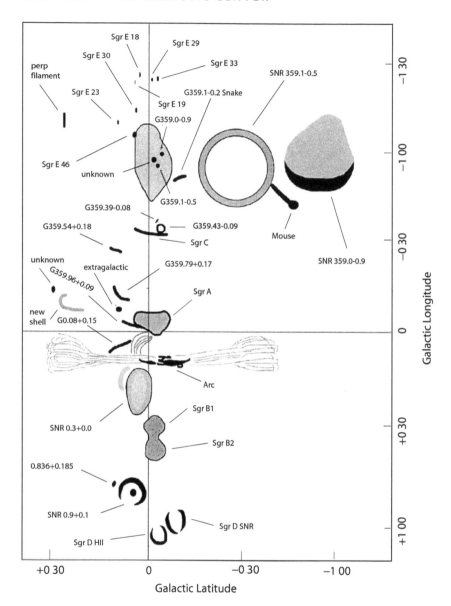

Figure 1.1 This is a schematic diagram of the extended sources shown in the 90 cm image of the galactic center in color plate 1. The perspective has been rotated so that the galactic plane is vertical in this representation. (From LaRosa, Kassim, and Lazio 2000)

that Sagittarius A West is seen in absorption against the background of Sagittarius A East, indicating that the latter must lie behind the former. Sagittarius A clearly contains detail within that is not evident in color plate 1; magnified views of the principal subcomponent sources are shown at higher frequency in color plates 2 and 3.

Some 15' to 20' (or 50 pc in projection) north of Sagittarius A is located the galactic-center arc. First resolved into a large number of narrow filaments by Yusef-Zadeh, Morris, and Chance (1984), they show strong polarization with no line emission and are therefore nonthermal synchrotron sources, probably magnetic flux tubes flushed with relativistic electrons. The fact that several HII Regions appear to be interacting with the filaments in this arc suggests that particles are being accelerated in situ via magnetic reconnection. Several other (isolated) filaments within the central half degree also contribute to the nonthermal magnetic structure, and most are oriented perpendicular to the galactic plane.

Other features of note in this image are supernova remnants (such as Sgr D and SNR $0.9 + 0.1$) and giant molecular clouds (such as Sgr B1 and Sgr B2). Although stars are not visible, the drama of their collective births and deaths is manifest throughout the galactic center. These clouds are in fact regions of star formation and become discernible when newborn stars heat the surrounding gas and make it shine in the radio. All in all, this radio continuum view, together with observations at mm, infrared, and X-ray wavelengths (see below), points to the galactic center as constituting a weak, Seyfert-like nucleus that sometimes also displays mild outbursts of active star formation, as we shall see.

The bright central source in color plate 1 may be magnified further by tuning the receivers to a higher frequency and therefore a better resolution. The region bounded by a box in color plate 1 is shown at 20 cm in color plate 2, a radio continuum image spanning the inner 50 pc × 50 pc portion of the Galaxy. On this level, the distribution of hot gas within the Sagittarius A complex displays an even richer morphology than at 90 cm, with the evident coexistence of both thermal and nonthermal components.[12]

Sagittarius A East is the diffuse ovoid region to the lower right in color plate 2, surrounding (in projection) a spiral-like pattern in red, which is Sagittarius A West. The central spot in this structure identifies Sagittarius A*, which is coincident with the concentration of dark matter inside 0.015 pc. The arc becomes apparent as a set of radio-emitting streamers

---

[12] See also Yusef-Zadeh and Morris (1987) and Pedlar et al. (1989).

interacting with the Sagittarius A complex and together with the other filamentary structures within a few hundred light-years of the center are believed to trace the large-scale magnetic field in the region.

Sagittarius A East appears to be a supernova remnant (perhaps a bubble driven by several supernovae), based primarily on recent *Chandra* X-ray observations that point to a young ($\sim 10^4$ yr) member of the metal-rich, mixed-morphology class of remnants (Maeda et al. 2002). Observations of this source also show it to be associated with a prominent ($50\,km\,s^{-1}$) molecular cloud near the galactic center (see below).

At a wavelength of 6 cm (see color plate 3), Sagittarius A West shines forth as a three-armed spiral consisting of highly ionized gas radiating a thermal continuum. Each arm in the spiral is about 3 light-years long, but one or more of these may be linked to the overall structure merely as a superposition of gas streamers seen in projection. At a distance of 3 light-years from the center, the plasma moves at a velocity of about $105\,km\,s^{-1}$, requiring a mass concentration of just over $3.5 \times 10^6\,M_\odot$ inside this radius. Again, the hub of the gas spiral corresponds to the very bright radio source Sagittarius A*, the dynamical center of our Galaxy.

The central 2 light-year × 2 light-year portion of Sagittarius A West is shown at 2 cm in color plate 4. This is to be compared with the corresponding infrared image of this field in color plate 5, a crowded infrared photograph of unprecedented clarity produced recently with the 8.2-meter VLT Yepun telescope at the European Southern Observatory in Paranal, Chile. (Each of the four telescopes in the Very Large Telescope [VLT] array has been assigned a name based on objects known to the Mapuche people, who live in the area south of the Bío-Bío River, some 500 kilometers from Santiago, Chile. Yepun, the fourth telescope in this set, means *Venus*, or evening star.) The sharpness of the image we see here was made possible with the use of adaptive optics, in which a telescope mirror moves constantly to correct for the effects of turbulence in Earth's atmosphere.

Sagittarius A West probably derives its heat from the central distribution of bright stars evident in color plate 5, rather than from a single point source, such as Sagittarius A*.[13] Some hot, luminous stars are thought to have been formed as recently as a few million years ago.[14] It is

---

[13] An early discussion of this inference was made by Zylka et al. (1995), Gezari (1996), Chan et al. (1997), and Latvakoski et al. (1999).

[14] See Tamblyn and Rieke (1993), Najarro et al. (1994), Krabbe et al. (1995), and Figer et al. (1999).

not surprising, therefore, to see a sprinkling of several IR-bright sources throughout Sagittarius A West that are probably embedded luminous stars. It is not known yet whether these particular stars formed within the gas streamer or just happen to lie along the line of sight.

On a slightly larger scale ($\sim$3 pc), Sagittarius A West orbits about the center within a large central cavity, surrounded by a gaseous and dusty circumnuclear ring.[15] Color plate 6 shows a radio-wavelength image of ionized gas at 1.2 cm (due to free-free emission) superimposed on the distribution of hydrogen cyanide (HCN), which traces the molecular gas. The picture that emerges from a suite of multiwavelength observations such as these is that this molecular ring, with a mass of more than $10^4 \, M_\odot$, is clumpy and is rotating around a concentrated cluster of hot stars, known as IRS 16 (see color plate 5), with a velocity of about $110 \, \mathrm{km \, s^{-1}}$, according to Güsten et al. (1987) and Jackson et al. (1993).

Most of the far infrared luminosity of the circumnuclear ring (or disk) can be accounted for by this cluster of hot, helium emission line stars, which bathe the central cavity with ultraviolet radiation, heating the dust and gas up to 8 pc from the center of the Galaxy. The IRS 16 complex consists of about two dozen blue stellar components at 2 μm and appears to be the source of a strong wind with velocity on the order of $700 \, \mathrm{km \, s^{-1}}$ and an inferred mass loss rate of $4 \times 10^{-3} \, M_\odot \, \mathrm{yr^{-1}}$. These blue stars are themselves embedded within a cluster of evolved and cool stars with a radial density distribution r^{-2} from the dynamical center. However, unlike the distribution of evolved cluster members, which extend over the central 500 pc of the galactic bulge, the hot stars in IRS 16 are concentrated only within the inner parsecs.[16]

It should be pointed out that, whereas the stars orbit randomly about the galactic center, the ionized gas is part of a coherent flow with a systematic motion that is decoupled from the stellar orbits. Identifying kinematics of the ionized gas is complicated by our incomplete view of its three-dimensional geometry; in addition, the orbiting gas may be subject to nongravitational forces, for example, from collisions with the winds produced by the central cluster of hot, mass-losing stars. Even so, Yusef-Zadeh, Roberts, and Biretta (1998) have recently reported some progress

[15] A more detailed description of this structure may be found in Becklin, Gatley, and Werner (1982) and Davidson et al. (1992).

[16] See Hall, Kleinmann, and Scoville (1982), Geballe et al. (1987), and Allen, Hyland, and Hillier (1990).

in mapping the motion of the interstellar medium by combining the transverse velocities measured over nine years with the radial velocities measured for the ionized gas.

The predominant motion projected in the plane of the sky is from east to west for most of the gaseous features (see color plate 4), with the exception of only a few cases where the velocity of the ionized gas is anomalously large, possibly due to an interaction with the stellar winds. In addition, velocity gradients exceeding $600\,\mathrm{km\,s^{-1}\,pc^{-1}}$ seem to be produced by the strong gravitational potential associated with the dark matter at the location of Sagittarius A*.

1.3 X-Ray Morphology of the Central Region

A rather different—though no less interesting—view of the galactic center emerges with progressively sharper images of this region in the X-ray band. X-ray emission has been observed on all scales, from structure extending over kiloparsecs down to a fraction of a light-year, with contributions from thermal and nonthermal, pointlike and diffuse sources. In figure 1.2, which shows the $1.5\,\mathrm{keV}$ map produced with ROSAT (Snowden et al. 1997), we detect evidence for a large-scale outflow of hot gas from the nucleus. Resembling the morphology seen in nearby galaxies with active nuclear star formation, the hollow-cone-shaped soft X-ray feature on either side of the galactic plane points to the efflux of plasma as the agent accounting for much of the diffuse soft X-ray background in the Milky Way. The presence of various spectral features, particularly the $6.7\,\mathrm{keV}$ Fe XXV Kα line detected with *ASCA*, suggests further that a large fraction of this gas is so hot[17] that confinement due to gravity is not feasible, though *Chandra* has more recently forced us to refine this global conclusion (see below).

The magnified view of the central $3° \times 3°$ shown in figure 1.3 reveals additional evidence for the expulsion of hot matter from the nucleus, in the form of a prominent, bright soft X-ray plume that apparently connects the galactic center to the large-scale X-ray structure hundreds of parsecs above and below the galactic plane. A direct comparison of the ROSAT and IRAS 100-micron images of this region suggests that any gaps in the X-ray emissivity are likely due to X-ray shadowing by foreground interstellar, dusty gas.

---

[17]The ASCA Fe line observations apparently require a temperature as high as $\sim 10^8$ K. See Koyama et al. (1996).

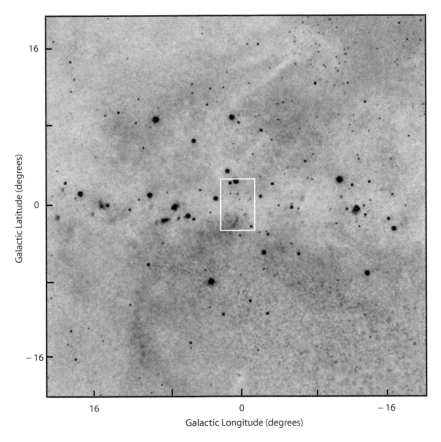

Figure 1.2 ROSAT all-sky survey of the inner $40° \times 40°$ region of the Galaxy in the $\sim 1.5\,keV$ band ($1° \approx 144\,pc$). The hot gas emanating from the central region may be responsible for the diffuse soft X-ray background—for example, the large hollow-cone-shaped features seen on either side of the galactic plane. Resembling the morphology in nearby galaxies with active nuclear star formation, this distinct structure suggests a hot gas outflow from the nucleus. The central box marks the region targeted by ROSAT-pointed observations and is shown magnified in figure 1.3. (Image courtesy of S. L. Snowden at the Goddard Space Flight Center, and NASA)

Without a doubt, however, the most detailed X-ray view of the galactic center has been provided by *Chandra's* Advanced CCD Imaging Spectrometer (ACIS) detector, which combines the wide-band sensitivity and moderate spectral resolution of *ASCA* and *Beppo*SAX with the much higher spatial resolution ($\sim 0.5''$–$1''$) of *Chandra's* High-Resolution Mirror Assembly (HRMA). The central rectangular box oriented along

11

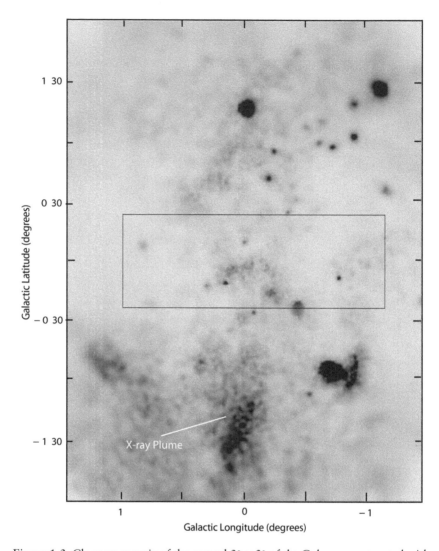

Figure 1.3 Close-up mosaic of the central 3° × 3° of the Galaxy, constructed with ROSAT PSPC observations in the highest energy band (0.5–2.4 keV). The bright, soft X-ray plume apparently connects the central region to the southern large-scale X-ray cone (see figure 1.2) some 300 pc away from the plane. The plume is the most prominent and coherent vertical diffuse soft X-ray feature seen at the galactic center; it may represent the hot gas outflow from the nucleus into the surrounding halo. The central rectangular box oriented parallel to the galactic plane outlines the field mapped out by the more recent *Chandra* survey, shown in color plate 7 and figure 1.4. (Image courtesy of L. Sidoli at INAF-IASF Milano, T. Belloni at INAF-Osservatorio di Brera, and S. Mereghetti at INAF-IASF Milano)

the galactic plane in figure 1.3 outlines the field mapped out in the 1–8 keV range by the most complete *Chandra* survey to date. This study consists of thirty separate pointings, all taken in July 2001; a mosaic of these observations is shown in color plate 7, covering a field of view ~2° × 0.8° centered on Sagittarius A. The saw-shaped boundaries of this map, plotted in galactic coordinates, result from a specific roll angle of the observations.[18]

The high spatial resolution of the *Chandra* X-ray Observatory (see color plate 7 and figure 1.4) allows for a separation of the discrete sources from the diffuse X-ray components pervading the galactic-center region. This analysis has led to a detection of roughly 1,000 discrete objects within the inner 2° × 0.8°, very few of which were known prior to this survey. Their number and spectra indicate the presence of numerous accreting white dwarfs, neutron stars, and solar-size black holes. Based on a comparison with the source density in another (relatively blank) region of the galactic plane,[19] one can estimate that as many as half of these discrete objects could be luminous background active galactic nuclei. Most of the other sources have a luminosity $\sim 10^{32}$–10^{35} ergs s^{-1} in the 2–10 keV band.

One of the fundamental questions that motivated the *Chandra* survey concerns the relative contribution of the point-source and diffuse components to the overall X-ray emission from the center of the Milky Way. For example, earlier observations with *ASCA*[20] implied that the ubiquitous and strong presence of the He-like Fe Kα line (at ~6.7 keV) throughout the central region required the existence of large quantities of $\sim 10^8$ K gas—a situation that is very difficult to explain on physical grounds.

A direct comparison of the accumulated point-source spectrum within the central region to that of the diffuse emission (see figure 1.5) reveals a distinct emission feature centered at ~6.7 keV (with a Gaussian width of ~0.09 keV) in the former but not the latter. The characteristics of this feature agree with those inferred previously with *ASCA*; that is, the high-resolution *Chandra* measurements seem to have resolved the issue of how the He-like Fe Kα line is produced—this emission is typical of X-ray binaries containing white dwarfs, neutron stars, or black holes,

[18] From Wang, Gotthelf, and Lang (2002).
[19] These observations were reported by Ebisawa et al. (2001).
[20] See Tanaka et al. (2000).

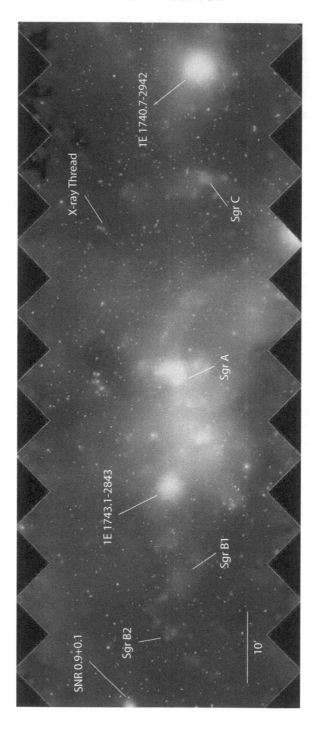

Figure 1.4 This image is the same as that in color plate 7, except here it shows the uncolorized intensity for the purpose of identifying the principal X-ray sources within the inner 2° of the Galaxy. At the distance to the galactic center, 10' is approximately 24 pc.

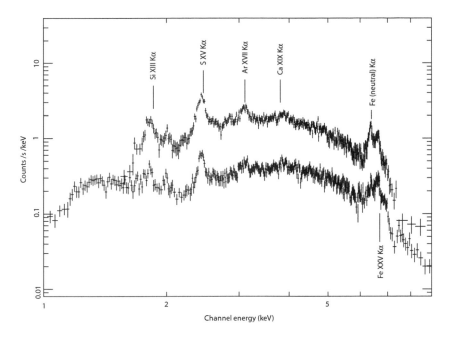

Figure 1.5 The *Chandra* spectrum of the diffuse X-ray flux enhancement above the surrounding background (upper curve) is shown in comparison with that of the accumulated point-source radiation (lower curve), both centered on Sagittarius A* and oriented along the galactic plane. The latter excludes regions around the two brightest sources (1E 1740.7−2942 and 1E 1743.1−2843; see figure 1.4) in order to minimize spectral pileup. This comparison seems to settle the issue of how the He-like Fe line is produced (see text). (From Wang, Gotthelf, and Lang 2002)

particularly during their quiescent state.[21] Rather than being attributed to the diffuse emission, the He-like Fe Kα line is instead found largely due to these discrete X-ray source populations.

On the other hand, the line emission from ions such as S XV, Ar XVII, and Ca XIX is quite prominent in the diffuse X-ray spectrum, which together with the weaker He-like Fe line, now points to the presence of an optically thin thermal plasma with a characteristic temperature of ∼10^7 K—typical of young supernova remnants.

Still, the overall spectrum of the diffuse X-ray emission (figure 1.5) is considerably harder than one would expect for a thermal component

[21]Sample spectra of these sources have been reported by Barret et al. (2000) and Feng et al. (2001).

15

alone. Nearly half of the detected diffuse emission in the 5–8 keV band is due to the Fe 6.4 keV line, part of which is likely due to the fluorescent radiation from discrete sources. The intensity profile shown in color plate 7 and figure 1.4, when compared with maps of HCN and CO in the central 630 pc of the Galaxy,[22] does in fact show that the distribution of the line emission tends to be correlated with lumpy dense molecular material (see figure 1.6). The problem is that the known population of bright X-ray objects in the galactic center region is not sufficient to produce this fluorescence.[23] Instead, it is likely that certain X-ray sources—possibly even the supermassive black hole itself—may have varied greatly in the past, so that their averaged luminosity was several orders of magnitude higher than today. Much of the present 5–8 keV diffuse emission could then be due to this past discrete-source irradiation of the molecular clouds, producing the scattered/fluoresced photon field that we observe now.

The *Chandra* observations also indicate that the 4–6 keV X-ray band, lacking any prominent emission line, differs considerably in profile compared to the distribution of 6.4 and 6.7 keV line flux. According to Wang, Gotthelf, and Lang (2002), the softer X-ray emission may be due to a combination of thermal hot gas, scattered point-source radiation, and the additional contribution of bremsstrahlung processes associated with nonthermal cosmic ray electrons. Perhaps surprisingly, the most prominent nonthermal radio filaments (see color plate 2) do not produce an enhanced X-ray flux, suggesting that inverse Compton scattering of the microwave background radiation is not an important contributor to the observed diffuse X-ray emission.

In fact, the scant correlation between diffuse X-ray and radio features extends even beyond the nonthermal filaments. But this situation is particularly acute for them because out of the eight most prominent cases known, only one has a direct X-ray counterpart, G359.54 + 0.18, shown in figure 1.7. A comparison between the radio and X-ray maps therefore provides a particularly useful perspective on the interplay among the various discrete and diffuse components in this portion of the Galaxy.

---

[22] See Jackson et al. (1996) and Price et al. (2001).

[23] This point is made observationally by Murakami, Koyama, and Maeda (2001) and theoretically by Fromerth, Melia, and Leahy (2001).

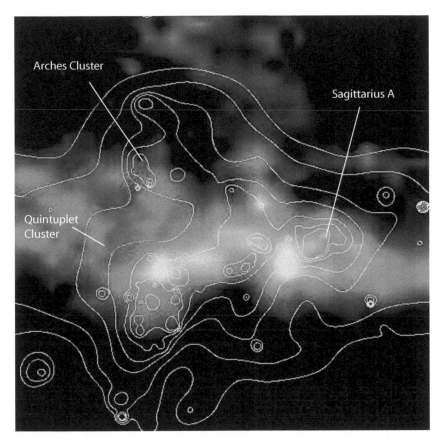

Figure 1.6 This composite image shows the 6.4 keV line intensity contours superimposed on the HCN $J = 1 \rightarrow 0$ emission map of the inner $\sim \times 25 \, pc^2$ region of the Galaxy. A continuum contribution in the 4–6 keV and 7–9 keV bands has been subtracted. This comparison illustrates the fact that X-rays in different energy bands arise in different regions, suggesting diverse origins. The 6.4 keV emission may be globally correlated with dense molecular gas tracers but not on scales smaller than a few arcminutes. (X-ray image courtesy of Q. Daniel Wang at the University of Massachusetts, Amherst, and NASA; the HCN map is from Jackson et al. 1996)

1.4 SAGITTARIUS A EAST

Let us now turn our attention to the X-ray glow from the inner parsecs of the Galaxy. We cannot help but notice first the high-energy shroud encasing the 20 pc $\times$ 20 pc region surrounding Sagittarius A* (see figure 1.8). The primary origin of these X-rays appears to be

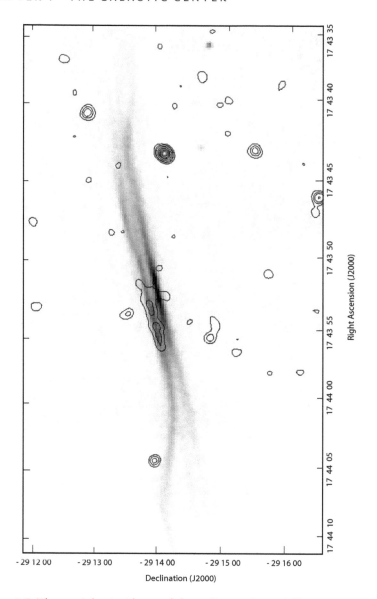

Figure 1.7 The spatial coincidence of the radio nonthermal filament G359.54 + 0.18 (continuum image) and a *Chandra*-discovered X-ray "thread" (shown with contours) argues convincingly for a physical association between these two features. The X-ray thread is about 1′ long and has a width that is not adequately resolved on an ~1″ scale. It also displays a flat spectrum, consistent with its inferred nonthermal origin. (X-ray contour image courtesy of Q. Daniel Wang et al. at the University of Massachusetts, Amherst, and NASA; radio continuum image from Yusef-Zadeh, Wardle, and Parastaran 1997)

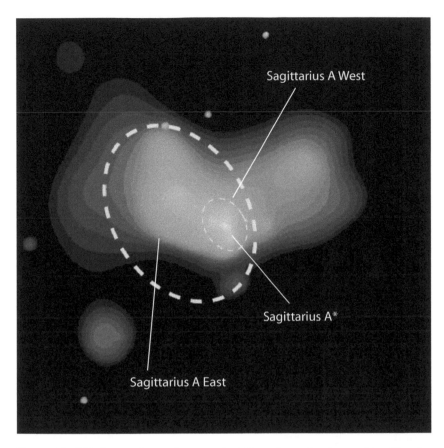

Figure 1.8 This image shows the smoothed X-ray intensity detected by *Chandra* in the 1.5–3.0 keV band from the inner $8'.4 \times 8'.4$ ($\sim$20 pc $\times$ 20 pc) of the Galaxy, centered on Sagittarius A*. The large and small white dashed ellipses identify the Sagittarius A East nonthermal shell and the outer boundary of Sagittarius A West, respectively. Compare with color plates 2 and 3. (Image from Maeda et al. 2002)

Sagittarius A East, a nonthermal radio source with a supernova-like morphology located near the galactic center (see color plate 2). Its elliptical structure is elongated along the galactic plane with a major axis of length 10.5 pc and a center displaced from the apparent dynamical nucleus by 2.5 pc in projection toward negative galactic latitudes. The actual distance between Sagittarius A* and the geometric center of Sagittarius A East has been estimated at $\sim$7 pc.[24]

[24] See Yusef-Zadeh and Morris (1987) and Pedlar et al. (1989).

Broadband radio observations of Sagittarius A East have placed it among the supernova remnants detected at 1,720 MHz, the transition frequency of OH maser emission.[25] In general, the detection of this line establishes the presence of shocks at the interface between the supersonic outflow and the dense molecular cloud environment with which the remnants are known to be interacting. In the case of Sagittarius A East, several maser spots with velocities $\approx 50\,\mathrm{km\,s^{-1}}$ have been resolved in the region where this remnant is interacting with the dense molecular cloud known as M–0.02–0.07, at the southeastern boundary. An additional spot has been observed near the northern arm of Sagittarius A West[26] (the $\sim 6\,\mathrm{pc}$ minispiral structure of ionized gas orbiting about the center; see color plate 3) at a velocity of $134\,\mathrm{km\,s^{-1}}$. The detection of these OH masers is a principal reason behind the identification of Sagittarius A East as the remnant of a powerful explosion.

At least one of these OH maser lines shows Zeeman splitting, from which a magnetic field strength of 2–4 mG has been estimated within the remnant's nonthermal radio shell. The remnant's intricate physical properties have received additional clarification with X-ray observations early in the *Chandra* mission,[27] as illustrated in figure 1.9. The smoothed broadband X-ray intensity map is here overlaid with radio contours from a 20 cm VLA image of the same region. We shall see shortly that the existence of relativistic particles within this unusually strong magnetic field makes Sagittarius A East a unique particle accelerator in the Galaxy, contributing significantly to the cosmic ray and gamma ray flux emerging from the nucleus.

Figures 1.8 and 1.9 provide evidence that no significant X-ray continuum or line emission is occurring at the location of the radio shell itself. Instead, the source of this diffuse X-ray flux appears to be associated with a hot, optically thin thermal plasma located within the cavity. Also, a division of the X-ray emissivity into soft and hard energy bands shows that the morphology is spectrally dependent (compare figure 1.8 with 1.10), characterized by a half-power radius $\sim 20''$ at 6–7 keV, compared with $\sim 30''$ at lower energies. Overall, the X-ray-emitting region appears to be concentrated within the innermost 2 pc of this remnant.

---

[25] A survey of supernova remnants W28, W44, and IC 443, detected at 1,720 MHz, is given by Claussen et al. (1997).

[26] See Yusef-Zadeh et al. (1996).

[27] This was reported by Maeda et al. (2002).

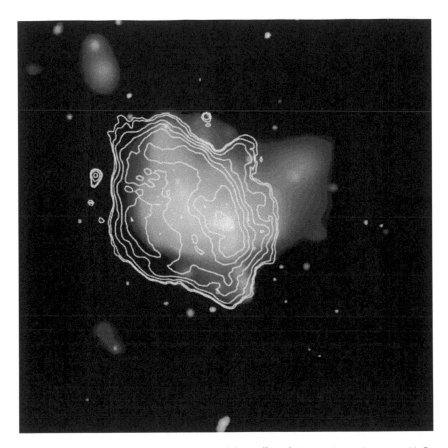

Figure 1.9 In this image, the smoothed broadband X-ray intensity map (1.5–7.0 keV) is overlaid with radio contours from a 20 cm VLA image of Sagittarius A. The outer oval-shaped radio structure is associated with synchrotron emission from the shell-like nonthermal radio source Sagittarius A East. (Image from Maeda et al. 2002)

The X-ray spectrum of Sagittarius A East contains strong $K\alpha$ lines from highly ionized ions of S, Ar, Ca, and Fe, for which a simple isothermal model yields an electron temperature $\sim$2 keV. The inferred metallicity is overabundant by a factor of four compared with solar values, concentrated toward the middle. Maeda et al. (2002) conclude from this that Sagittarius A East is probably the result of a Type II supernova explosion, with a 13–20 $M_\odot$ main-sequence progenitor, and that the combination of its radio and X-ray properties classifies it as a metal-rich "mixed morphology" remnant. However, the size of the Sagittarius A East shell is the smallest known for this category of sources,

21

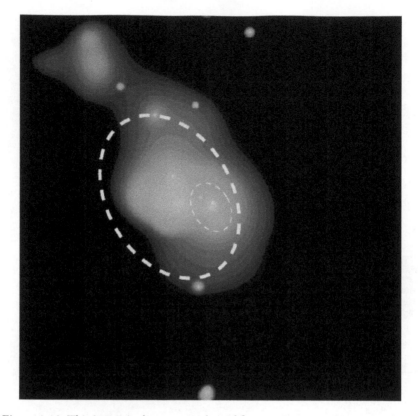

Figure 1.10 This image is the same as that of figure 1.8, except this image shows only the X-ray intensity within the 6.0–7.0 keV band. A comparison between this figure and figure 1.8 shows that the structure of Sagittarius A East is spectrally dependent. The half-power radius of the emission is ~20″ in the 6.0–7.0 keV band, compared to ~30″ at lower energies. The hard emission is concentrated toward the center. (Image from Maeda et al. 2002)

implying that the ejecta have been expanding into a uniquely dense interstellar medium. For a 10,000-year-old structure, the implied ambient density is ~10^3 cm^{-3}, fully consistent with the observed properties of the 50 km s^{-1} M–0.02–0.07 molecular cloud, into which Sagittarius A East is apparently expanding.

The identification of Sagittarius A East as a supernova remnant has been further strengthened by the EGRET detection of a ~30 MeV–10 GeV continuum source (3EG J1746–2852) within 1° of the galactic center[28]—a notable development because supernova remnants detected

[28] See Mayer-Hasselwander et al. (1998).

at 1,720 MHz also tend to be clearly associated with EGRET sources.[29] This connection, however, is subject to two important caveats. First, nonthermal radio emission observed from Sagittarius A East at 6 cm and 20 cm is characterized by a spectral index of $\sim$1, which requires an underlying population of nonthermal leptons (primarily electrons and positrons) with a power-law distribution index $p \sim 3$.[30] In contrast, typical supernova remnants display radio emission characterized by a spectral index $\sim$0.5 and an attendant lepton index $p \sim 2$. Second, the implied gamma ray luminosity of Sagittarius A East ($\sim 2 \times 10^{37}$ ergs s^{-1}) is roughly two orders of magnitude greater than that of the other remnants detected by EGRET.

Thus, although the *Chandra* observation may have resolved the mystery surrounding the birth of Sagittarius A East, it may have created another in terms of its relatively large gamma ray luminosity. Of course, it is quite plausible that 3EG J1746–2852 is not associated with Sagittarius A East at all. For example, the primary source of gamma rays may be the filaments in the arched magnetic field structure to the north of the galactic center (see figure 1.7).[31]

Given the striking similarity between the gamma ray spectrum of 3EG J1746–2852 and that of the EGRET supernova remnants, however, the association of this high-energy source with Sagittarius A East is probably real. In that case, the mechanism responsible for producing Sagittarius A East's broadband spectrum would be the decay of neutral and charged pions created in collisions between shock-accelerated protons and the ambient medium.[32] In this scenario, the neutral pions decay directly into two photons ($\pi^o \to \gamma\gamma$), while the charged pions decay into muons and subsequently into "secondary" relativistic electrons and positrons ($\pi^\pm \to \mu^\pm \nu_\mu$, with $\mu^\pm \to e^\pm \nu_e \nu_\mu$). In this fashion, not only does the cascade initiated by the shock-accelerated protons produce Sagittarius A East's gamma ray spectrum, but it also self-consistently accounts for its radio emission via leptonic synchrotron processes involving the decay products.[33]

---

[29] This category of sources is described fully by Esposito et al. (1996), Combi, Romero, and Benaglia (1998), and Combi et al. (2001).

[30] See Pedlar et al. (1989).

[31] This model was developed by Pohl (1997) and argued on the basis of circumstantial observational evidence by Yusef-Zadeh et al. (2002).

[32] See Gaisser, Protheroe, and Stanev (1998) and Melia et al. (1998).

[33] This scenario is described by Fatuzzo and Melia (2003).

But this is where the kinship between Sagittarius A East and the other EGRET supernova remnants ends, for the element that binds them—the interaction between their supersonically expanding ejecta with a dense molecular cloud environment—at the same time renders Sagittarius A East unique in the Galaxy. Its singularly high gamma ray luminosity, as well as its unusually steep radio spectral index, appears to be due to the high density ($\sim 10^3 \, cm^{-3}$) and strong magnetic field (~ 2–$4 \, mG$) of the surrounding medium. While the greater density enhances the particle collision rate—and hence the luminosity—the intense magnetic field facilitates the acceleration of particles to energies ($\sim 10^{18} \, eV$) not seen in any other remnant.[34]

Yet the bremsstrahlung emission due to the shock-accelerated leptons is at best $\sim 10^{32} \, ergs \, s^{-1}$, roughly an order of magnitude weaker than the X-ray luminosity inferred by *Chandra*. So whereas the radio and gamma ray photons from Sagittarius A East are produced primarily in its shell, the X-rays are evidently emitted instead by a thermal plasma within its cavity (see figures 1.8 and 1.10).

The environment occupied by Sagittarius A East is about as close as we can get to the nucleus without beginning to sense the influence of dark matter concentrated within. And so we will end our brief survey of the galactic center here. In chapter 2, we shall start to focus on the nature of strong gravity pervading the inner 0.01 pc of the Galaxy—a region 1,000 times smaller than the unusual remnant we have been exploring in this section.

[34] An early treatment of this acceleration mechanism in the presence of strong magnetic fields was made by Jokipii (1992). The most recent analysis of relativistic particle acceleration in Sagittarius A East may be found in Crocker et al. (2005).

CHAPTER 2

The Radio Source Sagittarius A*

The identification and localization of Sagittarius A* have been woven inextricably into the braided history that chronicles the discovery of our Galaxy's structure and its central region. The story begins in the early decades of the nineteenth century, prior to which the Milky Way was thought to encompass the entire universe, with the solar system at its nucleus. But the Copernican revolution, which removed Earth from the cosmic center, would eventually also displace the Sun from this exalted position and subdue even the Milky Way's aspirations of being the measure of all things.

The earliest indication that the center of the Milky Way might be far from the solar system was a peculiarity observed in the distribution of globular clusters—gravitationally bound aggregates of thousands to millions of stars, spread over a volume several hundred light-years in diameter. Globular clusters are characterized by high central densities and tend to be extremely round. They apparently formed early in the history of the universe. Systems of globular clusters surround all bright galaxies, providing a fossil record of the dynamical and chemical conditions at the time when the hosts were forming.

John Herschel (1792–1871) noticed in the 1830 s that a large number of these clusters occurred in a relatively small portion of the sky, mainly in the direction of Sagittarius. (The current census shows that 77 globular clusters, out of a total of 150, are contained within this constellation and two of its neighbors, Scorpius and Ophiuchus.) Almost 100 years later, Harlow Shapley (1885–1972) correctly interpreted this unusual distribution while studying what he thought were Cepheid variables sprinkled among the aggregated stars.

A Cepheid is a young star of several solar masses and roughly 10,000 solar luminosities, whose brightness changes periodically. Singly ionized helium in its atmosphere may become doubly ionized by the outwardly streaming radiation, rendering the stellar envelope more opaque. The net effect of this is to heat the atmosphere, pushing it outward, thereby increasing the star's size and luminosity. But the expansion then cools the gas, allowing the ionized helium to regain its electron, which lowers the transparency, causing the atmosphere to shrink again. The period of this repetitive behavior is related to the Cepheid's intrinsic luminosity. Thus, measuring the repetition cycle also provides an unmistakable determination of its absolute brightness, making the Cepheid a reliable standard candle for assessing the cosmic distance scale.

Assuming a symmetric spatial distribution of globular clusters about the galactic center, Shapley could use the distances he had estimated to infer the overall size of the system and the displacement of its center from the Sun. He concluded that the globular clusters formed a halo around a flat disk-shaped body with a diameter of 300,000 light-years and a centroid some 50,000 light-years away. It was realized only later that the standard candles he had been observing were not Cepheids at all but rather RR Lyrae variables—stars similar to Cepheids, though fainter. The Milky Way is now thought to be 100,000 light-years across, and the distance to its center is only 20,000–30,000 light-years.

Revolutionary ideas such as this are bound to incur critical appraisal, and for Shapley, this took form during the now-famous "Great Debate," in which Heber D. Curtis (1872–1942) challenged his estimates and the implied cosmological model.[1] Their lectures on "The Scale of the Universe" were delivered on April 26, 1920, in Washington, D.C. The thirty-four-year-old Shapley, who was then a staff member at Mount Wilson Observatory, spoke first. In support of his numbers, he argued that the spiral nebulae seen through the largest telescopes of the day were part of the Milky Way. They were thought to be solar systems, like our Sun and its planets, in the process of formation.

Speaking after Shapley, Curtis agreed with the notion that the globular clusters were outside the disk of our Galaxy, but he believed they were much closer. Curtis argued that the Galaxy overall was smaller than the size proposed by Shapley. Curtis proposed that the spiral nebulae were objects similar to our own Milky Way, based in part on the fact that the light spectrum of a spiral nebula is indistinguishable from that of the Galaxy. He described them as large collections of stars, comparable in size to the Milky Way and located well beyond the latter's boundary.

At the time of the debate, Curtis was the director of the Allegheny Observatory and had already established himself as an outstanding

---

[1]There are many accounts of this event. A useful summary may be found in Hoskin (1976).

speaker. Shapley, who was hoping to become the new director of the Harvard College Observatory, faced a daunting task. Surprising no one, Curtis won the debate, and Shapley was not offered the position, though he did eventually become the director at Harvard. It is said that Henry Norris Russell, who attended the debate, suggested to Shapley afterwards that he teach a course to hone his speaking skills.

Today, Curtis is also remembered for his correct conclusion that spiral nebulae were enormous aggregates of stars separate from the Milky Way. They became known as "extragalactic nebulae," then as "island universes," and finally simply as other "galaxies."

Many of the attempts to improve Shapley's localization of the galactic center and its distance from us have involved improvements in the calibration of the standard candles. Besides RR Lyrae stars, clusters are also now known to contain Mira variables, among others, providing greater confidence in the determination of the distance scale using optical means.[2] One of the most recent attempts in this regard[3] has concluded that the best estimate for the distance to the center of our Galaxy is 7.9 ± 0.3 kpc.

2.1 POSITION OF SAGITTARIUS A*

This story took a significant turn in the 1960s, when it became increasingly clear that the location of the galactic center could not be discussed in isolation from the dominant radio source in this region, known as Sagittarius A. The strong maximum of radio continuum emission in the constellation Sagittarius was first recognized as a discrete source in 1951 by Jack H. Piddington (1910–1997) and Harry C. Minnett. Because the bright emission originated from the center of the Milky Way, as indicated by optical observations, it was assumed that Sagittarius A should indeed be at the galactic nucleus. For this reason, Sagittarius A was subsequently used to define the zero of longitude in the revised system of galactic coordinates.[4]

---

[2]Reid (1993) has written a review of the galactic distance scale and the methods used for its determination in other regions of the electromagnetic spectrum, including the X-ray, radio, and infrared bands.

[3]See McNamara et al. (2000).

[4]See Blaauw et al. (1960). For a more detailed set of arguments used in the early support for adopting Sagittarius A as the center of the Milky Way, see Downes and Maxwell (1966).

As we saw in chapter 1, continued refinements in radio imaging have brought us to where we are today. Sagittarius A is now recognized as a complex field of diffuse and pointlike emission, from both thermal and nonthermal plasma. Increasingly, Sagittarius A* is viewed as the dominant object in the central region, due in part to some very clever observations that have allowed astronomers to determine its location with unprecedented precision.

This statement immediately begs the question, relative to what is Sagittarius A*'s position being determined? For example, how does one know where sources seen on a radio image should be placed relative to stars seen on an infrared map or relative to Shapley's distribution of globular clusters? While Sagittarius A* is a very bright radio source, it is very dim in the infrared, so locating it on infrared images is extremely difficult given the high projected number density of stars at the galactic center and by the absence of any obvious infrared counterpart to its radio emission.

In the mid-1990s, Karl Menten, then at the Harvard-Smithsonian Center for Astrophysics, made the intriguing suggestion that the 43 GHz SiO maser emission from Mira variables ought to be detectable at the galactic center. Such a measurement would be remarkable because Mira stars would then be seen as both radio and infrared sources, instantly permitting the two maps containing these objects to be cross-registered with breathtaking accuracy. For a typical Mira variable, SiO maser emission occurs at a radius of about 8 AU inside its atmosphere, which corresponds to about 1 mas (or roughly 10^{-4} light-year) at the ≈ 8 kpc distance to the galactic center. Since the infrared emission presumably peaks at the projected center of this star, one can therefore hope for a positional accuracy of 1 mas or so in the determination of Sagittarius A*'s location relative to other objects in the field.

Karl Menten and Mark Reid, also at the Harvard-Smithsonian Center for Astrophysics, pitched this idea to Reinhard Genzel and his group at the Max-Planck-Institut für Extraterrestrische Physik in Garching, Germany, who had been busy acquiring high-resolution infrared images of the galactic center with the hope of measuring stellar proper motions near the heavy concentration of dark matter in this region. Preoccupied with their own equally exciting work, about which we shall have more to say in chapter 5, the German group could not easily be distracted. So in November 1995, Menten and Reid jump-started the collaboration by visiting Garching to demonstrate the viability of this project and to move things along. Just as they arrived in Munich, however, the

U.S. government shut down because of the wranglings in Congress. As government workers, Menten and Reid were told they should return home. But they ignored that request, and the creation of a landmark paper was underway.

Using the MPE SHARP camera on the European Southern Observatory's 3.5-meter New Technology Telescope in Chile, Genzel and his collaborators produced a diffraction-limited 2.2 μm image of the galactic center that Menten and his group could then compare with their 43 GHz map of the same region produced with the Very Large Array.[5] The simultaneous identification of several SiO (and, as it turns out, H_2O) maser and infrared sources produced a localization for Sagittarius A* accurate to 60 mas. And for the first time, it was possible to constrain its infrared emission (given the absence of a counterpart on the 2.2 μm map) to a level of 9 mJy (dereddened). As we shall see later in this book, the nondetection of Sagittarius A* at infrared wavelengths provides a severe constraint on possible models of the physical environment producing the cm radiation.

In chapter 5, we will show that it is now possible to actually observe gravitational acceleration near Sagittarius A* directly, by tracing the orbital curvature of stars in the dense cluster surrounding the nucleus. Some of these stellar motions exceed $1,000 \, \mathrm{km \, s^{-1}}$ at a projected distance of 0.015 pc from the middle; indeed, one of these stars, known as S2, is apparently traveling on an elliptical orbit about the center with a speed exceeding $5,000 \, \mathrm{km \, s^{-1}}$ at a distance of only 124 AU—approximately 16 mas—at pericenter passage (see figures 2.1 and 2.2). So even the unprecedented 60 mas accuracy of Sagittarius A*'s position is inadequate for probing the gravitational potential in the now-accessible inner ~ 0.01–0.1 pc of the Galaxy.

A more recent application of this technique has produced a determination of Sagittarius A*'s position with significantly higher precision. Using a combination of Very Large Array and Very Long Baseline Array (VLBA) observations, Reid et al. (2003) have acquired greatly improved positions and, for the first time, proper motions of stars with circumstellar SiO masers in the galactic center region, while the new instrumentation on one of the 8.2-meter components of the Very Large Telescope (VLT) in Chile produced a deep ($K_s = 18$ mag) infrared map of this field. The application of this technique using seven bright stars

[5] See Menten et al. (1997).

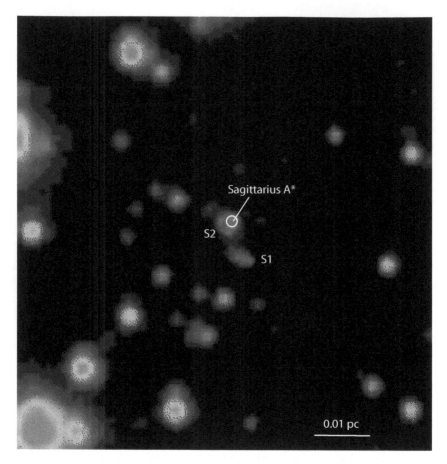

Figure 2.1 The position of Sgr A* is indicated by the center of the small circle on this May 2, 2002, infrared image of the inner 2″ of the Galaxy. This circle, with a radius of 10 mas (or roughly 0.0004 pc), corresponds to a 1 σ uncertainty in Sgr A*'s position. Though it appears on this image that Sgr A* is coincident with an infrared counterpart, in fact the infrared emission is due to the fast-moving star S2 projected only 16 mas from the center. Orbiting Sgr A*, this star was close to pericenter when this image was taken (Schödel et al. 2002). (Image from Reid et al. 2003)

within 15″ of Sagittarius A* has resulted in position and proper motion accuracies of ~ 1 mas and ~ 1 mas yr^{-1}, respectively.

The position of Sagittarius A* on the May 2, 2002, image is shown in figure 2.1. Reid et al. (2003) estimate that the small differences between the radio and fitted infrared positions account for a 1σ uncertainty in Sagittarius A*'s infrared localization of about 10 mas. It should be noted

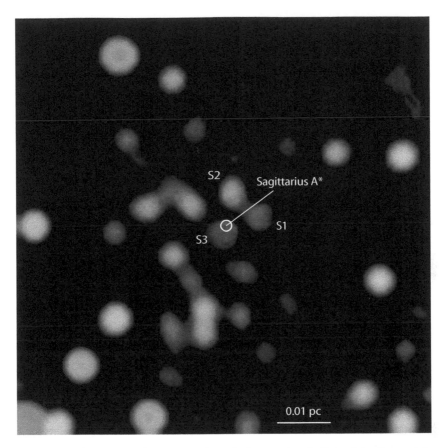

Figure 2.2 This image shows a region similar to that indicated in figure 2.1, though this one was taken in July 1995. The circle representing the location of Sgr A* has a radius of 15 mas, corresponding to the estimated $1\,\sigma$ uncertainty. This time, the infrared emission close to Sgr A* is produced by the star S3, projected about 30 mas from the center. (Image from Reid et al. 2003)

that the star S2 had moved to within about 16 mas of Sagittarius A* on this date, so the infrared emission detected near the center is not due to Sagittarius A* itself.

A comparison between figure 2.1 and the corresponding image produced in July 1995 (see figure 2.2) demonstrates the significant proper motion of stars orbiting within a mere $\sim 2\text{--}5 \times 10^{-3}$ light-year of the nucleus. In chapter 5, we shall use data such as these to probe the concentration of dark matter in this region. In the meantime, Reid et al. (2003) have combined the orbits of several stars to show that the infrared

31

motions, corrected to zero mean motion, are consistent with them being in a reference frame tied to Sagittarius A*, with an uncertainty of ~ 1 mas yr^{-1}, or roughly 40 km s^{-1}. In addition, the current best limit on the steady infrared emission from Sagittarius A* based on these measurements is ≈ 2 mJy.

2.2 PROPER MOTION

It is also possible—indeed, desirable—to determine Sagittarius A*'s position relative to an external frame of reference. Distant objects, such as quasars, are immobile on a timescale of years compared to points of reference within the Galaxy. However, it is not so much the absolute position of the galactic center that matters, since that has little dynamical value, but, rather, its proper motion. The reason for this is that fluctuations in the Hubble expansion could account for an apparent motion of Sagittarius A* relative to background extragalactic sources of at most ~ 0.01 mas yr^{-1}, a number undetectable using current technology. A motion closer to what is now detectable (i.e., 1 mas yr^{-1} or more) would then signify local dynamics and possibly produce a measurement of the local gravity.

The first program to determine Sagittarius A*'s proper motion began shortly after its discovery, using the National Radio Astronomy Observatory's Green Bank interferometer.[6] Interest and motivation in carrying out this measurement have been driven by the expectation that a stellar-size object would be buzzing around the central gravitational potential well with a velocity commensurate with that of gas clouds and other stars in this region. If Sagittarius A* were instead a supermassive object, then one might reasonably expect it to be at rest in the center. Of course, one should keep in mind that the residual motion of a black hole relative to other sources in the galactic nucleus also depends on its formation history and the local galactic dynamics.

The best determination of Sagittarius A*'s proper motion to date has been made from 8 epochs of VLA observations spanning the years 1982 to 1998,[7] followed more recently with even higher resolution VLBA measurements, which confirm and improve upon the earlier results.[8] The

[6] See Backer and Sramek (1982).
[7] Reported by Backer and Sramek (1999).
[8] See Reid et al. (1999).

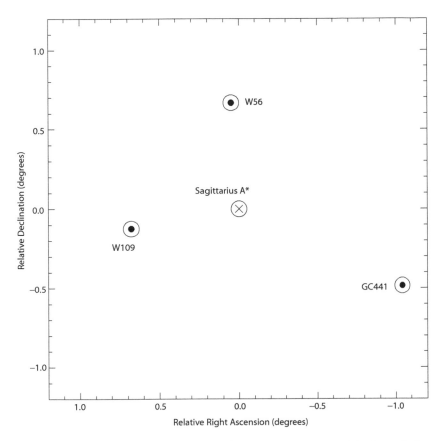

Figure 2.3 A map of the sky surrounding Sagittarius A* with the relative locations of three background extragalactic sources with sufficiently strong fluxes (> 25 mJy) and compact structure (< 1″) to be used as positional references. While these sources have not yet been identified as quasars or active galactic nuclei, their radiative characteristics strongly suggest that they are extragalactic. (Image from Backer and Sramek 1999)

interferometric observations were carried out at 4.885 GHz using the VLA in its 36-kilometer (A) configuration, with three reference sources (W56, W109, and GC441) whose strong flux (>25 mJy) and compact structure (< 1″) made them appropriate stationary background beacons of light at this frequency (see figure 2.3).

Of course, any observation from the solar system of an object at the galactic center relative to the extragalactic firmament is subject to the secular parallax produced by the rotation of the Galaxy itself, plus the small additional contribution due to the Sun's motion relative to the local standard of rest (LSR). For an object at rest at the galactic center, this secular parallax is given, in galactic coordinates, by

$$[\mu_l, \mu_b]_\Pi = [\mu_l, \mu_b]_{GR} + [\mu_l, \mu_b]_\odot$$
$$= -[(A - B), 0] - [V_\odot/R_0, W_\odot/R_0], \qquad 2.1$$

where A and B are Oort's constants expressed in angular terms, $V_\odot$ and $W_\odot$ are the components of the solar motion relative to the LSR in the directions $l = 90°$ and $b = 90°$, respectively, and R_0 is the distance to the galactic center. By now, several determinations of the constant $(A - B)$ have been made,[9] consistent with the value

$$[\mu_l, \mu_b]_{GR} = [-5.57 \pm 0.42, 0.0] \text{ mas yr}^{-1}. \qquad 2.2$$

And using the solar motion derived from observations with *Hipparcos*,[10] assuming a distance to the galactic center $R_0 = 8$ kpc (see above), we determine that

$$[\mu_l, \mu_b]_\odot = [-0.12 \pm 0.02, -0.16 \pm 0.01] \text{ mas yr}^{-1}. \qquad 2.3$$

So the total secular parallax is

$$[\mu_l, \mu_b]_\Pi = [-5.69 \pm 0.42, -0.16 \pm 0.01] \text{ mas yr}^{-1}. \qquad 2.4$$

Based on their 8 epochs of VLA measurements, Backer and Sramek (1999) reported an observed proper motion for Sagittarius A* of

$$\mu_{l,*} = [-6.18 \pm 0.19] \text{ mas yr}^{-1},$$
$$\mu_{b,*} = [-0.65 \pm 0.17] \text{ mas yr}^{-1}. \qquad 2.5$$

[9]Hanson (1987) used data from the Lick Northern Sky Proper Motion program to obtain 25.2 ± 1.9 km s^{-1} kpc^{-1}; Feast and Whitelock (1997) concluded from a *Hipparcos* study of Cepheid stars that $(A - B) = 27.2 \pm 1.0$ km s^{-1} kpc^{-1}; Olling and Merrifield (1998) used a more complete model of the galactic mass distribution to get 25.2 ± 1.9 km s^{-1} kpc^{-1}; and Feast, Pont, and Whitelock (1998) analyzed the Cepheid period-luminosity zero point from radial velocities and Hipparcos proper motions to update their previous result to 27.23 ± 0.86 km s^{-1} kpc^{-1}. The 1984 IAU adopted value of 26.4 ± 1.9 km s^{-1} kpc^{-1} (Kerr and Lynden-Bell 1986) is a reasonable intermediate value for all of these.

[10]See Dehnen and Binney (1998).

Subtracting the expected secular parallax from these measurements therefore leaves a peculiar motion for Sagittarius A* of

$$\Delta\mu_{l,*} = [-0.49 \pm 0.46] \text{ mas yr}^{-1},$$

$$\Delta\mu_{b,*} = [-0.49 \pm 0.17] \text{ mas yr}^{-1}. \qquad 2.6$$

At a distance of 8 kpc, this translates into a peculiar velocity of

$$v_{l,*} = [-18 \pm 18] \text{ km s}^{-1},$$

$$v_{b,*} = [-18 \pm 7] \text{ km s}^{-1}. \qquad 2.7$$

In principle, VLBA observations of Sagittarius A*, phase referenced to extragalactic radio sources, should improve considerably on this accuracy, given that its global baseline greatly exceeds that ($\sim$36 km) of the VLA. However, one must contend with the blurring effects of interstellar scattering in the dense, turbulent plasma surrounding the galactic center, about which we will have more to say in the next section. In addition, one cannot avoid having to carefully model atmospheric effects due to the low source elevation.

The effects of interstellar scattering diminish with increasing frequency, so the VLBA must be pushed to its limit of sensitivity at or below 7 mm. From the highest frequency VLBI observations, we infer an upper limit to the size of Sagittarius A* (corresponding to its scatter-broadened image) of about 1 AU at 86 GHz.[11] At the 8 kpc distance to the galactic center, this corresponds to about 0.1 mas, which should therefore be the effective upper limit to the precision with which the VLBA can improve on the above results.

Between 1995 and 1997, Reid and his collaborators (1999) observed the galactic center with the VLBA in the late-night and early-morning periods (to diminish high water vapor turbulence in the atmosphere) and estimated a peculiar motion for Sagittarius A* of 0 ± 15 km s^{-1} toward positive galactic longitude. After subtracting the small solar motion (7.17 km s^{-1}) out of the galactic plane, they also estimated Sagittarius A*'s peculiar motion toward the north galactic pole to be 15 ± 11 km s^{-1}. For both components, this estimated peculiar motion is statistically indistinguishable from a null result and appears to be

---

[11] See Rogers et al. (1994).

consistent with the earlier VLA measurements, though the errors may have been underestimated in one or both cases.

These measurements of the proper motion in Sagittarius A* therefore provide a conservative upper limit of around $20\,\mathrm{km\,s^{-1}}$ in and out of the galactic plane. Stars in the central cluster,[12] however, move at speeds in excess of $1,000\,\mathrm{km\,s^{-1}}$. Thus, Sagittarius A* must be much more massive than a typical star in this region, because it would otherwise incur gravitational Brownian motion with respect to the galactic barycenter in response to the influence of the uneven momentum distribution of objects surrounding it. N-body simulations of stars orbiting a $2.6 - 3.6 \times 10^6\ M_\odot$ black hole (which appears to be the case for Sagittarius A*; see chapter 5) show that after only 10,000 years, a quasi-steady state condition is achieved, in which the motion of the central object is never more than $\sim 0.1\,\mathrm{km\ s^{-1}}$ in each coordinate. Sagittarius A*'s mass would have to be lower than $\sim 3,000\ M_\odot$ before its orbital speed could exceed $20\,\mathrm{km\,s^{-1}}$.[13]

One could argue that Sagittarius A*'s motion just happens to be along the line of sight to the galactic center, in which case this restriction on mass is fictitious. But Backer and Sramek (1999) seem to have ruled this out as well, since Sagittarius A* had no measurable acceleration between the years 1982 and 1998. Suppose Sagittarius A* were simply a random object orbiting the concentration of dark matter known to be present at the galactic center from the measured proper motion of stars in the central parsec (see chapter 5). Its acceleration[14] within this potential well would then be roughly $1.35\ \mathrm{mas\ yr^{-2}}$. However, the upper limit to the acceleration inferred from the full set of VLA observations is only $0.3\,\mathrm{mas\ yr^{-2}}$. Thus, if the galactic center harbors a supermassive black hole, the radio source Sagittarius A* must be its radiative manifestation, and its heft must be greater than a few thousand solar masses.

---

[12] See Eckart and Genzel (1997) and Ghez et al. (1998).

[13] The force acting on an object embedded within a stellar cluster consists of two distinct components: the first originates from the "smoothed-out" average distribution and varies slowly with position and time; the second arises from discrete encounters with individual stars and fluctuates more rapidly. The magnitude of the velocity associated with the ensuing gravitational Brownian motion scales inversely with black hole mass and has a value less than $\sim 20\,\mathrm{km\,s^{-1}}$ only for $M > 3,000\ M_\odot$. Recent numerical simulations that show this effect directly have been reported by Reid et al. (1999), Chatterjee, Hernquist, and Loeb (2002), and Laun and Merritt (2006).

[14] See Gould and Ramirez (1998).

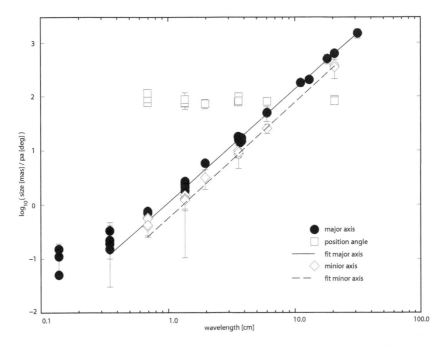

Figure 2.4 The major source axis (filled circles) of Sagittarius A*, the minor source axis (open diamonds), and the position angle (PA) of the major axis (open squares), as measured by VLBI, plotted versus wavelength. (Image from Krichbaum, Witzel, and Zensus 1999)

2.3 Structure as a Function of Frequency

An important element that makes Sagittarius A* unique among galactic nuclei is its proximity to Earth, which allows radio observers to resolve it with VLBI. Its apparent angular diameter ($\propto \lambda^{+2.0}$) depends strongly on wavelength, consistent with angular broadening by the scattering of its radio waves in the intervening plasma.[15] This broadening is very similar to that of OH masers within 0.5° of the galactic nucleus, implying that the diffuse thermal plasma within the central 140 pc (in longitude) is sufficiently turbulent to produce the observed scattering.[16] The data available prior to 1999 are displayed in figure 2.4.

---

[15] Some of the key papers on this thread include Lo et al. (1981, 1985), Backer (1988), Jauncey et al. (1989), Lo et al. (1993), Alberdi et al. (1993), Yusef-Zadeh et al. (1994), Backer (1994), Rogers et al. (1994), Bower and Backer (1998), and Lo et al. (1998).

[16] See van Langevelde et al. (1992) and Frail et al. (1994).

Most of the imaging observations have been carried out in the Northern Hemisphere, where the VLBI baseline coverage of the galactic center—a southern source—is relatively poor. Even so, the multitude of experiments have shown convincingly that Sagittarius A*'s scatter-broadened appearance is elliptical with an axial ratio $\sim 2 : 1$ and a position angle (PA) $\sim 80°$. This structure has been interpreted as being due mostly to an anisotropic scattering medium, probably caused by a relatively uniform magnetic field whose stress dominates the thermal and turbulent pressure of the gas.

The scattered image also appears to be very stable. In 1974, the size reported at 8.4 GHz was 17 ± 1 mas,[17] with a principal resolution in the east-west direction. Measurements in 1983 had sufficient uv coverage to provide elliptical source parameters,[18] which yielded a size of 15.5 ± 0.1 mas, an axial ratio of 0.55 ± 0.25, and a PA of $98° \pm 15°$. In 1998, the VLBA-determined size was 18.0 ± 1.5 mas, with a ratio of 0.55 ± 0.14 and a PA of $78° \pm 6°$. Over these twenty-three years of study, the source size thus appears to have changed very little—whether via random or secular variation—by no more than 5%–10%. This translates into a rate of expansion or contraction at 8.4 GHz of less than ~ 0.07 mas yr^{-1}.

VLBA observations at 7 mm have confirmed the hypothesis that the image of Sagittarius A* is a resolved elliptical Gaussian, with a measured major axis at this wavelength of 0.76 ± 0.04 mas.[19] The expected size, extrapolated from longer wavelengths, is 0.67 ± 0.03 mas. The 7 mm image of Sagittarius A* shown in figure 2.5 also reveals that its axial ratio is 0.73 ± 0.10, with a PA of $77° \pm 7°$.

Recently, an interesting refinement to these measurements was made possible by the near-simultaneous VLBA mapping of Sagittarius A* at five different wavelengths. By permitting the determination of *both* the major and minor axis diameters versus wavelength, this effort eliminated much of the remaining uncertainty associated with the source-size variability.[20] Its diameter along the major axis is evidently

$$\theta_{\mathrm{major}} = (1.43 \pm 0.02)\,(\lambda/1\ \mathrm{cm})^{1.99 \pm 0.03}\ \mathrm{mas.} \qquad 2.8$$

Within the quoted accuracy, the index in this expression is indistinguishable from 2, so the major axis diameter appears to follow the previously

---

[17]See Lo et al. (1981).
[18]See Lo et al. (1985).
[19]These observations were reported by Bower and Backer (1998).
[20]See Lo et al. (1998).

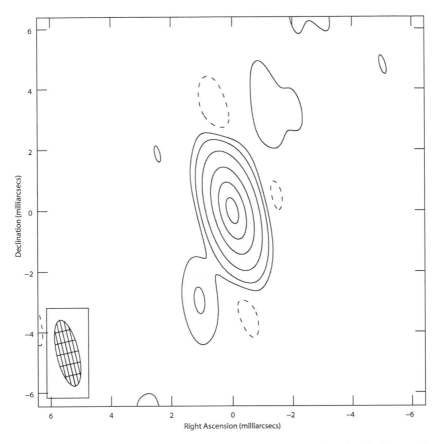

Figure 2.5 Uniformly weighted image of Sagittarius A* produced with the VLBA at 7 mm. The beam appears in the lower left-hand corner of the diagram. The contours shown here are −0.01, 0.01, 0.03, 0.10, 0.30, 0.60, and 0.90 times the peak intensity of 0.87 Jy/beam. North is up and east is left in this image. The galactic plane runs from the upper left-hand corner to the lower right. (Image from Bower and Backer 1998)

inferred λ^2-dependence over a wide range of wavelengths, from 7 mm to 6 cm. It is now understood that, when the size follows this law, the scattering is dominated by fluctuations in the electron density of the interstellar medium. The power spectrum of these fluctuations goes as $k^{-\beta}$, where k is the wavenumber of the irregularities.[21] The scattering angle scales as $\lambda^{1+2/(\beta-2)}$, and $[1+2/(\beta-2)] = 2$ when $\beta = 4$.

[21] See Romani, Narayan, and Blandford (1986).

However, a similar fit for all five wavelengths sampled along the minor axis by this VLBA study produced the following result:

$$\theta_{\text{minor}} = (1.06 \pm 0.10) \, (\lambda/1 \text{ cm})^{1.76 \pm 0.07} \text{ mas}, \qquad 2.9$$

which appears to be at odds with the effects of interstellar scattering.

The fit is somewhat closer to the expected λ^2-dependence if only points corresponding to $\lambda \geq 1.35$ cm are included, in which case the minor axis diameter is then $(0.87 \pm 0.23) \, (\lambda/1 \text{ cm})^{1.87 \pm 0.16}$ mas. A possible interpretation of this disparity is that the deviation at 7 mm between the measured minor axis diameter θ_{minor} and the scattering angle θ_{sc} is in fact the emergence of the intrinsic source diameter θ_{int}, which affects the observed size in quadrature:[22] $\theta_{\text{minor}} = (\theta_{\text{int}}^2 + \theta_{\text{sc}}^2)^{1/2}$. The implication is that at 7 mm Sagittarius A*'s intrinsic size is $\theta_{\text{int}} = (0.45 \pm 0.11)$ mas (along the minor axis) with PA $= -10°$, nearly in the north-south direction.

At the same time, the measured diameter of 0.7 ± 0.01 mas along the major axis (PA $= 80°$; essentially east-west) and the extrapolated scattering size of 0.69 ± 0.01 mas imply that the intrinsic size along this direction has to be ≤ 0.13 mas. Together, these results suggest that at 7 mm Sagittarius A* may be elongated along an essentially north-south direction, with an axial ratio of <0.3.

Given the importance of a direct measurement of the intrinsic size of Sagittarius A* as a discriminant between various emission mechanisms, the 7 mm result needs to be confirmed, preferably at shorter wavelengths, which are less susceptible to the effects of interstellar scattering (see figure 2.4). Though the reported statistical significance of the 7 mm elongation is five standard deviations, there appear to be additional systematic errors in the calibration. In 1999, Doeleman et al. (2001) therefore reobserved Sagittarius A* at 86 GHz using a six-station array, including the VLBA antennas at Pie Town, Fort Davis, and Los Alamos, the 12-meter antenna at Kitt Peak, and the mm arrays at Hat Creek and Owens Valley. The motivation for this was that intrinsic structure of Sagittarius A*, if visible at 7 mm, should be readily observable at 3.5 mm as long as its size does not increase faster than λ^2.

At 86 GHz, assuming the scattering and intrinsic structure add in quadrature, the limits on the intrinsic size of Sagittarius A* in a PA of $-10°$ and $80°$ are 0.25 mas and 0.13 mas, respectively. The

---

[22] See, e.g., Narayan and Hubbard (1988).

data are best modeled by a circular Gaussian brightness distribution with a full width at half maximum (FWHM) of 0.18 ± 0.02 mas. The 3.5 mm observations are therefore consistent at the 2σ level with the elliptical scattering model of a point source. They are also consistent at the 3σ level with a model in which the elliptical scattering disk is combined with an intrinsic structure extrapolated from 7 mm, assuming a $\lambda^{0.9}$-dependence. Doeleman et al. (2001) report that increasing the spectral index of this power law—thus decreasing Sagittarius A*'s intrinsic size—improves agreement with the 3.5 mm data.

In summary, the 3.5 mm imaging of Sagittarius A* is consistent with the interstellar scattering of an unresolved source, assuming an elliptical Gaussian structure. These observations, however, do not yet exclude an extrapolation of Sagittarius A*'s size at 7 mm, as long as its intrinsic size evolves faster than $\lambda^{0.9}$.

The situation is significantly clearer at mm to sub-mm wavelengths, where refractive scintillation measurements set a *lower* limit to Sagittarius A*'s size. Unlike the multidish interferometric techniques employed at longer wavelengths, refractive scintillation is based on observations carried out with single-dish telescopes, under the assumption that inhomogeneities in the intervening medium are well understood. To see how this works, imagine staring at a lightbulb, while passing an opaque object across your line of sight. You would see a smaller variation in the intensity of the light as you reduce the size of the eclipsing material. We know that the clumps in the interstellar medium are about 10^{12} cm wide, and so the absence of refractive scintillation in Sagittarius A* at 1.3 and 0.8 mm implies that the intrinsic source must be at least this large (i.e., roughly 0.1 AU) at these wavelengths. [23]

---

[23] These measurements were carried out over a decade ago, by Gwinn et al. (1991).

CHAPTER 3

Sagittarius A*'s Spectrum

3.1 THE RADIO SPECTRUM

We shall see now why the measurement of Sagittarius A*'s size is of paramount importance, beyond simply giving us a sense of what is there. A source of radio emission is often characterized by the *brightness* temperature T_b instead of its frequency-dependent intensity I_ν. Though its physical meaning is somewhat limited, T_b is nevertheless a valuable tool for estimating the conditions in the radiating gas, whose volume and temperature together determine the overall luminosity.

The brightness temperature is the temperature of a blackbody whose radiant intensity

$$B_\nu(T_b) = \frac{2h\nu^3/c^2}{\exp(h\nu/kT_b) - 1} \qquad 3.1$$

in the range of frequencies between ν and $\nu + d\nu$ is the same as that of the observed source. This is the Planck function written in terms of the Planck constant h, the Boltzmann constant k, and the speed of light c. However, the vast majority of cosmic radio emitters are not thermal, so the brightness temperature for a given source may vary widely with frequency. For many of them, including Sagittarius A* as we shall see, the inferred brightness temperature is so high that equation (3.1) may be simplified with the limited expansion $\exp(h\nu/kT_b) \approx 1 + (h\nu/kT_b)$. The result is the Rayleigh-Jeans form of the blackbody law, defined by

$$T_b = \frac{1}{2k} \left(\frac{c}{\nu}\right)^2 I_\nu, \qquad 3.2$$

when the observed intensity I_ν is equated to $B_\nu(T_b)$.

Now, in order to produce a certain luminosity, the brightness temperature of a radio source must be higher the smaller its radiating surface area. In chapter 2, we saw that Sagittarius A*'s intrinsic size at 7 mm is roughly 0.1 mas, or ∼1 AU, at the ∼8 kpc distance to the galactic center.

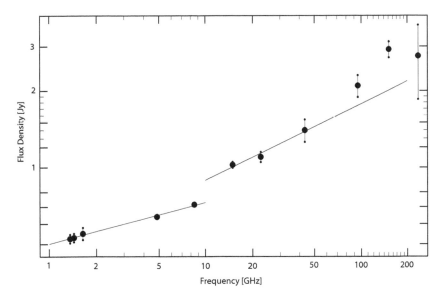

Figure 3.1 The spectrum of Sagittarius A*—flux density versus frequency—from radio to mm wavelengths (1 Jy = 10^{-23} ergs cm^{-2} s^{-1} Hz^{-1}). The data shown here were averaged over the week of observation in late October 1996. The error bars indicate variability (one standard deviation) during this period. The flux densities at neighboring frequencies were also combined from the various mm telescopes. The solid lines represent power-law fits to guide the eye in the low- and high-frequency VLA portions of the spectrum. An unmistakable excess emerges at mm wavelengths. (Image from Falcke et al. 1998)

Thus, given its luminosity at this wavelength, calculated from the flux density shown in figure 3.1, the corresponding brightness temperature must be greater than $\sim 10^{10}$ K.

Its minimum size of ~ 0.1 AU corresponds to an upper limit on the brightness temperature (at 0.8 mm) of about 0.5×10^{12} K. Sagittarius A* therefore appears to be within the range of typical radio cores in active galactic nuclei[1] and shines below the so-called Compton limit (at approximately 10^{12} K).[2] The particles that emit the synchrotron photons may also energize them further via inverse Compton scattering. The Compton limit corresponds to the brightness temperature above which the radiation is so heavily Comptonized that most of it emerges at frequencies well beyond the gigahertz range.

[1]See, e.g., Readhead (1994).
[2]For an early discussion of this phenomenon, see Kellermann and Pauliny-Toth (1969).

The Compton limit represents the maximum possible brightness temperature of incoherent synchrotron emission from an electron plasma.

These two limits on T_b imply that, regardless of whether the emitting particles are Maxwellian, they must attain relativistic energies in order to produce Sagittarius A*'s radio spectrum. This is well known in the case of nonthermal particles,[3] but it is straightforward to see even for a thermal population of leptons of mass m_e, given that $(\gamma - 1)m_e c^2 \sim (3/2)kT_b$. Evidently the Lorentz factor γ needs to be $\sim 10\,(T_b/6 \times 10^{10}$ K$)$ or greater to account for Sagittarius A*'s measured brightness.

A closer inspection of figure 3.1 reveals that, although Sagittarius A*'s radio spectrum compiled from published data may be described as a $\sim \nu^{1/3}$ power law,[4] simultaneous multifrequency VLA observations such as these show it to be bumpy with a spectral index at gigahertz frequencies varying between $\alpha = 0.1$ and $\alpha = 0.4$ (where the flux density is given as $S_\nu \propto \nu^\alpha$).[5] The spectrum appears to have a low-frequency turnover around 1 GHz,[6] which is probably due to free-free absorption, self-absorption, or simply an error arising from the fact that Sagittarius A*'s scattered size is largest in this portion of the spectrum. There is no question that at low frequencies ($\sim$330 MHz) the entire Sagittarius A region suffers from free-free absorption.[7]

Sagittarius A*'s radio spectrum also drops off steeply at sub-mm wavelengths, for which the most likely explanation is a decrease in the source emissivity, that is, a thinning of the medium.[8] This faintness in the infrared is somewhat out of character for an active galactic nucleus,[9] which may be indicative of a peculiarity in the emitting region of weakly active nuclei, such as Sagittarius A*, compared with those of their more distant and powerful brethren. However, the steep sub-mm dropoff in Sagittarius A*'s spectrum is fully consistent with its X-ray emissivity, as we shall see later in this chapter.

---

[3] A classic text on the physics of radio sources is that by Pacholczyk (1970).

[4] See, e.g., Duschl and Lesch (1994).

[5] A representative sample of papers addressing this point includes the following: Brown and Lo (1982), Wright and Backer (1993), Morris and Serabyn (1996), and Falcke et al. (1998).

[6] See Davies, Walsh, and Booth (1976).

[7] A full analysis is provided in Pedlar et al. (1989).

[8] This is simply due to the fact that, as the frequency-dependent emissivity drops toward shorter wavelengths, the integrated intensity and, correspondingly, the optical depth along the line of sight both drop with increasing frequency. An early analysis of this important spectral feature in Sagittarius A* may be found in Melia (1992a, 1994).

[9] This was first noted by Rieke and Lebofsky (1982).

The mm to sub-mm portion of Sagittarius A*'s spectrum also happens to display one of the most interesting and potentially probative features seen within its broad radiative output. Although the spectrum at cm wavelengths is adequately described by a power law with spectral index $\alpha \sim 0.1$–0.4, the index increases to $\alpha = 0.52$ in the mm range, peaking at an even higher level ($\alpha \sim 0.76$) at 2–3 mm. It is safe to conclude that there probably exists a significant mm excess—a mm bump— or possibly even a separate component at these frequencies. So far, no bright source of confusion has been found in high-resolution mm wave maps that could account for such an excess as being extrinsic to Sagittarius A*.[10]

The mm bump is somewhat reminiscent of an extreme ultraviolet feature seen in the spectrum of Seyfert galaxies. Called the Big Blue Bump (BBB), this excess emerges above an underlying nonthermal power-law component in the ultraviolet to soft X-ray portion of the spectrum.[11] Early theoretical work[12] on the BBB spectrum focused on the role of optically thick emission from a hypothesized accretion disk surrounding the central engine. More recent analysis has cast some doubt on this simple interpretation, though it appears that an accretion disk must be present to produce the BBB, possibly via thermal, optically thin free-free radiation rather than a blackbody.[13]

Of course, the range in brightness temperature inferred for Sagittarius A* precludes the possibility that the same exact mechanism is at work in this source, and anyway, the mm bump emerges at substantially lower frequencies than the BBB. However, as we shall see in chapters 7 and 8, Sagittarius A*'s size of ~ 0.1–1 AU at mm wavelengths corresponds to an extent merely ~ 2–20 times the size of the central supermassive black hole. In addition, the shape of the spectrum shown in figure 3.1 suggests that the mm bump is produced in a self-absorbed region, probably via synchrotron emission.

A radiating medium undergoes self-absorption at a given frequency when the emissivity is so high that the exiting intensity approaches (and possibly exceeds) that of a blackbody at the same frequency. Since blackbody radiation corresponds to a distribution of photons in complete

---

[10] See, e.g., Zhao and Goss (1998).

[11] See, e.g., Sanders et al. (1989).

[12] The list of relevant papers on this topic is extensive, but see Shields (1978), Malkan and Sargent (1982), Czerny and Elvis (1987), and Laor and Netzer (1989).

[13] See Antonucci and Barvainis (1988) and Nayakshin and Melia (1997).

equilibrium with the matter, the so-called blackbody limit could only be violated if a mechanism were present to permit the rate of photon emission to exceed the rate of absorption, even in the presence of physical conditions for which equilibrium would otherwise have been maintained. A radio spectrum rising indefinitely toward lower frequencies would eventually exceed the blackbody limit. A telltale signature of self-absorption is therefore a spectral turnover, with a decreasing flux density toward lower photon energy—as indeed appears to be the case below $\sim$150 GHz in figure 3.1. So if this hot, radiating gas is in orbit about the central object, it constitutes an analog of the BBB-producing accretion disks seen in Seyfert galaxies.

3.2 LINEAR AND CIRCULAR POLARIZATION

The electric $\mathbf{E} = (E_1, E_2, E_3)$ and magnetic $\mathbf{B} = (B_1, B_2, B_3)$ field vectors of an electromagnetic wave oscillate in a plane perpendicular to the wavevector $\mathbf{k}$. The surface containing $\mathbf{E}$ and $\mathbf{k}$ is often called the *plane of polarization*; by superposing waves corresponding to two such oscillations perpendicular to each other, we can construct the most general state of polarization for a wave of given $\mathbf{k}$ and angular frequency ω.

Since $\mathbf{E}$, $\mathbf{B}$, and $\mathbf{k}$ form an orthogonal set, the most general way to write the electric field vector is

$$\mathbf{E}(\mathbf{x}, t) = (\mathbf{E}_{01} + \mathbf{E}_{02}) \exp\{i(\mathbf{k} \cdot \mathbf{x} - i\omega t)\}$$

$$\equiv (E_{01}\hat{\epsilon}_1 + E_{02}\hat{\epsilon}_2) \exp\{i(\mathbf{k} \cdot \mathbf{x} - i\omega t)\}, \qquad 3.3$$

where $\hat{\epsilon}_1$, $\hat{\epsilon}_2$, and $\mathbf{k}$ constitute an alternative set of orthogonal vectors. The benefit of this may not be immediately obvious; it seems that we can just rotate the vectors $\hat{\epsilon}_1$ and $\hat{\epsilon}_2$ arbitrarily, and we haven't gained anything. But this is only true when E_{01} and E_{02} are real or have the same phase. In general,

$$E_{01} = |E_{01}| \exp(i\phi_1), \qquad 3.4$$

$$E_{02} = |E_{02}| \exp(i\phi_2), \qquad 3.5$$

so that

$$\mathbf{E}(\mathbf{x}, t) = [\hat{\epsilon}_1 |E_{01}| + \hat{\epsilon}_2 |E_{02}| \exp\{i(\phi_2 - \phi_1)\}]$$

$$\times \exp\{i\mathbf{k} \cdot \mathbf{x} - i\omega t + i\phi_1\}. \qquad 3.6$$

But the overall phase ϕ_1 of the field represents just a constant rotation in the complex plane, which does not affect the physics. We can therefore drop it and effectively write the field as

$$\mathbf{E}(\mathbf{x}, t) = [\hat{\epsilon}_1 |E_{01}| + \hat{\epsilon}_2 |E_{02}| \exp\{i(\phi_2 - \phi_1)\}]$$
$$\times \exp\{i\mathbf{k} \cdot \mathbf{x} - i\omega t\}. \qquad 3.7$$

The measurable electric field is the real part $\Re(\mathbf{E})$ of equation (3.7). We see that when $\phi_2 - \phi_1 = 0$ the wave is *linearly polarized*, since $\Re(\mathbf{E}_1) = |E_{01}| \cos(\mathbf{k} \cdot \mathbf{x} - \omega t)$ and $\Re(\mathbf{E}_2) = |E_{02}| \cos(\mathbf{k} \cdot \mathbf{x} - \omega t)$. For such a wave, $\chi \equiv \arctan[\Re(\mathbf{E}_2)/\Re(\mathbf{E}_1)]$ remains constant as the field evolves in space and time, so $\mathbf{E}$ oscillates along a fixed direction, making an angle χ with respect to the unit vector $\hat{\epsilon}_1$. However, when $\phi_2 - \phi_1 \neq 0$, the wave is instead *elliptically polarized* because the tip of the electric field vector traces an ellipse as it rotates about $\mathbf{k}$. The easiest way to see this is to note that the phase difference between $\mathbf{E}_{01}$ and $\mathbf{E}_{02}$ causes the field component in the $\hat{\epsilon}_1$ direction to pass through its nodes at different spatial locations and/or times compared with the other. They do not change proportionately, and the net effect of this is a rotation in the $\hat{\epsilon}_1 - \hat{\epsilon}_2$ plane, usually with a phase-dependent amplitude. According to equation (3.7), $\mathbf{E}$ sweeps around once every $2\pi/\omega$ seconds, and so the angular frequency of this rotation must be ω. In the special case where $|E_{01}| = |E_{02}|$ with $\phi_2 - \phi_1 = \pm\pi/2$, the amplitude of the electric field $|\mathbf{E}|$ is constant, and $\Re(\mathbf{E}_1) = |E_{01}| \cos(\mathbf{k} \cdot \mathbf{x} - \omega t)$ and $\Re(\mathbf{E}_2) = \mp |E_{01}| \sin(\mathbf{k} \cdot \mathbf{x} - \omega t)$. Clearly, $\mathbf{E}$ here rotates about $\mathbf{k}$ with a constant magnitude and angular frequency ω. This special case therefore constitutes *circular polarization*.

An alternative general expression for $\mathbf{E}$ employs the vectors

$$\hat{\epsilon}_\pm \equiv \frac{1}{\sqrt{2}}(\hat{\epsilon}_1 \pm i\hat{\epsilon}_2). \qquad 3.8$$

In this instance,

$$\mathbf{E}(\mathbf{x}, t) = (E_{0+}\hat{\epsilon}_+ + E_{0-}\hat{\epsilon}_-) \exp\{i(\mathbf{k} \cdot \mathbf{x} - i\omega t)\}. \qquad 3.9$$

Following our discussion in the previous paragraph, it is apparent that $\hat{\epsilon}_+$ represents a wave with positive helicity; that is, the rotation of this field component is in the right-hand sense because the $\hat{\epsilon}_2$ piece leads the other (by a phase exactly equal to $\pi/2$). For analogous reasons, $\hat{\epsilon}_-$ represents a wave with negative helicity. We can see how this sense of rotation comes about more fundamentally by considering the two components E_1 and E_2 in $\mathbf{E} = E_0(\hat{\epsilon}_1 + i\hat{\epsilon}_2)\exp\{i(\mathbf{k}\cdot\mathbf{x} - \omega t)\}$, which corresponds to an electric field with $E_{0-} = 0$. Taking the real part of $\mathbf{E}$, with $\mathbf{k} = k\hat{z}$, we infer that

$$\Re(E_1) \propto \cos(kz - \omega t), \qquad\qquad 3.10$$

whereas

$$\Re(E_2) \propto -\sin(kz - \omega t). \qquad\qquad 3.11$$

Thus, for fixed z, increasing t for an electric field vector in the fourth quadrant results in an increasing $\Re(E_1)$, while $\Re(E_2)$ approaches zero from negative values. That is, the electric field vector is rotating clockwise when seen from behind the x-y plane in the direction of $\mathbf{k}$.

Is it possible to determine the polarization of a field vector experimentally and therefore to interrogate the radio signal from Sagittarius A* at a level beyond merely its intensity? Clearly, we can describe the "internal" structure of the electric field with these two elegant basis vector systems, but unless we have a way of measuring the various components, this capability would not be of much practical interest. Stokes parameters are quantities defined in terms of the projected amplitudes along each of the basis directions, such that together they allow us to isolate the various dependences of $\mathbf{E}$ on the phase and component amplitudes. Written in the $(\hat{\epsilon}_1, \hat{\epsilon}_2)$ basis, the four Stokes parameters are[14]

$$S_0 = |\hat{\epsilon}_1 \cdot \mathbf{E}|^2 + |\hat{\epsilon}_2 \cdot \mathbf{E}|^2 = |E_{01}|^2 + |E_{02}|^2, \qquad\qquad 3.12$$

$$S_1 = |\hat{\epsilon}_1 \cdot \mathbf{E}|^2 - |\hat{\epsilon}_2 \cdot \mathbf{E}|^2 = |E_{01}|^2 - |E_{02}|^2, \qquad\qquad 3.13$$

$$S_2 = 2\,\mathrm{Re}\,[(\hat{\epsilon}_1 \cdot \mathbf{E})^*(\hat{\epsilon}_2 \cdot \mathbf{E})] = 2|E_{01}||E_{02}|\cos(\phi_2 - \phi_1), \qquad 3.14$$

$$S_3 = 2\,\mathrm{Im}\,[(\hat{\epsilon}_1 \cdot \mathbf{E})^*(\hat{\epsilon}_2 \cdot \mathbf{E})] = 2|E_{01}||E_{02}|\sin(\phi_2 - \phi_1)\,. \qquad 3.15$$

Alternative names for the Stokes parameters, often seen in the literature, are $I = S_0$, $Q = S_1$, $U = S_2$, and $V = S_3$.

Notice that

$$S_0^2 = S_1^2 + S_2^2 + S_3^2 \qquad\qquad 3.16$$

[14]These are based on the notation of Born and Wolf (1970).

(or $I^2 = Q^2 + U^2 + V^2$), so not all of them are independent variables. S_0 is nonnegative and represents the total energy flux or intensity of the wave. One often chooses a single proportionality factor in all of the definitions in equations (3.12)–(3.15) to force S_0 to be precisely the flux or intensity. The parameter S_1 contains information regarding the preponderance of $\hat{e}_1$-linear polarization over $\hat{e}_2$-linear polarization, and S_2 and S_3 provide phase information. For example, the wave has positive or negative helicity depending on the sign of S_3, while $S_3 = 0$ is the condition for linear polarization. Circular polarization leads to the condition $S_1 = S_2 = 0$. Knowing these parameters (e.g., by passing a wave through perpendicular polarization filters) is sufficient for us to determine the amplitude and relative phase of the field components.

Of course, one must be careful with the practical use of the Stokes parameters, which strictly apply to monochromatic waves. Beams of radiation, even if quasi-monochromatic, actually consist of a superposition of finite wave trains and therefore, by Fourier's theorem, contain a range of frequencies. The measurable Stokes parameters must then be treated as averages over some time interval. One consequence of the averaging process is that they end up satisfying the inequality $S_0^2 \geq S_1^2 + S_2^2 + S_3^2$, rather than the equality in equation (3.16).

The early papers that discussed the radio emission from the galactic center reported an absence of any significant polarization from Sagittarius A's complex amalgamation of thermal and nonthermal sources. The peak flux from this morphologically rich region, which in the early 1970s was not well resolved in 5 GHz Westerbork observations, was thought to have an upper limit of ~1% linear polarization.[15] This result was later confirmed with the VLA,[16] which also failed to detect any significant linear polarization from Sagittarius A* itself.[17]

There are several reasons why Sagittarius A* may exhibit a low degree of linear polarization at cm wavelengths (but see the entirely different situation at mm wavelengths discussed below). If the cause is not intrinsic, then we may be seeing the effects of a scattering screen between the galactic center and Earth. We know scattering occurs in the intervening medium because OH masers within 0.5° of the

---

[15] See Ekers et al. (1975).

[16] See Yusef-Zadeh and Morris (1987).

[17] This conclusion was somewhat surprising in view of the situation with AGNs, which typically exhibit linear polarization at the level of a few percent. See Hughes, Aller, and Aller (1985) and Marscher and Gear (1985).

galactic nucleus—not to mention Sagittarius A* itself (see figure 2.4)—are broadened according to a $\lambda^{+2.0}$ law. Initially polarized radiation may become depolarized in transit through such a screen due either to differential Faraday rotation or to a considerable variation in the Faraday rotation at different points within the screen. The differential Faraday rotation of a homogeneous medium may be so high that within the bandwidth of the observation the polarization vector is rotated by more than 180° and is therefore largely canceling itself. If the intervening medium is inhomogeneous, then rays reaching the observer may get rotated unevenly, which reduces the overall polarization significantly.

Faraday rotation arises in an ionized, magnetized medium, where waves with positive and negative helicity have different indices of refraction. Imagine a circularly polarized wave propagating down a magnetic field line. The charges it encounters along the way are accelerated more easily in a clockwise or counterclockwise sense, depending on their sign. Thus, unless the plasma has equal densities of electrons and positrons, one of these waves propagates faster than the other because it encounters "less resistance." According to equation (3.9), a linearly polarized signal may be written in terms of the basis vectors $\hat{\epsilon}_\pm$, with $|E_{0+}| = |E_{0-}|$. Uneven indices of refraction introduce a growing relative phase shift between the $\hat{\epsilon}_+$ and $\hat{\epsilon}_-$ components of $\mathbf{E}$ and thereby induce a rotation of the position angle χ of the linearly polarized wave.

Fortunately, there is a direct observational technique for determining whether differential, or uneven, Faraday rotation is behind the absence of measurable linear polarization at cm wavelengths in Sagittarius A*. It is not difficult to show that the change in position angle of a linearly polarized signal is

$$\Delta\chi = \text{RM}\,\lambda^2, \qquad\qquad 3.17$$

where

$$\text{RM} \equiv 0.8\, n_e\, B_\parallel\, L \quad \text{rad m}^{-2} \qquad\qquad 3.18$$

is the rotation measure, written in terms of the electron number density n_e (in cm^{-3}), the component of magnetic field $B_\parallel$ (in μG) parallel to the line of sight, and the length L (in pc) over which scattering occurs.[18] Since $\Delta\chi$ varies as λ^2 for a given line of sight, one can determine the RM by measuring the position angle at several different wavelengths for the same source.

---

[18]A straightforward derivation of this result may be found in Rybicki and Lightman (1979).

Now, a linearly polarized signal is significantly depolarized within an observing bandwidth $\Delta \nu$ if χ rotates by more than about 1 rad across that band. And so, from equation (3.17), we see that any rotation measure greater than

$$\text{RM}_\text{depol} \equiv \frac{1}{2\lambda^2} \frac{\nu}{\Delta \nu} \qquad\qquad 3.19$$

dilutes the polarized signal to undetectable levels. This means that for a typical VLA bandwidth of 50 MHz at 4.8 GHz the critical rotation measure RM_depol is $\sim 10^4$ rad m^{-2}, which is deemed to be within reasonable values for the galactic center. To sample larger values of RM, one must therefore split the bandwidth into smaller channels. For example, splitting the band into 256 channels increases RM_depol by two orders of magnitude to 3.5×10^6 rad m^{-2}.

Toward the end of the 1990s, these considerations led to a reappraisal of the older VLA results and a new suite of measurements of the linear polarization fraction in Sagittarius A*; the task was now to determine what value of RM_depol was needed in order to attribute the lack of a measurable signal in this source to simple interstellar scattering effects. The results of continuum polarimetry at 4.8 GHz and spectro-polarimetry at 4.8 and 8.4 GHz indeed confirmed the absence of linear polarization in Sagittarius A* with an even lower upper limit of < 0.1% fractional polarization.[19] But more important, a Fourier transform of the spectro-polarimetric data to sample multiple rotations of the polarization vector across the entire band showed that Faraday rotation could be *excluded* as the cause of this tiny polarization fraction for RM values up to 10^7 rad m^{-2}. The observed RM in the galactic center—as well as its variations—is therefore many orders of magnitude below what is required to depolarize Sagittarius A* by either differential or uneven Faraday rotation.[20] It appears that the cause of Sagittarius A*'s low degree of linear polarization at cm wavelengths is intrinsic to the source.

Imagine how surprising it was, then, to hear within a few years of these observations[21] that the linear polarization of Sagittarius A* at mm and sub-mm wavelengths had been measured at a level of

---

[19] These results were reported in a series of papers from 1999 through 2003 by Geoffrey Bower and his collaborators.

[20] See, e.g., Yusef-Zadeh, Wardle, and Parastaran (1997).

[21] See Aitken et al. (2000).

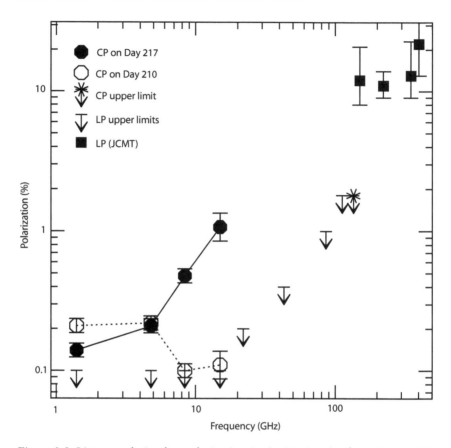

Figure 3.2 Linear and circular polarization in Sagittarius A* from 1.4 to 400 GHz. The *down* arrows indicate upper limits for the linear polarization measurements. The filled squares represent the 10%+9%−4% detection of linear polarization at 0.75, 0.85, 1.35, and 2 mm by Aitken et al. (2000). Note that the position angle changed by ~ 80° (not shown here) between the mm and sub-mm portions of the spectrum. Circular polarization measurements are shown as *open* octagons (from the VLA on July 28, 1999) and *filled* octagons (from the VLA on August 5, 1999). (Image from Bower et al. 2002)

~10%. (A summary of these data, including the circular polarization we shall discuss next, is provided in figure 3.2.) From the lack of linear polarization at longer wavelengths, it is natural to conclude that these large measured values at higher frequencies must arise in a separate component, strengthening the view espoused in the previous section that the mm/sub-mm spectral excess suggests the presence of two separate species of emitting particles. These observations also point to the tantalizing result that the position angle χ changes considerably (by about 80°) in the transition from mm to sub-mm wavelengths, which one would think must surely have something to do with the fact that the emitting gas becomes transparent in this portion of the spectrum.

Still, the mm/sub-mm polarization measurements were taken with the SCUBA camera at the James Clerk Maxwell Telescope (JCMT), with a relatively poor resolution (22″ at 220 GHz). The possibility of confusing Sagittarius A*'s flux with that from dust emission in the surrounding circumnuclear disk (color plate 6) and from the minispiral (color plate 3) in Sagittarius A West means that a substantial fraction of polarized and unpolarized radiation must be subtracted from the overall flux detected in this beam. Indeed, it appears that as much as 3.55 Jy of (unpolarized) free-free emission and 0.75 Jy of dust emission at 220 GHz contributes to the central 22″ beam. This leaves 2.2 ± 0.5 Jy for Sagittarius A*, and so the ~10% linear-polarization fraction based solely on these results should be viewed as tentative at best.

However, there is no longer any doubt that the ~10% linear-polarization fraction in Sagittarius A* at mm/sub-mm wavelengths is real. This is based on follow-up interferometric observations by the BIMA array.[22] With a resolution of 3.6×0.9 arcsec^2 at 230 GHz, these measurements have detected an unambiguous polarized signal at the level of $7.2\% \pm 0.6\%$, clearly ruling out any contribution from dust and affirming the synchrotron nature of the source associated with Sagittarius A*. The puzzle, though, is why the position angle ($139° \pm 4°$) measured by BIMA does not agree with that ($89° \pm 3°$) inferred earlier with the JCMT. Polarized dust emission cannot account for this discrepancy, since it would take ~40% of the estimated 0.75 Jy flux density to be polarized in a position angle of 80°. This high polarization fraction is inconsistent with other measured polarization fractions at the galactic center, leaving variability and observational error as the sole explanations.

Interestingly, there does in fact exist some evidence that the small disk feeding the supermassive black hole at the galactic center may be precessing on a timescale of about 100 days—presumably the consequence of a very small black hole spin. (This topic will be covered fully in chapters 4 and 9.) Thus, if the mm/sub-mm radiation originates from within this disk or some structure coupled to it, a time-dependent position angle

---

[22] The latest interferometric observations have been reported by Bower et al. (2003).

associated with Sagittarius A*'s mm/sub-mm linearly polarized flux may be a tracer of that motion.[23]

Circular polarization, on the other hand, is not well understood in AGNs, rendering it a more difficult tool than linear polarization to use with Sagittarius A*. When detected, the degree of circular polarization is typically <0.1%, rarely approaching 0.5%. It peaks near 1.4 GHz and decreases rapidly with increasing frequency.[24] It was therefore quite surprising to find that the degree of circular polarization in Sagittarius A* is quite high, $-0.37\% \pm 0.04\%$ at 4.8 GHz and $-0.26\% \pm 0.06\%$ at 8.4 GHz, with an average spectral index $\alpha = -0.6 \pm 0.3$ between these two frequencies.[25]

It may be that the circular polarization is intrinsic to the synchrotron emission itself.[26] Alternatively, circular polarization may result from one or more propagation-related mechanisms, including circular *repolarization*, a process that converts linear to circular polarization in either a cold plasma or an electron-positron-pair–dominated plasma.[27] Circular polarization may also be induced by scintillation in the interstellar medium, which, however, predicts a very steep spectrum ($\alpha = -4$ or steeper) that does not appear to be consistent with the measured spectral index.[28]

By now it should already be evident that discerning how Sagittarius A* produces its polarized light constitutes an essential step in our theoretical study of this object. However, to conduct this investigation effectively, we need to have a better idea of where Sagittarius A*'s various spectral components originate. Most of chapter 7 is dedicated to this topic. We will therefore now shift our attention to observations of Sagittarius A* at other wavelengths, and we will return to the polarization problem at the appropriate juncture four chapters hence.

---

[23] The suggestion that Sagittarius A*'s long-term variability at cm wavelengths may be due to a disk precession, which itself is due to a small black hole spin, was made by Liu and Melia (2002b).

[24] See, e.g., Weiler and de Pater (1983) and Rayner, Norris, and Sault (2000).

[25] These results were first obtained with the VLA and reported by Bower et al. (1999a); they were confirmed soon thereafter by Sault and Macquart (1999), using the Australia Telescope Compact Array at 4.8 GHz.

[26] See Legg and Westfold (1968) and Epstein (1973).

[27] The former is discussed in Pacholczyk (1973), the latter in two papers by Jones and O'Dell (1977a, 1977b).

[28] Scintillation was discussed toward the end of chapter 2. This effect, which requires a scattering region with a fluctuating rotation measure gradient, has been studied in detail by Macquart and Melrose (2000).

3.3 INFRARED OBSERVATIONS

Unlike the situation at wavelengths longward of ~1 mm that we have just considered, Sagittarius A* has been rather difficult to detect in the infrared. In fact, a confirmed measurement of its ~1–8 μm flux has been reported only as recently as 2003. Prior to that, the tightest limits on Sagittarius A*'s infrared emissivity had been culled from high-resolution, near-infrared 2 μm observations of the galactic center conducted over a period of six years starting in 1995, using both speckle and adaptive optics imaging techniques on the Keck 10-meter telescopes.[29]

These techniques allow the observer to circumvent the image distortions produced by turbulence in Earth's atmosphere. We touched briefly on the adaptive optics method in §1.2. Speckle imaging is an alternative way of producing diffraction-limited images. Earth's atmosphere usually limits the resolution attainable on astronomical sources to about 0."5, whereas the theoretical limit on the image resolution is set by the size of the telescope's aperture. For example, the diffraction limit at the Keck Telescope is ≈ 0."05, about ten times better than may be achieved without image reconstruction. Speckle imaging requires taking many short exposures that freeze out the effects due to atmospheric turbulence. The individual exposures are then Fourier transformed and averaged in Fourier space to preserve their high-resolution information. The Fourier amplitudes (from the power spectrum) and Fourier phases may be calculated separately and then recombined into the final (relatively undistorted) image.

On the basis of this search, which involved the identification and removal of all the stars in the crowded inner ~ 0."6 × 0."6 of the Galaxy, an upper limit to Sagittarius A*'s 2 μm emission emerged for several epochs of observation. The quiescent emission from Sagittarius A* at this wavelength was constrained to ≤ 0.09 mJy (2.0 mJy, dereddened), whereas any possible variable component was determined to be fainter than 0.8 mJy (19 mJy, dereddened) at the 2σ confidence level.

This experiment was repeated in 2003, using strictly adaptive optics (as opposed to speckle imaging) on the VLT.[30] One of the shortcomings with speckle imaging is that integration times of ≥ 10 minutes (a significant

---

[29]See Hornstein et al. (2002).

[30]These observations, utilizing the newly commissioned VLT adaptive optics imager known as the Nasmyth Adaptive Optics System (NACO), were reported by Genzel et al. (2003a).

fraction of the apparent variation timescale seen subsequently in this source) are required to confidently detect Sagittarius A* at the now-measured flux densities. It is also evident that Sagittarius A* may have become more active in the infrared over the past decade. Whatever the cause (or causes) for the disparity between the two sets of measurements, the latest round of observations with the VLT has finally determined an infrared flux for Sagittarius A*, not only in quiescence and not only in K_s-band (2.15 μm), but also in a flaring state and also in H-band (1.65 μm) and L'-band (3.76 μm).

The quiescent infrared source coincident (to within 20–30 mas) with the position of Sagittarius A* is spatially unresolved and exhibits a twenty- to thirty-minute scale, small-amplitude (20%–60%) variability. Wire-grid K-band polarimetry also indicates that this quiescent source is significantly more polarized than the average star in its immediate vicinity, though more calibration needs to be carried out before a precise polarization fraction and position angle can be determined.

The inferred 1.6–3.8 μm spectral energy density (SED) for Sagittarius A* has a power-law index of -2.1 ± 1.4, placing it approximately on the extrapolation of the mm/sub-mm bump toward high energies (as we shall see in the next section). The actual dereddened flux densities measured in the 2003 VLT observations are 2.8 mJy (H-band), 2.7 mJy (K_s-band), and 6.4 mJy (L'-band). Adopting these values, one must remember that at near-infrared wavelengths (1–2 μm) a clear detection of an associated point source is always made difficult by the potential confusion with stellar sources. In the 3–5 μm window, on the other hand, the stellar and dust emission (which dominates at midinfrared wavelengths ≥ 8 μm) are both declining, potentially facilitating the unambiguous detection of a radiative counterpart in this portion of the spectrum.

This was the principal reason in fact why in its own 2003 retry of the experiment to measure Sagittarius A*'s infrared flux the UCLA group chose to conduct the search primarily in the L'-band.[31] This set of observations, made with the Keck II 10-meter telescope's adaptive optics system, uncovered a variable point source coincident to within 6 mas with the supermassive black hole. In earlier observations, the L' counterpart to Sagittarius A* had apparently blended in with the star S0-2, which had experienced its closest approach to the nucleus,[32] with a

---

[31] See Ghez, Wright et al. (2004).
[32] See, e.g., Schödel et al. (2002).

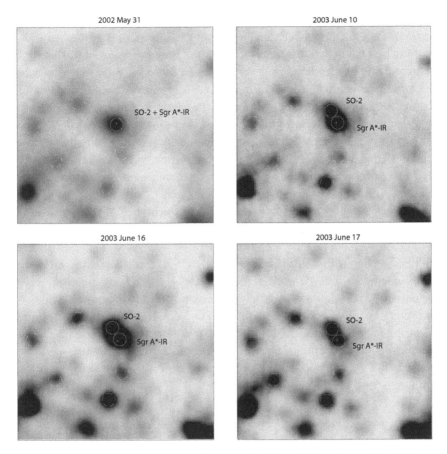

2002 May 31 2003 June 10

SO-2 + Sgr A*-IR

SO-2

Sgr A*-IR

2003 June 16 2003 June 17

SO-2

Sgr A*-IR

SO-2

Sgr A*-IR

Figure 3.3 L' (3.8 μm) images of the central 1.″2 × 1.″2 region of the Galaxy obtained on the W. M. Keck II 10-meter telescope between May 31, 2002, and June 17, 2003. The cross denotes the dynamically determined position of the central black hole (see chapter 5) and its uncertainties. An L'-source, called Sgr A*-IR, is coincident with this position at all times. The five comparison stars are circled with dashed lines. (Image from Ghez et al. 2004)

projected separation of a mere 14 mas and a proper motion of 170 mas yr^{-1}, or equivalently 6, 600 km s^{-1} (see figure 3.3).

The object S0-2 is one of a handful of stars whose orbits are so tight about Sagittarius A* that their trajectories have been mapped with startling precision. We shall revisit them in chapter 5, where we will see how their orbital kinematics may be used to determine not only Sagittarius A*'s location but also its mass with unprecedented accuracy. In 2003, S0-2 was 87 mas away from the dynamical center of the Galaxy,

so the identification of a new L′ source at the location of Sagittarius A*
is unambiguous. It has been given the name Sagittarius A*-IR.

Matched to the five comparison stars identified in figure 3.3, this new
source distinguishes itself in several important ways. First, its projected
location is coincident with Sagittarius A* to within 6 mas. It appears to
be stationary, with an upper limit to its transverse motion of 300 km
s^{-1}, significantly lower than the proper motions ($> 6,000$ km s^{-1}) of
any of the other sources that have come within 10 mas of the dynamical
center (see chapter 5). It is also a variable source, with large flux-density
changes observed on timescales as short as forty minutes. Sagittarius A*-
IR's measured L′ flux density in June 2003 varied between 4 and 17 mJy
(dereddened) and was therefore consistent with the VLT observations
made during the same period.

The most exciting aspect of these infrared observations is clearly the
implied strong variability of Sagittarius A*. Indeed, we have not even
mentioned yet that flares detected during the VLT monitoring were
apparently modulated with a characteristic period of about seventeen
minutes. For a black hole mass of 2.6–$3.6 \times 10^6 \, M_\odot$, this places the
emitting region precariously close to the event horizon. We shall devote a
major section of chapter 4 to a discussion of Sagittarius A*-IR's variable
nature, and we will address the strong-field implications of the seventeen-
minute period in chapter 8.

3.4 X-ray Observations

The earliest X-ray observations of the central region surrounding Sagittarius A* were carried out with rocket- and balloon-borne instruments,
but it was left to *Einstein*, the first satellite equipped with X-ray imaging
optics, to conduct the initial detailed study.[33] Its Imaging Proportional
Counter (IPC) produced images with an angular resolution of $\sim 1'$
and revealed a dozen discrete sources within the central square degree
centered on Sagittarius A*. It was duly noted at the time that the
brightest of these sources (labeled 1E 1742.5–2859) was centered only
20″ from the position of Sagittarius A*, though no variability had been
detected in its emission over a six-month period. The first estimate
of what was thought to be Sagittarius A*'s X-ray power—that associated with 1E 1742.5–2859—was made assuming an absorbed thermal

---

[33] A review of the early X-ray missions may be found in Skinner (1989), and a description
of the *Einstein* observations is provided in Watson et al. (1981).

bremsstrahlung spectrum with $kT = 5$ keV and a corresponding column density $N_H = 6 \times 10^{22}$ cm^{-2}, which yielded an absorption-corrected 0.5–4.5 keV luminosity of 9.6×10^{34} ergs s^{-1}.

But as the angular resolution of X-ray instruments improved over the next two decades, it became increasingly clear that source confusion at the galactic center persisted as the most serious detriment toward a clear identification of Sagittarius A*'s X-ray counterpart. In 1994, ROSAT re-observed the galactic center[34] with a much better spatial resolution (10″–20″ FWHM) and this time resolved 1E 1742.5–2859 into three separate sources, one of which—RX J1745.6–2900—is coincident with the radio position of Sagittarius A* to within 10″. Unfortunately, ROSAT was designed to measure source intensity in the soft energy band (0.1–2.5 keV), where line-of-sight absorption toward the galactic center is particularly severe. These measurements could provide only a very limited capability for determining Sagittarius A*'s X-ray spectrum and luminosity.

The situation is much better in the hard X-ray (2–30 keV) portion of the spectrum, where the interstellar medium becomes optically thin. In the mid-1990s, several new satellites took advantage of this feature to refine our X-ray view of the galactic nucleus. For example, the ART-P telescope[35] on GRANAT uncovered a long-term variable source near the position of Sagittarius A* with a measured 3–10 keV luminosity of $(2–10) \times 10^{35}$ ergs s^{-1}. But to render this result consistent with the ROSAT spectrum, one had to assume a significant absorption of the soft X-rays, which requires a column density $N_H \approx (1.5–2) \times 10^{23}$ cm^{-2}. This is roughly three times larger than the value inferred from IR observations of nearby stars, calling into question the tentative identification of a common source for both the soft and hard X-rays.

This problem with source confusion emerged again when the Japanese satellite *ASCA* imaged the central region up to 10 keV with a spatial resolution of 3′. Part of the difficulty is the relatively strong diffuse thermal emission ($kT \approx 10$ keV) pervading the central square degree (see figure 1.10). Indeed, the *ASCA* measurements[36] showed that the 2–10 keV luminosity of the hot gas within the $2' \times 3'$ elliptical shell of Sagittarius A East is $\sim 10^{36}$ ergs s^{-1}, after correcting for the measured absorption corresponding to a column density $N_H \approx 7 \times 10^{22}$ cm^{-2}.

---

[34] See Predehl and Trümper (1994).

[35] See Pavlinsky, Grebenev, and Sunyaev (1994).

[36] The galactic center was observed twice with *ASCA*, first in 1993 (see Koyama et al. 1996) and then again in 1994 (see Maeda et al. 1996).

Without subtracting for any variable local background, *ASCA* could therefore only place an upper limit of $\sim 10^{36}$ ergs s^{-1} to Sagittarius A*'s hard X-ray luminosity.

To complicate matters further, the *ASCA* observations revealed the presence of a new low-mass X-ray binary named AX J1745.6–2901, a mere 1'.3 away from Sagittarius A*. The appearance of X-ray bursts and eclipses with a period of 8.356 hours from the hard component in this binary established it as an accreting neutron star, which at the time was producing a variable X-ray flux between 1×10^{-11} and 4×10^{-11} ergs cm^{-2} s^{-1}, roughly the same range of variability reported earlier from the Sagittarius A* region.

It is clear, then, that any X-ray fluxes measured by instruments with insufficient spatial resolution (i.e., certainly those without the capability to resolve sources a few arcseconds apart) and attributed to Sagittarius A* are subject to significant contamination by AX J1745.6–2901 and other nearby objects.

This situation changed dramatically in 1999 following the space shuttle launch of the *Chandra* X-ray Observatory, which has without a doubt had a profound influence on our understanding of the galactic center. Formerly known as the Advanced X-ray Astrophysics Facility (AXAF), this state-of-the-art detector received its latest appellation in honor of the late Indian American Nobel laureate Subrahmanyan Chandrasekhar. In Sanskrit, the nickname Chandra means "Moon" or "luminous," a very fitting name for this mission in view of Chandrasekhar's tireless devotion to the pursuit of truth. His career spanned many decades, during which he made fundamental contributions to the theory of black holes and other phenomena now being studied by the *Chandra* X-ray Observatory. He is widely regarded as one of the foremost astrophysicists of the twentieth century, winning the Nobel Prize in 1983 for his theoretical work on the physical processes that govern the structure and evolution of stars. *Chandra's* unprecedented ability to distinguish features barely one-twentieth of a light-year across at the distance to the galactic center is providing X-ray images that are fifty times more detailed than those of any previous mission. At almost 14 meters in length and weighing more than 45 metric tons, it is also one of the largest objects ever placed in Earth's orbit by the space shuttle.

The X-ray map of the inner $1' \times 1'$ of the Galaxy produced by *Chandra*[37] is shown in figure 3.4. Each pixel's intensity is based on

[37] See Baganoff et al. (2003).

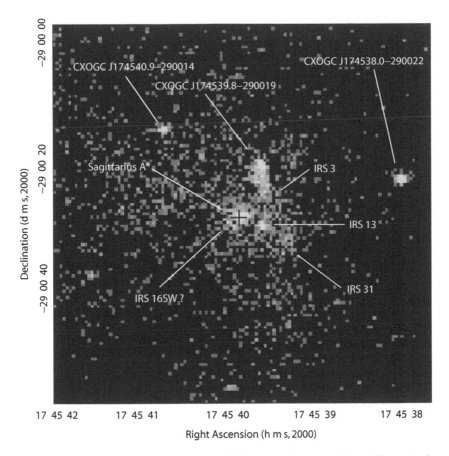

Figure 3.4 *Chandra* 0.5–7 keV image of the central $1' \times 1'$ of the Galaxy. Each pixel subtends a solid angle of $0.''5 \times 0.''5$ on the sky. The radio position of Sagittarius A* is marked with a black cross, which lies superposed on the X-ray counterpart. Also shown in this image are the tentative identifications of IR sources. The question mark next to IRS 16SW indicates the possible detection of an X-ray source that may simply be coincident with IRS 16SW within $1''$–$2''$. Two of the brightest sources in this field—CXOGC J174540.9–290014 and CXOGC J174538.0–290022—appear to be new X-ray binaries. (Image from Baganoff et al. 2003)

counts within the 0.5–7 keV band. At the very center of this image, the source CXOGC J174540.0–290027 is coincident with the radio-interferometric position of Sagittarius A* within $0.''35$, corresponding to a maximum projected distance of 16 light-days. Based on the number of sources (i.e., 143) detected by *Chandra* within a circle of radius $8'$ at the galactic center, one can estimate a probability of 5.6×10^{-3} that a randomly detected absorbed X-ray source as bright or brighter than CXOGC J174540.0–290027 would lie within any given circle of radius 0.″35. Given this unlikelihood, CXOGC J174540.0–290027 must almost certainly be the X-ray counterpart to Sagittarius A*.

A smoothed, false-color version of figure 3.4 is shown in the 1.′3 × 1.′5 close-up of the region surrounding Sagittarius A* (the red dot at $17^h45^m40.0^s$, $-29°00'28''$) in color plate 8, overlaid with VLA 6 cm contours of the radio intensity in color plate 3. The western boundary of the brightest diffuse X-ray emission (shown in green) coincides very well with the shape of the Western Arc of thermal emission from Sagittarius A West, whereas the emission on the eastern side continues smoothly into the heart of Sagittarius A East (compare this image with figure 1.8). The Western Arc is thought to be the ionized inner edge of the circumnuclear ring of molecular material orbiting about the galactic center, so the morphological similarities between the X-ray and radio features strongly suggest that the brightest X-ray-emitting plasma is confined by the western portion of this ring. In §7.1 we shall discuss the likely origin of this diffuse X-ray emission being wind-wind collisions within the inner few parsecs of the Galaxy, and we will analyze the implications this may have on the accretion of matter onto Sagittarius A*.

It is important to note that spatial analysis of the morphology of the source coincident with Sagittarius A* and two nearby point sources (CXOGC J174538.0–290022 and CXOGC J174540.9–290014) indicates that the X-ray emission from Sagittarius A* may be slightly extended (see figure 3.4). *Chandra's* on-axis high-resolution mirror assembly has a FWHM point-spread function ≈ 0.″5, whereas the apparent intrinsic size of CXOGC J174540.0–290027 is ≈ 1″ or about 0.04 pc at the distance to the galactic center. As we shall see in §7.1, structure in Sagittarius A* on this scale may be consistent with the radius (1″–2″) at which matter is captured by the black hole and begins its hydrodynamic infall toward the center.[38]

Extracting a spectrum from the X-ray photon flux produced by Sagittarius A* is difficult even for *Chandra* because of the unusually low luminosity of this source. A total of only 258 counts were detected between 0.5 and 9 keV from a 1.″5-radius circle centered on CXOGC J174540.0–290027. This aperture is large enough, given the point-spread function, to encircle more than 85% of the energy radiated by a point

---

[38] See also Melia (1992a, 1994) and Quataert (2002).

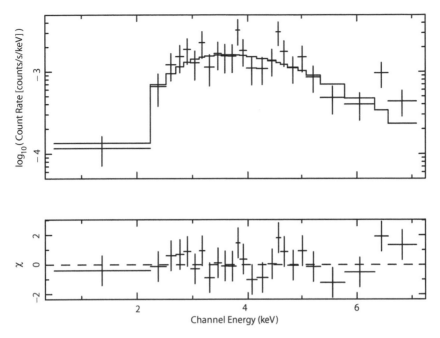

Figure 3.5 The X-ray spectrum of Sagittarius A* obtained with *Chandra*. The solid line in the upper panel represents the best-fit absorbed power-law model, with $N_H = 9.8^{+4.4}_{-3.0} \times 10^{22}$ cm^{-2} and a photon spectral index $\Gamma = 2.7^{+1.3}_{-0.9}$. The 2–10 keV absorption-corrected luminosity is $2.4^{+3.0}_{-0.6} \times 10^{33}$ ergs s^{-1}. The χ^2/d.o.f. for this fit is 19.8/22, and the residuals are shown in the lower panel. (Image from Baganoff et al. 2003)

source at its center, yet small enough to minimize contamination from several nearby sources. The latter is particularly important in view of the difficulty encountered by all earlier missions to clearly identify Sagittarius A*'s X-ray flux.

With such a low count rate, finding a unique spectral model is not feasible. In this case, both a power-law spectrum and an absorbed optically thin thermal plasma model fit the data rather well. Figure 3.5 compares the observed energy-dependent count rate with that expected for an absorbed power-law distribution, characterized by a photon index $\Gamma = 2.7^{+1.3}_{-0.9}$ (where $N[E] \propto E^{-\Gamma}$ ph cm^{-2} s^{-1} keV^{-1}) and an inferred column density $N_H = 9.8^{+4.4}_{-3.0} \times 10^{22}$ cm^{-2}. However, a fit of similar quality (specifically, χ^2/d.o.f. $= 16.5/22$) is obtained for an optically thin bremsstrahlung emitter in which the abundances are twice solar and $kT = 1.9^{+0.9}_{-0.5}$ keV, with $N_H = 11.5^{+4.4}_{-3.1} \times 10^{22}$ cm^{-2}. In either case,

Sagittarius A*'s absorption-corrected luminosity in the 2–10 keV band is $2.4^{+3.0}_{-0.6} \times 10^{33}$ ergs s^{-1}.

We can now begin to understand why Sagittarius A* has been so difficult to study at high energy. Its X-ray luminosity is significantly fainter than that expected of a solar mass neutron star, let alone a four million solar mass behemoth. The maximum rate at which a compact object may accrete is dictated by the efficiency with which dissipated gravitational energy is converted into radiation. The interaction between these escaping photons and the infalling plasma retards the flow and can—given the right circumstances—actually lead to a temporary expulsion of the matter rather than its accretion. Of course, geometry plays a role as well, but one may infer the following characteristic value assuming simple spherical symmetry.

When the infalling gas is fully ionized, the radiative force on the electrons greatly exceeds that on the protons because of the significant disparity in their scattering cross sections. However, Coulomb forces between the charges strongly couple these particles, so the effective system interacting with the radiation may be thought of as an (unbound) electron-proton pair (assuming mostly hydrogen abundance). The gravitational force on this couplet is then

$$f_{\text{grav}} = -\frac{GM(m_p + m_e)}{r^2}, \qquad 3.20$$

where M is the mass of the compact object and m_p and m_e are, respectively, the proton and electron masses.

As we shall see in chapter 6, a photon with energy ϵ carries a momentum ϵ/c, so the momentum flux of radiation with energy flux F is

$$\Pi \equiv \frac{F}{c}. \qquad 3.21$$

Thus, the radiative force on an electron-proton pair is

$$f_{\text{rad}} = \Pi \times \sigma_{\text{T}} = \left(\frac{F}{c}\right) \times \sigma_{\text{T}}, \qquad 3.22$$

where we have explicitly used the Thomson scattering cross section (ignoring the cross section due to the proton) to represent the area through which the momentum flux Π must pass as the radiation impacts the particles. And so, for a point source of radiation with luminosity

$L = 4\pi r^2 F$, we may also write

$$f_{\text{rad}} = \frac{\sigma_T L}{4\pi \, cr^2}.$$ 3.23

Presumably, the maximum rate at which a compact object may accrete is set by the condition

$$f_{\text{grav}} + f_{\text{rad}} = 0.$$ 3.24

It should not surprise us that both f_{grav} and f_{rad} scale as r^{-2}; after all, any influence emanating from a point source drops off inversely as the surface area of a sphere in three-dimensional space. The limit on L, known as the Eddington luminosity, is therefore independent of radius:

$$L_{\text{Ed}} \equiv \frac{4\pi c \, GM \, (m_p + m_e)}{\sigma_T}.$$ 3.25

For a neutron star (with $M \approx 1 \, M_\odot$), $L_{\text{Ed}} \approx 1.3 \times 10^{38}$ ergs s^{-1}, already far greater than Sagittarius A*'s X-ray luminosity ($\approx 2 \times 10^{33}$ ergs s^{-1}) estimated by *Chandra*. Indeed, the black hole at the galactic center appears to be radiating at a rate $\sim 10^{10}$ times below the Eddington (i.e., maximal) value for an object with mass $M \approx 3 \times 10^6 \, M_\odot$. We will revisit this surprising observational conclusion later, particularly in chapter 7.

3.5 OBSERVED HIGH-ENERGY CHARACTERISTICS

In the spring of 2003, the then recently deployed International Gamma-Ray Astronomy Laboratory (INTEGRAL) began its deep exposure of the galactic center, focusing on the inner 10′–15′ region. The maps shown in figures 3.6 and 3.7 were constructed from more than 500 individual "snapshots" taken over the course of almost one million seconds of observation. In the first of these images—a 20–40 keV intensity map—we see what appear to be seven distinct sources, including one coincident with Sagittarius A*. Of these, 1E 1740.7−2942 is a black hole candidate and microquasar,[39] KS 1741−293 and 1A 1742 − 294 are neutron-star

---

[39] As the name suggests, this source produces twin jets of relativistic plasma reminiscent of quasars. Like many of its more distant and powerful brethren, 1E 1740.7−2942 also possesses features in its overall radio intensity map that move across the sky superluminally. See, e.g., Mirabel and Rodriguez (2002).

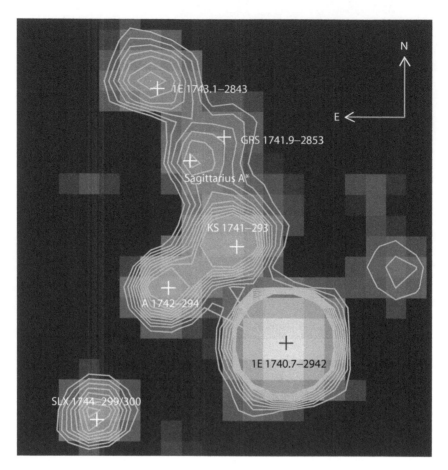

Figure 3.6 A 20–40 keV image of the central 2° × 2° of the Galaxy seen by the IBIS/ISGRI instrument on INTEGRAL. Each image pixel size is equivalent to about 5 arcmin or roughly 12 pc. The ten contour levels mark isosignificance linearly from about 5σ to 15σ. Of the seven distinct sources seen here, one appears to be coincident with Sagittarius A*, though its contours are elongated toward the nearby object GRS 1741.9–2853, which is probably contributing to the overall 20–40 keV flux seen from the center. (Image from Bélanger et al. 2004)

Low-Mass X-ray Binary (LMXB) burster systems, SLX 1744–299/300 are, in fact, two LMXBs that cannot be resolved with INTEGRAL's angular resolution of 12′ (FWHM), and 1E 1743.1–2843 is an X-ray source whose nature is still uncertain.

In the 20–40 keV energy band, contours of the central emitting region peak at the position of Sagittarius A* but are elongated toward

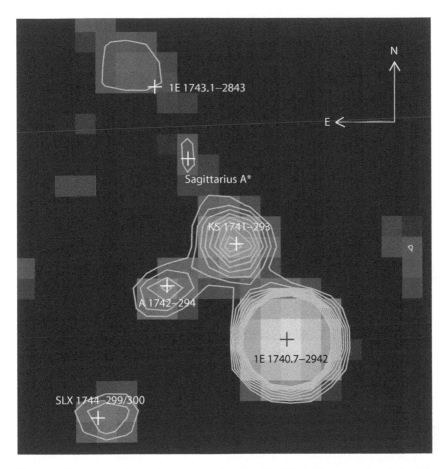

Figure 3.7 Same as figure 3.6, except for the 40–100 keV energy band. In this case, the central source coincident with Sagittarius A* is marginally visible at a level of 5.3σ but without any contribution from the direction of GRS 1741.9–2853. (Image from Bélanger et al. 2004)

GRS 1741.9–2853, suggesting that the latter may be contributing to the former's overall high-energy emission. On the other hand, the source is also marginally visible in the 40–100 keV band, though without any apparent contribution from the direction of GRS 1741.9–2853.

Unfortunately, it is not yet clear how much—if any—of the high-energy radiation seen from the central source, called IGR J17456 − 2901, is actually associated with Sagittarius A* itself. The best fit position for IGR J17456 − 2901 is $17^{\mathrm{h}}45^{\mathrm{m}}38.^{\mathrm{s}}5$, $-29°01'15''$, which lies within

0.'9 (roughly 2 pc) of Sagittarius A*, though one must remember that INTEGRAL's angular resolution is 12' (FWHM). Its 20–100 keV luminosity is $(2.89 \pm 0.41) \times 10^{35}$ ergs s^{-1}, two orders of magnitude greater than Sagittarius A*'s X-ray power in the 2–10 keV band.

A known point source lying only 1.'3 from Sagittarius A* that could be contributing to this high-energy signal is AX J1745.6 − 2901, the eclipsing binary and burster detected by ASCA in the mid-1990s. Though it was seen in a high state of emissivity only once, an extrapolation of its flux then would have been compatible with that of IGR J17456 − 2901. However, AX J1745.6 − 2901 had not been detected by either XMM-*Newton*[40] or *Chandra* in the three years prior to the INTEGRAL observation, in spite of their much greater sensitivity compared with ASCA. In addition, the INTEGRAL data did not reveal any statistically significant modulation in IGR J17456 − 2901's 20–40 keV flux, contrary to what might be expected on the basis of AX J1745.6 − 2901's 8.356-hour period. It is therefore unlikely that IGR J17456 − 2901's flux can be fully attributed to AX J1745.6 − 2901.

Nonetheless, it is difficult to avoid the conclusion that IGR J17456 − 2901 is probably a superposition of several sources lying within an arcminute or two of Sagittarius A*. To appreciate this point, one needs only to realize that on at least one occasion (June 19, 2003) during INTEGRAL's long exposure, *Chandra* detected three or four sources with a combined 2–8 keV intensity more than thirty times that of Sagittarius A* itself. As of now, we have no choice but to wait for future 20–100 keV observations of the galactic center with improved spatial resolution before we can say anything definitive regarding Sagittarius A*'s spectrum in this energy range.

At higher energies still, the latest observations of Sagittarius A* are perhaps even more tantalizing than those at $\sim$100 keV. To date, the

---

[40]Launched in 1999 by the European Space Agency, the XMM-*Newton* mission was so named because of its X-ray Multi-Mirror design and to honor Sir Isaac Newton, the person who first introduced spectroscopy to the scientific community. Formally known as the High Throughput X-ray Spectroscopy Mission, it was designed to to detect X-rays in abundance and quickly in order to acquire accurate spectra. Its energy coverage is 0.1–12 keV, and its barrel-shaped Mirror Modules, each containing fifty-eight "Wolter-type I" wafer-thin concentric mirrors, produce a spatial resolution of 5" (FWHM). In addition to confirming each other's principal observational results, XMM-*Newton* and *Chandra* work in a complementary fashion, the latter providing exquisite imaging power, while the former produces spectra with excellent energy resolution.

galactic center has been identified by three air Cerenkov telescopes[41] in the $\sim$ TeV range: Whipple, CANGAROO, and most recently and significantly, HESS. The Hegra ACT instrument[42] has also put a (weak) upper limit on galactic center emission at 4.5 TeV, and the Milagro water Cerenkov extensive air-shower array[43] has released a preliminary finding of a detection at similar energies from the region defined as $l \in \{20°, 100°\}$ and $|b| < 5°$.

HESS detected a signal from the galactic center in observations conducted over 2 epochs (June–July 2003 and July–August 2003) with a $\sim 6\sigma$–9σ excess evident over this period. These data can be fitted by a power law with a spectral index of 2.21 ± 0.09 and normalization of $(2.50 \pm 0.21) \times 10^{-8}$ m^{-2} s^{-1} TeV^{-1} with a total flux above the instrument's 165 GeV threshold of $(1.82 \pm 0.22) \times 10^{-7}$ m^{-2} s^{-1}. There is also a 15%–20% error from energy resolution uncertainty.

In chapter 1, we discussed the EGRET source 3EG J1746–2851 and considered whether Sagittarius A East might be the origin of these gamma rays at GeV energies. It would be difficult to reconcile the HESS detection with this object for several reasons. The first is that a careful analysis of data from the EGRET 3EG catalog[44] has shown that the galactic center may be excluded at the 99.9% confidence limit as the true position of 3EG J1746–2851. The HESS source, on the other hand, is coincident within $\sim 1'$ of Sagittarius A*, though its centroid is displaced roughly $10''$ (corresponding to ~ 0.4 pc) to the east of the galactic center. In addition, an extrapolation of the EGRET spectrum into HESS's range overpredicts (by a factor of about 20) the TeV gamma ray flux of the GC source. We are apparently dealing with two separate emission regions. Perhaps both are associated with the supernova remnant Sagittarius A East, but given

---

[41] Air Cerenkov telescopes incorporate Earth's atmosphere into the overall detector design. Energetic photons create electron-positron pairs at high altitude; these continue to interact on their way down, emitting secondary energetic photons via bremsstrahlung and Compton scattering. The ensuing cascade grows until the charges eventually run out of energy. Particles in this shower travel faster than the speed of light in air and consequently emit a faint, bluish light known as Cerenkov radiation, after the Russian physicist who made comprehensive studies of this phenomenon. The photons in this glow typically form a pancakelike front, some 200 meters in diameter and only about 1 meter thick. The mirrors in the telescope on the ground capture this light and feed it to a central detector. See Kosack et al. (2004), Tsuchiya et al. (2004), and Aharonian et al. (2004).

[42] See Aharonian et al. (2002).

[43] See Fleysher and the Milagro Collaboration (2002).

[44] See Hartman et al. (1999).

its angular proximity to Sagittarius A*, the TeV source could somehow be associated with the black hole itself.

Toward the end of chapter 7, we will examine the likelihood that these TeV photons are being produced by relativistic particles escaping the stochastic acceleration site near Sagittarius A*. Proton-proton collisions in the medium surrounding the black hole induce neutral pion decays, which, subject to certain observationally motivated restrictions, can produce TeV gamma rays in numbers detectable by HESS.

CHAPTER 4

Variability

On **Wednesday, September 5, 2001**—six days before the attack on the World Trade Center in New York City—representatives of the media gathered at the Washington Plaza Hotel for a press conference on an unusual and exciting discovery made by *Chandra* during the previous year's set of observations. This date was chosen to coincide with the release of *Nature*'s cover story that week,[1] headlined "Galactic Centre Flares Up."

The event reported in this article was the detection of a short-duration highly variable X-ray flare produced by Sagittarius A*. This transient phenomenon was over in a couple of hours, but while it lasted, about forty-five times as many X-rays were being emitted per second as are usually produced in Sagittarius A*'s quiescent state (see figure 4.1; see also figure 3.5 for Sagittarius A*'s spectrum in the nonflaring state). Even more startling was the realization that, during the flare, the X-ray output dropped abruptly by a factor of five in less than ten minutes, only to recover just as quickly. This type of variability is rarely seen in the emission from multimillion solar mass objects. It is far more common in solar-size stars accreting gas from their orbiting companion. The implication—that the region where the X-rays were produced is no bigger than 10 light-minutes across, roughly the distance between Earth and the Sun—has impacted our theoretical understanding of this object in several important ways; these will command our attention in the rest of this chapter and several others to follow.

4.1 SHORT-TERM VARIABILITY IN THE IR AND X-RAYS

Very soon after *Chandra*'s first detection of an X-ray flare in Sagittarius A*, XMM-*Newton* followed with its own confirming measurements,

---

[1] See Baganoff et al. (2001) and the accompanying News and Views article by Melia (2001a).

including the discovery of an even more powerful burst several years later. None of the previous X-ray satellites had the sensitivity and spatial resolution to identify such short-duration low-luminosity events at the distance to the galactic center. Now, *Chandra* and XMM-*Newton* were detecting them at a rate of about one per day when pointed in that direction.

The best-fit photon index during the *Chandra* event (figure 4.1) was $\Gamma = 1.3^{+0.5}_{-0.6}$ (where $N[E] \propto E^{-\Gamma}$ ph cm^{-2} s^{-1} keV^{-1}), representing a flattening (i.e., a $\Delta\Gamma$) of about 1 compared to Sagittarius A*'s spectrum in the quiescent state (figure 3.5). The hardness ratio, defined as the hard-band (4.5–8 keV) count rate divided by the soft-band (2–4.5 keV) count rate, increased during the flare (panel c in figure 4.1); the spectrum "hardened" (i.e., became flatter and extended more strongly to higher energies) during the burst. This evolution away from the quiescent state signals either a spectral shift across the X-ray band, possibly due to a change in temperature of the radiating gas, or a variable blending of different emission mechanisms.

In addition, the rise/fall timescale of a few hundred seconds and the ten-minute variability seen within the burst are consistent with the light-crossing and dynamical timescales of an accretion disk within a mere 10 Schwarzschild radii of an object with mass $M \approx 3 \times 10^6 M_\odot$. (We shall define the radius $2GM/c^2$ of a black hole's event horizon more precisely in chapter 6.) Later in this book, we will explore in detail the implications of this constraint and conclude that a "conventional" AGN disk extending out to hundreds of thousands of Schwarzschild radii could not be present in this system.

Yet this is only the first of several important ways in which Sagittarius A*'s X-ray variability distinguishes it from AGNs. It is indeed true that strong, variable X-ray emission is a generic property of galactic nuclei, where we often measure factors of two to three variations in emissivity on timescales of minutes to years. Moreover, moderate- to high-luminosity AGNs (i.e., Seyfert galaxies and quasars) often display a general trend of increasing variability with decreasing luminosity.[2] But this trend does not extend to low-luminosity AGNs, which show little or no significant variability on timescales of less than a day. Sagittarius A* stands out as an oddity, given that it is the nearest and least luminous member of the known AGN community. However, one must remember that flares of similar strength to that seen at the galactic center would be

---

[2] See, e.g., Nandra et al. (1997).

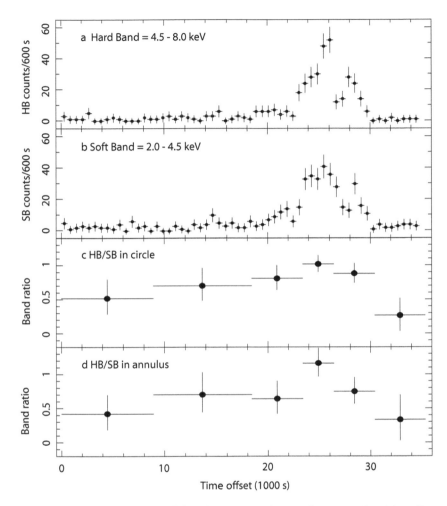

Figure 4.1 X-ray lightcurve of the photon arrival times from a circle with radius 1.″5 in the direction of Sagittarius A* on October 26 and 27, 2000. The x-axis shows the time offset from the beginning of the observation, which lasted about twenty-one hours. Panel (a) shows the count rate in the 4.5–8 keV band, and panel (b) is the corresponding count rate in the 2–4.5 keV band. Panel (c) is the ratio of these two lightcurves, whereas panel (d) is this ratio for counts received from an annulus with inner and outer radii 0.″5 and 2.″5, respectively, rather than from within a circle of radius 1.″5. (Image from Baganoff et al. 2001)

undetectable by *Chandra* (and XMM-*Newton*) in the nucleus of even the nearest spiral galaxy M31 in the constellation of Andromeda. In Sagittarius A*, we may be learning about the behavior of supermassive

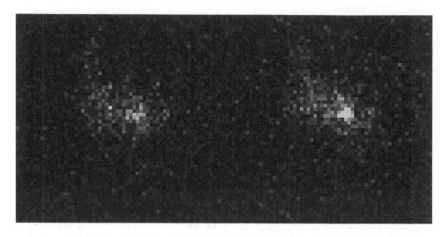

Figure 4.2 Two 2–10 keV images of the central $5' \times 5'$ region surrounding Sagittarius A*, taken by XMM-*Newton* on September 4, 2001. The left panel shows events integrated over the 1,000 seconds before the flare, whereas the right panel shows events integrated over the subsequent 1,000 seconds, which includes the flare. The pixels are binned to a size of $5.''5 \times 5.''5$. The radio source Sagittarius A* is positioned right in the middle of the central bright pixel visible in the flare image (right). (Image from Goldwurm et al. 2003)

black holes accreting at a rate well below that required to convert them into high- or even low-luminosity AGNs.

Of course, one cannot strictly rule out an unrelated contaminating source as the origin of the flare—for example, an X-ray binary along the line of sight—but alternative explanations are unlikely given the size of the stellar cluster surrounding Sagittarius A*. The stars are distributed over $\sim 5''$–$20''$, whereas the source of the *Chandra* flare could be pinpointed to within $1/3''$ of the center. Since this cluster contains up to 1 million solar masses of stars and remnants, it seems highly improbable that it would contain only one very unusual flaring member that would, in addition, be fortuitously superposed right on top of Sagittarius A*.

An association between the flare and Sagittarius A* has since been reaffirmed by the discovery of several other such events, at least one of which has been linked causally to this object's variability at other wavelengths. Within a year of the first discovery, XMM-*Newton* began its own observational campaign of the galactic center and, at the end of a 26-kilosecond exposure, detected a second brightening of the central source, though with somewhat poorer spatial resolution. Figure 4.2 shows images of the central $5' \times 5'$ region integrated over the 1,000 seconds before the flare and the last 1,000 seconds including the event.

An analysis of the bursting source position showed that it is compatible with Sagittarius A* within 1.″5 (i.e., within the residual systematic uncertainties).[3] Since the nearest X-ray point source identified by *Chandra* (the infrared star IRS 13 in figure 3.4) is 4″ away, contamination of the flare by objects other than Sagittarius A* may be excluded.

Interestingly, though, not all the flares fit neatly into a single category. Already with the second flare the inferred spectrum was harder—$\Gamma = 0.7 \pm 0.6$, meaning flatter—than that of the first, and its maximum luminosity was $\sim 4 \times 10^{34}$ ergs s^{-1} versus 1×10^{35} ergs s^{-1} for the first.

One year later (on October 3, 2002), Sagittarius A* surprised the XMM-*Newton* observers yet again, producing the brightest flare seen thus far, with a peak 2–10 keV luminosity of $\approx 4 \times 10^{35}$ ergs s^{-1}. This was an unprecedented factor of 160 greater than the quiescent value, reflecting an enormous change in Sagittarius A*'s accretion profile. There was clearly something very different about this flare compared to the earlier ones, because its spectrum was much softer, with $\Gamma = 2.5 \pm 0.3$. Evidently, the radiative process emitting the burst flux was indistinguishable from that producing the quiescent spectrum, only with a much greater efficiency.

The October 3, 2002, flare in Sagittarius A* was also the first to demonstrate variability correlated with emission at other wavelengths. For several decades now, the VLA has been conducting a weekly monitoring program of Sagittarius A* at ~ 1 cm. An investigation of twenty years of data has found marked outbursts with an amplitude around 0.4 Jy at 23 GHz, with a characteristic timescale of less than twenty-five days.[4] (To gauge the significance of this amplitude, see Sagittarius A*'s overall radio spectrum in figure 3.1.) A separate study, based on 540 daily observations of Sagittarius A* at 2.3 and 8.3 GHz using the Green Bank[5] Interferometer, found a peak-to-peak variability of 0.25 Jy with an RMS (i.e., modulation index) of 6% and 2.5% at 8.3 and 2.3 GHz, respectively. The median spectral index between the two observed frequencies for the whole period was $\alpha = 0.28$ ($S_\nu \propto \nu^\alpha$), varying between 0.2 and 0.4. The spectral index tends to become larger when the flux density in both bands increases, implying that the radio outbursts in Sagittarius A* are more pronounced at higher frequencies.

---

[3] See Goldwurm et al. (2003).
[4] This work was carried out by Zhao, Bower, and Goss (2001).
[5] See Falcke (1999).

One observation[6] of the weekly VLA monitoring program was co-incidentally carried out just half a day after XMM-*Newton* observed the X-ray flare on October 3, 2002. Figure 4.3 shows the flux density measurements at 2 cm, 1.3 cm, and 7 mm, spanning two months centered on October 4, 2002. The flux density of 1.9 ± 0.2 Jy measured at 7 mm exceeds the mean value (1.00 ± 0.01 Jy) by a factor of about 2 (or 4σ), one of the two highest increases observed during the previous three years. The variations in flux density at 1.3 cm and 2 cm were 3σ and $< 1\sigma$, respectively. Given the VLA's sampling period of about one week and the fact that Sagittarius A*'s measured flux density a week before and a week after the outburst was consistent with its mean value, the timescale of the outburst at 7 mm is inferred to be shorter than two weeks. Correspondingly, there were no significant variations in flux density over the approximately one-hour observation, so the radio outburst must have lasted at least this long.

But what can such a coordinated observation at several different wavelengths tell us about the emitting region and the underlying source? The analysis can be quite revealing, and Sagittarius A* will not disappoint us. We won't have the tools to fully address this question until we have developed the relativistic framework in chapter 6. Even now, however, several tantalizing conclusions may be drawn. The October 3, 2002, X-ray flare observed with XMM-*Newton* displayed variability on a 200-second timescale, implying an emitting region no bigger than ~ 5–10 Schwarzschild radii. The radio outburst, on the other hand, varied not at all during the one-hour observation. In addition, the delay between the two events may have been as long as 13.5 hours, this being the temporal gap between the XMM-*Newton* and VLA exposures. It is natural to suppose that the burst started close to the black hole, signaled by an outpouring of intense X-radiation, followed by an expulsion of energetic matter into a larger volume, where it subsequently produced the radio flare.

As we shall see later in this book, there is a telling piece of evidence that strongly favors this interpretation. The burst spectrum measured by the VLA was a power law with index $\alpha \approx 2.4^{+0.3}_{-0.6}$. Had α been 2.0, we could have argued simply on the basis of equation (3.2) that the emitter must have been a blackbody radiating in the Rayleigh-Jeans limit. But even though α was not quite 2.0, it was suggestively close to 2.5, which is the corresponding spectral index produced by an optically thick, nonthermal synchrotron emitter. (This is quite easy to show,

---

[6] See Zhao et al. (2004).

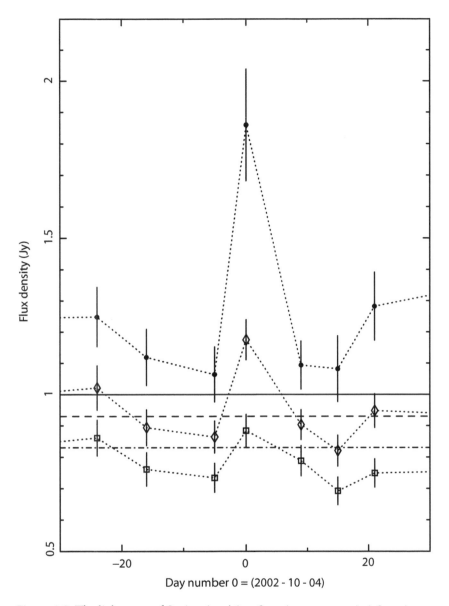

Figure 4.3 The lightcurve of Sagittarius A* at 2 cm (open squares), 1.3 cm (open diamonds), and 7 mm (solid dots) covering a sixty-day span that included the October 3, 2002, X-ray flare detected by XMM-*Newton*. The horizontal lines mark the mean values at 7 mm (solid), 1.3 cm (dashed), and 2 cm (dot-dashed). Note that day number 0 occurred on October 4, 2002. (Image from Zhao et al. 2004)

but we will need the language of special relativity to do so.) The key descriptor here is *optically thick*. For such a source, the luminosity scales directly with the surface area of the emitter, and a region only 5 Schwarzschild radii across is simply too small to have produced the enormous radio flux observed during the burst. Following this train of thought, we will identify the number and type of particle ejected during the event *and* the likely mechanism that accelerated them in the first place. Surely and deliberately, we will unravel the mystery that is the highly energetic environment hovering precariously close to Sagittarius A*'s event horizon.

But before we engage ourselves more intently on the physics behind this fascinating phenomenon, we will consider what is ostensibly the most tantalizing result emerging thus far from Sagittarius A*'s variable light profile. We touched briefly on the infrared observations in §3.4, but we have yet to expound on the apparent detection of a periodic signal from this source. Theorists have long anticipated that a cyclic (or quasi-cyclic) modulation in Sagittarius A*'s sub-mm, infrared, and possibly X-ray luminosity ought to be detectable, should the basic picture of how it produces its radiative flux be correct.[7]

A powerful H-band flare[8] was observed by the Yepun telescope at the VLT on May 9, 2003 (see figure 4.4). Sagittarius A*'s 1.65 μm flux rose by a factor of six in about five minutes, subsiding back to its quiescent value some thirty minutes later. Additional infrared flares, this time in the K-band (2.16 μm), were detected during a second observing run in June 2003, one of which is shown in figure 4.5. The K-band flux rose to a factor of three above the quiescent level and lasted for approximately eighty-five minutes. Remarkably, both the May 9, 2003, and June 16, 2003, flares (figures 4.4 and 4.5) showed several major peaks spaced by thirteen to twenty-eight minutes, resembling a 15%–40% quasi-periodic modulation in the lightcurve. Based on the combined infrared exposure of the galactic center with Yepun, it appears that Sagittarius A* undergoes this type of activity between two and six times per day, at least twice the rate of X-ray flares detected via the *Chandra* and

---

[7]The earliest discussion on the detection of a modulated infrared signal from Sagittarius A* appears in Hollywood and Melia (1995), Hollywood et al. (1995), and Hollywood and Melia (1997). This subject was revisited several years later, with a reference to a possible cyclic modulation in Sagittarius A* at mm/sub-mm and X-ray wavelengths. See Melia et al. (2001) and Bromley, Melia, and Liu (2001).

[8]See Genzel et al. (2003a).

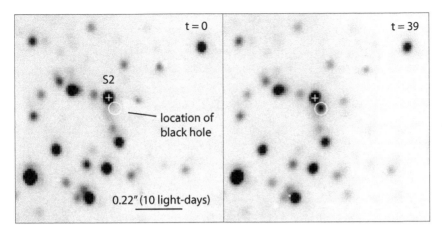

Figure 4.4 Detection of a near-IR flare from Sagittarius A*. Each panel is an H-band (1.65 μm) image of the central $\sim 1'' \times 1''$ of the Galaxy, obtained on the Yepun telescope at the VLT on May 9, 2003. The integration time for each image was sixty seconds, and the time (in minutes from the beginning of the set) appears in the upper right-hand corner of each panel. North is up and east is to the left. The left panel shows the map of this region in quiescence; the panel to the right shows the H-band intensity during the flare itself. Compare this also with figure 3.3, which shows the corresponding L' (3.8 μm) intensity maps of this region obtained on the W. M. Keck II 10-meter telescope. (Image from Genzel et al. 2003a)

XMM-*Newton* monitoring programs. It is not yet clear, however, whether this really points to a separate origin for the two types of flare, or whether (more likely) the difference in rates is an indication that the fainter X-ray flares are more difficult to identify.

The very short ($\sim$ minute-scale) rise and decay times of the infrared flares, not to mention the fine temporal substructure within the events (see figure 4.5), strongly point to emission from within a region no bigger than $\sim$ 5–10 Schwarzschild radii. In this regard, the H-band, K-band, and X-ray flares all require a consistent source geometry. However, the most tantalizing conclusion we may draw from the lightcurve shown in figure 4.5 is that the temporal substructure is a fundamental property of the flow—that it represents a relativistic modulation of the emitting gas orbiting in a prograde sense just outside the black hole's event horizon.[9] Interestingly, the period in figure 4.5 seems to decrease with time, from about seventeen to thirteen minutes (a phenomenon known

[9]We will return to this in §9.1. See also the discussion in Melia et al. (2001).

79

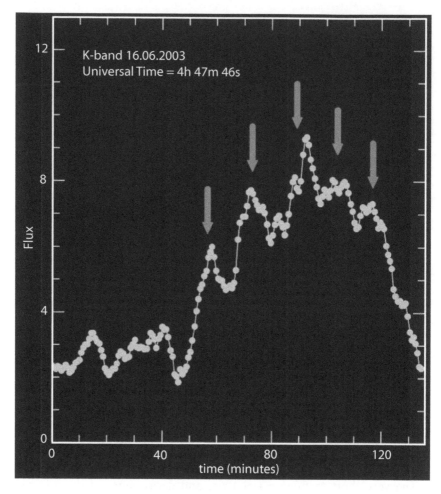

Figure 4.5 Sagittarius A*'s lightcurve of the June 16, 2003, K-band flare. The arrows mark the substructure peaks corresponding to an apparent seventeen-minute modulation. The vertical scale shows the dereddened flux density in mJy. The horizontal scale shows the time in minutes relative to the universal time shown in the figure. Note that the time structure of this flare has a tendency to chirp—periods decrease with time, perhaps suggesting a decaying Keplerian orbit. (Image from Genzel et al. 2003a)

as *chirping*). Though other explanations may be possible, this behavior suggests a decaying Keplerian orbit, adding some support to the orbital-period interpretation of the infrared modulation. We may be witnessing an instability in the accretion flow that drains inwards on a timescale of several periods.

4.2 Long-Term Variability in the Radio

The weekly VLA monitoring program that led fortuitously to a coordinated X-ray/radio observation of Sagittarius A* in 2002 has accumulated twenty years of variability data at 1.3, 2.0, 3.6, 6.0, and 20 cm wavelengths. The sampling within this period has been somewhat irregular.[10] Nonetheless, the power spectral density (PSD) analysis shown in figure 4.6 reveals a clear peak near 1×10^{-7} Hz, with a progressively smaller significance at longer wavelengths. This frequency corresponds to a periodic modulation of $\sim$100–120 days; the actual best-fit period extracted from the combined data sets is 106 days.

This result is at once intriguing and unsettling. The fact that a Keplerian period has been seen in Sagittarius A*'s infrared emission makes this cyclic modulation easier to accept. Yet the implied radio period ($\sim$ 106 days) contrasts sharply with the dynamical timescale (about seventeen minutes) associated with motion in the disk.

Perhaps the periodic radio signal is a false detection due to a combination of a random process and the irregular sampling pattern. However, Monte Carlo tests with data created from various sources of noise using this same sampling don't seem to bear this out. Regardless of the type of noise used in the simulations—including white noise, Gaussian noise around a mean, and a Poisson distribution of flares—the probability of false detection due to any such random process appears to be <5%.

Figure 4.7 shows the average profile of the 106-day modulation constructed by folding the cumulative data into the *inferred* 106-day period. The zero levels represent the mean flux densities observed in Sagittarius A* over the entire baseline, and zero phase corresponds to the reference day December 4, 1991. These panels provide evidence of another important trend—that both the absolute (ΔS) and fractional ($\Delta S/S$) amplitudes of the pulsed component increase toward shorter wavelengths.

The 106-day cycle evident primarily at 1.3 and 2.0 cm may be a valuable tool for probing Sagittarius A*'s inner workings should it truly have something to do with the source. High-resolution VLA observations have already ruled out the possibility that such a period might be produced by an orbiting emitting object.[11] The 106-day

---

[10] See Zhao, Bower, and Goss (2001).
[11] See Bower and Backer (1998).

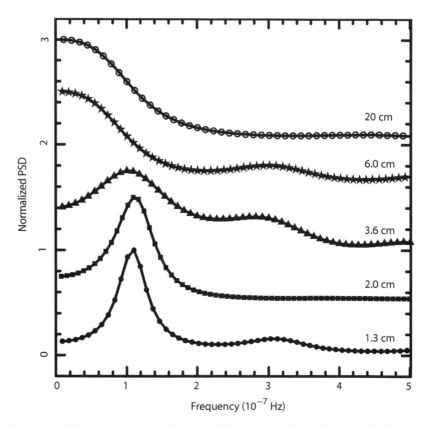

Figure 4.6 The power spectral density (PSD) derived from all the available data folded into six cycles of a 106-day period at each wavelength. Each profile is normalized by its peak value, and the curves are shifted vertically for clarity. A peak near 1×10^{-7} Hz is detected at 1.3–2 cm; the detection at 3.6 cm is marginal ($< 3\sigma$). Note also that no significant periodicities were seen at 6 and 20 cm. (Image from Zhao, Bower, and Goss 2001)

orbit of a companion to Sagittarius A* would have a radius ≈ 60 AU, corresponding to an angular separation of ≈ 8 mas at 8 kpc. A compact 0.2 Jy source separated from Sagittarius A* by this amount would have easily been detected with the VLBA at wavelengths shorter than 3.6 cm. The unlikelihood of Sagittarius A* having an orbiting companion is further supported by its observed lack of proper motion (see §2.2), which precludes any possible association with rapidly moving components. In addition, a stellar origin for such a source would fall well short of the power required to account for the measured radio emission. All in all, the

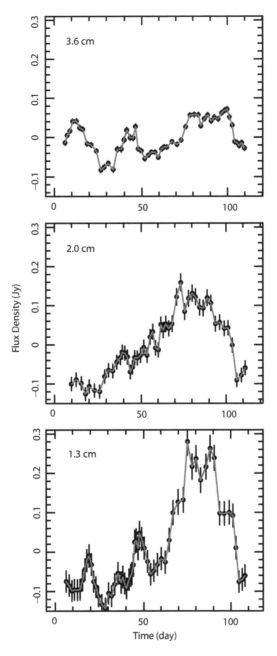

Figure 4.7 Mean (smoothed) lightcurves at 3.6, 2, and 1.3 cm obtained by folding the data into one single 106-day cycle. (Image from Zhao, Bower, and Goss 2001)

evidence seems to favor an interpretation in which the 106-day periodic variations are intrinsic to Sagittarius A* itself.

The characteristics of this 106-day cycle constrain the nature of its origin rather tightly. First, the observed period is independent of wavelength. We have already seen that the emission in Sagittarius A* at different frequencies is produced on different spatial scales,[12] so the period should be induced by a single process. Otherwise, we would expect to see different periods at different frequencies. What is required is something that can cause correlated fluctuations across a broad range of wavelengths. Second, the period is four orders of magnitude longer than the dynamical timescale in the inner disk surrounding Sagittarius A*. Could it be produced on a much larger spatial scale? According to figure 2.4, Sagittarius A*'s 2 cm emission is produced within a region no bigger than ~ 24 AU, for which the corresponding dynamical timescale is calculated to be about 1.5 days; so the answer is apparently no. Higher frequency emission is produced within still smaller regions, associated with even smaller timescales.

We may ask, then, whether this modulation could be produced by a corrugation wave in an accretion disk, which is used to account for the quasi-periodic oscillations (QPOs) seen in low-mass X-ray binaries. These waves have periods that are much longer than the corresponding dynamical timescale,[13] but they depend on radius and thus may not be able to account for the first feature described above. Moreover, Sagittarius A*'s lightcurves show quite stable periodic fluctuations, rather than the uncorrelated segments constituting QPOs. The evidence is pointing to a single process evolving in a relatively confined region, certainly no bigger than 100 Schwarzschild radii.

In chapter 9, we will investigate the leading candidate proposed thus far as the agent responsible for the long-term modulation in Sagittarius A*'s radio lightcurve. In a Kerr metric, the gravitational acceleration acquires a dependence on poloidal angle (relative to the black hole's spin axis). As we shall see, matter orbiting above or below the equatorial plane therefore experiences a restoring force toward the equator, which results in the precession of its angular momentum vector

---

[12] Observationally, this inference has been drawn on the basis of coordinated measurements, such as those discussed in connection with figure 4.3. We shall see later in this book that a "stratified" emission region is also expected from theory. (See, e.g., Melia, Jokipii, and Narayanan 1992).

[13] See Kato (1990).

about the black hole's axis of rotation. Under an appropriate set of circumstances, the precession period may be as long as 100 days or more, and the long-term radio modulation in Sagittarius A* may be closely related to its short-term X-ray and infrared variability after all—via the dynamical properties of the disk in a Kerr metric.

CHAPTER 5

The Central Star Cluster

To identify stars within the inner 1″ (≈ 0.12 lt-yr) of the Galaxy, one must separate out the small flux contribution of these (typically K = 14.5–15.0) sources from the much brighter neighboring IRS 16 cluster, whose members have K-magnitudes as low as 10. In recent years, significant progress has been made in this effort with near-infrared, high-spatial resolution imaging observations and spectroscopy of the central stars, which have provided us with new insights into their dynamics, evolution, and mass function.

The best current images resolve the near-infrared emission of the central parsec into thousands of stars with K-magnitudes up to 15 or 16, illustrated quite impressively with the photograph shown in color plate 5. At this sensitivity, all red and most blue supergiants, all red giants (including the brighter asymptotic giant branch, or AGB, stars) of spectral type later than K5, and all main sequence stars earlier than B2 are visible. Still, even images such as this only sample a fraction of the total stellar content of the cluster.

The most prominent feature near Sagittarius A* is a group of more than two dozen stars (the IRS 16 complex) centered $\sim 1″$–2″ east of the radio source (see color plate 5). Another compact group of bright stars (the IRS 13 complex) lies 3.″5 to the southwest, and the additional concentration of light within 1″ of the middle marks the presence of the so-called Sagittarius A* cluster, a grouping of stars orbiting within 0.12 light-year of the supermassive black hole. In the next section, we shall begin the process of unraveling the various stellar components at the galactic center, starting with a determination of their overall distribution out to ~ 100 pc but eventually working our way inwards to within a few light-days of the nucleus.

86

5.1 THE STELLAR CUSP SURROUNDING THE BLACK HOLE

Let us begin by identifying several key dynamical processes that shape a gravitating stellar system. These include both remote gravitational interactions, in which stars may exchange energy and momentum as "point particles," and local collisions involving an actual physical contact between them.

On the largest spatial scale, a star responds to the combined effects of the black hole's gravitational attraction and that of all the other stars in the system. The resultant gravitational potential may be approximated as a single smoothed function. On smaller scales, however, stellar motions may be randomized by two-body scattering events, in which the separation between two approaching stars is sufficiently small for their mutual interaction to temporarily dominate over that of the smoothed potential. Given sufficient time, these scattering events *relax* the system and redistribute the energy and angular momentum. As we shall see below, one may think of this process as a gradual erasure of the system's dependence on the initial conditions. Two-body relaxation forces the system's final configuration to depend only on the boundary conditions and the stellar mass function.

Usually, the largest characteristic scale is the so-called black hole's radius of influence r_h, within which the massive central object dominates the stellar dynamics. Beyond r_h, the black hole's impact is dwarfed by the gravitational force of the stars in the cluster. It is customary to define this radius as

$$r_h \equiv \frac{GM}{\sigma^2}, \qquad\qquad 5.1$$

where M is the black hole mass and σ is the one-dimensional (root mean square) velocity dispersion, though this is not always a precisely determined quantity since σ may depend on radius. Kinematically, the black hole potential dominates over that due to the cluster out to a radius beyond which its mass is less than that of the enclosed stars. As we shall see later in this chapter (see, e.g., figure 5.9), the measured enclosed mass at the galactic center suggests that r_h is approximately 3 pc (roughly 9 light-years) at that location.

The *collision* radius r_{col} is defined as the minimal distance from the black hole, where close gravitational encounters can produce large-angle deflections. The fact that the relative velocity $v_{rel}(r)$ $(\approx [2GM/r]^{1/2})$

of interacting stars increases with proximity to the black hole means that r_{col} cannot be zero, for otherwise v_{rel} would eventually exceed the escape velocity ($\equiv [2GM_*/R_*]^{1/2}$, in terms of the stellar mass M_* and radius R_*) at the stellar surface, and large-angle deflections would only occur for distances of closest approach less than R_*. However, when a star's orbital energy exceeds its internal binding energy, such a physical collision leads to complete disruption. Thus, for a black hole mass $M \sim 2.6$–$3.6 \times 10^6 \, M_\odot$ (see chapter 1 and later in this chapter), two-body relaxation gradually becomes inefficient at distances less than $r_{col} \approx 0.06$–0.08 pc (≈ 0.18–0.24 lt-yr). Of course, this does not mean that the region $r < r_{col}$ is devoid of stars, but, rather, that objects surviving this close to the central object will not have relaxed via two-body collisions.

When a star ventures even closer than r_{col} to the black hole, the difference between the latter's gravitational pull on the side of the star closest to the central object and that on the far side becomes a significant fraction of the star's own self-gravity. The magnitude of this effect depends on how close the intruder approaches the so-called tidal radius r_t, defined by the condition that the black hole's tidal field on the stellar surface is comparable to the field from the star itself, resulting in the expression

$$\left[-\frac{GM}{(r + R_*)^2} + \frac{GM}{(r - R_*)^2} \right]_{r_t} = \frac{GM}{R_*^2} \qquad 5.2$$

or

$$r_t = \left(\frac{4M}{M_*} \right)^{1/3} R_*. \qquad 5.3$$

We therefore infer that at the galactic center $r_t \sim 10^{13}$ cm or roughly 10^{-5} lt-yr. (Incidentally, in the next chapter, we will define the Schwarzschild radius—the size of the event horizon of a nonspinning black hole—as $r_S \equiv 2GM/c^2$. For Sagittarius A*, $r_S \approx 10^{12}$ cm, so a $\sim 1 \, M_\odot$ star would need to venture to within ~ 10 black hole radii of the center to be torn apart by tidal forces.)

How quickly the stellar cluster adjusts to the various dynamical processes depends on the characteristic timescales. For example, the *dynamical* time t_d is how long a star takes to cross the system. A convenient measure of this quantity is simply the orbital period (which, however, is a function of radius):

$$t_d \equiv 2\pi \left(\frac{r^3}{GM_{tot}(r)} \right)^{1/2},$$
5.4

where $M_{tot}(r)$ is the enclosed mass (black hole plus stars) at r. At r_b, this time is about 2×10^5 years, but it is only ~ 800 years at r_{col}.

Perhaps the most important timescale is that associated with two-body relaxation, which we will here label t_r. We will determine this quantity by considering the encounter between two stars and estimating the deflection in velocity δv incurred by one of them, subject to the assumptions that (1) $|\delta v|/v \ll 1$ and (2) the second star remains almost stationary (in our frame) during the event. Without an interaction, the straight-line path of the incoming star would take it to within a distance of closest approach b—the impact parameter—of the target.[1] The component of force $F_\parallel$ parallel to v is antisymmetric with respect to the point of closest approach, so its effect cancels out over the course of the interaction. However, the perpendicular component $F_\perp$ that gives rise to δv always points in the same direction, and we may integrate its cumulative effect to determine the magnitude of the perturbation.

Aligning our Cartesian coordinate system with $\hat{z}$ pointing along v and the origin and $t = 0$ at the point of closest approach, we have for the two equal-mass stars

$$F_\perp = \frac{GM_*^2 b}{(b^2 + z^2)^{3/2}} \approx \frac{GM_*}{b^2} \left[1 + \left(\frac{vt}{b} \right)^2 \right]^{-3/2}.$$
5.5

But according to Newton's laws of motion, $M_* \dot{v}_\perp = F_\perp$, which we may use in equation (5.5), with a change of variable $u = vt/b$, to integrate this expression with respect to time, yielding

$$|\delta v_\perp| \approx \frac{GM_*}{bv} \int_{-\infty}^{\infty} (1 + u^2)^{-3/2} \, du = \frac{2GM_*}{bv}.$$
5.6

Notice that $|\delta v_\perp|$ is roughly equal to the force at closest approach, GM_*/b^2, times the length of time ($\sim b/v$) over which this force acts.

---

[1] The precise amount δv by which the scattering deflects the velocity v of each star actually depends on the star's mass and speed, in addition to the impact parameter b. See Binney and Tremaine (1987) for the more detailed treatment of this two-body process.

In passing through a cluster with characteristic size L_c and average stellar number density n_c, such a star encounters a column density $\sim L_c n_c$ of scattering centers and thereby suffers $\sim L_c n_c 2\pi b\,db$ deflections with impact parameters in the range b to $b+db$. In so doing, the cumulative deflection it experiences, which we write as the sum over the square of the quantity on the left-hand side of equation (5.6) to avoid the cancellations that occur to first order in $|\delta\mathbf{v}_\perp|$, is

$$\delta v_\perp^2 \approx \left(\frac{2GM_*}{bv}\right)^2 L_c n_c 2\pi b\,db. \qquad 5.7$$

Clearly, our assumption that $|\delta\mathbf{v}|/v \ll 1$ breaks down when the impact parameter $b < b_{\min} \equiv GM_*/v^2$ (from equation 5.6). So to find the overall deflection suffered by the star, we should integrate equation (5.7) over all values of b between $b_{\min}$ and the largest possible impact parameter L_c. This gives

$$\Delta v_\perp^2 \equiv \int_{b_{\min}}^{L_c} \delta v_\perp^2\,db \approx 8\pi \left(\frac{GM_*}{v}\right)^2 L_c\,n_c\,\ln\Lambda, \qquad 5.8$$

where

$$\ln\Lambda \equiv \ln\left(\frac{L_c}{b_{\min}}\right) \qquad 5.9$$

is the Coulomb logarithm.

Using the one-dimensional velocity dispersion σ to represent the "typical" star in this system, we now estimate t_r by determining how many times it must cross the cluster before the scatterings produce a cumulative deflection with $\Delta v_\perp^2 \sim \sigma^2$. Evidently, this number is

$$\delta n \sim \frac{\sigma^2}{\delta v_\perp^2} \sim \frac{\sigma^4}{8\pi(GM_*)^2 L_c\,n_c\,\ln\Lambda}, \qquad 5.10$$

and the relaxation time is therefore

$$t_r \sim \frac{\delta n \times L_c}{\sigma} = \frac{\sigma^3}{8\pi(GM_*)^2\,n_c\,\ln\Lambda}. \qquad 5.11$$

Both σ^3 and n_c scale with radius in roughly the same way (somewhere between $\sim r^{1.5}$ and r^{-2}), as we shall see later in this chapter, so t_r is not a strong function of r. At $r = 1$ pc, $\sigma \sim 100\,\mathrm{km\,s^{-1}}$, and $n_c \sim 5 \times 10^5$ stars $\mathrm{pc^{-3}}$. In addition, $\ln \Lambda \sim 10$, and therefore, $t_r \sim 4 \times 10^8$ yr. This is easily much shorter than the age of the Galaxy (almost a Hubble time, $t_H \sim 13.7$ billion yr), and so we anticipate that the older stars in the central cluster, at least those beyond the collision radius r_{col}, will have relaxed to an equilibrium configuration by now.

Indeed, attempts to determine the global properties of the stellar population close to the black hole are based on the premise that there exists an underlying dynamically relaxed distribution accounting for most of the stars (and most of the mass). The young and bright stars (such as those within the IRS 16 and IRS 13 complexes) may still be unrelaxed or are perhaps susceptible to other environmental factors, such as stellar collisions, and are presumably not reliable tracers. Instead, it is the surface density and surface brightness of the old, faint stars that apparently trace the equilibrium distribution.

At the galactic center, the luminosity profile is consistent with a stellar volume density that follows an r^{-2} power law[2] from a projected radius $\sim 40'$ (≈ 100 pc) down to about $10''$ (≈ 0.4 pc). At least for radii $> r_h \sim$ 3 pc, the stars cannot be directly influenced by the central black hole. Instead, as we shall soon see, a stellar distribution following an r^{-2} power law is consistent with an *isothermal* profile, meaning that it may be characterized by a single energy variable equivalent to the temperature in a gaseous system.[3] (We will examine what happens below $\sim$3 pc shortly.)

An ingredient sought to characterize this distribution is the *core* radius r_c, at which the surface density falls to half its central value. Estimated to lie anywhere from 0.05 pc to $\sim$1 pc, this parameter is still unknown, due to differences between the various methods used in its determination, that is, whether the light distribution is measured or whether individual stars are counted, and whether the bright stars are included in the fits. As we have already seen, however, a radius of $\sim$1 pc falls within the black hole's sphere of influence, so an isothermal formulation for the stellar profile may be inappropriate there.

But we can still learn a great deal about the stars farther out by first examining the general properties of an isothermal distribution. At any

---

[2] Several independent studies have reached this conclusion, including those of Sellgren et al. (1990), Catchpole, Whitelock, and Glass (1990), and Serabyn and Morris (1996).

[3] However, note that such a power-law distribution is apparently realized in the nuclei of many galaxies, so it may also reflect how stars form in a typical nuclear environment, rather than simply requiring that they all attain a relaxed isothermal profile.

given time t, a full description of the state of the system is given by specifying the phase-space density $f(\mathbf{x}, \mathbf{v}, t)$, giving the number of stars per unit volume in coordinate space and per unit volume in velocity space. The quantity $f(\mathbf{x}, \mathbf{v}, t)$ is also known as the distribution function.

The behavior of stars moving under the influence of gravity is governed by Poisson's equation

$$\nabla^2 \Phi = 4\pi G \rho, \qquad\qquad 5.12$$

where Φ is the gravitational potential, G is the gravitational coupling constant, and

$$\rho = \int f(\mathbf{x}, \mathbf{v}, t) \, d^3 \mathbf{v} \qquad\qquad 5.13$$

is the (stellar) mass density, and we here ignore the black hole. In spherical systems, a stellar orbit depends only on the energy E and the angular momentum vector $\mathbf{L}$.[4] Thus, a spherical system in steady state can be described with a distribution function $f(E, \mathbf{L})$. If, in addition, the system is spherically symmetric, then f cannot depend on the direction of $\mathbf{L}$ but only on its magnitude L, and we can then write $f = f(E, L)$.

The fundamental equation governing spherical equilibrium stellar systems is therefore

$$\frac{1}{r^2} \frac{d}{dr} \left(r^2 \frac{d\Phi}{dr} \right) = 4\pi G \int f\left(\frac{1}{2} v^2 + \Phi, |\mathbf{r} \times \mathbf{v}| \right) d^3 \mathbf{v}, \qquad 5.14$$

though for convenience, one typically redefines the gravitational potential and the energy as

$$\Psi \equiv -\Phi + \Phi_0 \qquad\qquad 5.15$$

and

$$\varepsilon \equiv -E + \Phi_0 = \Psi - \frac{1}{2} v^2 \qquad\qquad 5.16$$

in terms of the constant Φ_0.

---

[4]Two-integral spherical systems have been discussed in the literature for many decades. See Jeans (1915) for one of the earliest treatments of such clusters.

The analog to a Maxwell-Boltzmann distribution for a stellar system may be written in the form

$$f(\varepsilon) = \frac{\rho_0}{(2\pi\sigma^2)^{3/2}} \exp(\varepsilon/\sigma^2), \qquad 5.17$$

where instead of using a temperature (as we would for a gas) we now characterize the population via the velocity dispersion σ. It is trivial to see that the density (from equation 5.13) is then

$$\rho = \rho_0 \exp(\Psi/\sigma^2). \qquad 5.18$$

Poisson's equation for this system reads

$$\frac{1}{r^2} \frac{d}{dr}\left(r^2 \frac{d\Psi}{dr}\right) = -4\pi G\rho \qquad 5.19$$

or, using equation (5.18),

$$\frac{d}{dr}\left(r^2 \frac{d\ln\rho}{dr}\right) = -\frac{4\pi G}{\sigma^2} r^2 \rho. \qquad 5.20$$

A simple solution to this equation is a power law

$$\rho(r) = \frac{\sigma^2}{2\pi G} \frac{1}{r^2} \qquad 5.21$$

known as the singular isothermal sphere (because its density would go to infinity at the origin). In reality, of course, $\rho(0)$ is finite, and anyway, we need to contend with the fact that the black hole's gravity cannot be ignored inside r_h. It turns out that if one integrates equation (5.20) numerically, starting with a *finite* density at $r = 0$, the solution eventually approaches equation (5.21) but is otherwise "flattened" near the origin. At the galactic center, the actual stellar surface number density at moderately bright magnitudes ($K \leq 15$) follows this behavior, going as r^{-2} at large radii but flattening toward the middle, and is well fit by the modified (or "flattened") isothermal distribution

$$\rho(r) = \frac{\rho_0}{1 + 3(r/r_c)^2}, \qquad 5.22$$

with a core radius $r_c \sim 0.4$ pc (though, as we said, this value is not unique, and its precise physical meaning is questionable, given that $r_c < r_h$) and a central density $\rho_0 = 4 \times 10^6 \, M_\odot \, \text{pc}^{-3}$.

However, the situation is expected to change for radii smaller than ~ 3 pc, where the black hole's influence may alter the impact of individual star-star collisions and, in principle, modify the distribution function given in equation (5.17). Dynamical models of a massive black hole's influence on the stellar population in its immediate vicinity generally predict the formation of a stellar *cusp* with a characteristic power-law density index that may reflect its formation history in cases where $t_r > t_H$. Still, one should keep in mind that observations of galactic nuclei often reveal a central stellar cluster with a cusp in the light distribution on scales too large to be related to the black hole itself. Evidently, stellar cusps may be created by a variety of dynamical means, so the existence of such a compact stellar distribution is not necessarily evidence for the presence of a supermassive pointlike object.[5]

We may understand how and why a cusp forms around the black hole by examining the equilibrium stellar distribution attained in a single-particle system,[6] where all the stars have the same mass M_* and $M_* \ll M$. For simplicity, we will adopt the essential assumptions and approximations needed to find a manageable, analytic solution to this problem.

Stars on bound orbits in the $\Phi = -GM/r$ Keplerian gravitational potential of the black hole diffuse from one bound orbit to another as a result of star-star gravitational scattering. For them, the distribution function $f(\mathbf{x}, \mathbf{v}, t)$ is spherically symmetric in coordinate space, though only approximately isotropic in velocity space. However, it is straightforward to show that even with star-star scattering events, $f(\mathbf{x}, \mathbf{v}, t)$ in the galactic center departs only slightly from isotropy, rendering it primarily a function of just the energy E (or ε) and time t. As before, we define the energy

$$\varepsilon = \frac{GM}{r} - \frac{1}{2}v^2 \qquad\qquad 5.23$$

(here with $\Phi_0 = 0$) to be the negative of the stellar energy per unit mass, with v the stellar velocity and r its distance from the black hole. To see

---

[5] See Phinney (1989).

[6] The pioneering work on this topic was that of Peebles (1972) and Bahcall and Wolf (1976).

why $f \approx f(\varepsilon, t)$, we first show that the time Δt_{col} between individual collisions is long compared with the orbital period t_d. Defining the stellar collision radius R_{col} with the condition

$$\frac{1}{2}v^2 = \frac{GM_*}{R_{col}}, \qquad 5.24$$

we estimate the individual collision cross section σ_{col} as πR_{col}^2, so that

$$\Delta t_{col} = \frac{1}{vn_c\sigma_{col}}, \qquad 5.25$$

where n_c (here equal to ρ/M_*) is the stellar density in the cluster. Thus, with the orbital period defined in equation (5.4), we find that

$$\frac{\Delta t_{col}}{t_d} \sim \frac{v^3 n_c^{-1}(GM_*)^{-2}}{rv^{-1}}. \qquad 5.26$$

For a typical radius, we adopt the value $r = r_h$ (see equation 5.1), which also functions as the "capture" radius in terms of the line-of-sight velocity dispersion $\sigma \equiv \langle \Delta v^2 \rangle^{1/2}$ in the core of the cluster. It is therefore also appropriate to take n_c equal to n_0 (here equal to ρ_0/M_*), the number density of stars in the core, and define the core mass M_c (in terms of the core radius r_c) accordingly:

$$M_c = \frac{4}{3}\pi n_0 r_c^3 M_*. \qquad 5.27$$

If we, in addition, take as a typical velocity $v \sim \langle \Delta v^2 \rangle^{1/2}$ and use the remaining relation

$$\frac{GM_c}{r_c} \sim \langle \Delta v^2 \rangle, \qquad 5.28$$

then equation (5.26) becomes

$$\frac{\Delta t_{col}}{t_d} \sim \frac{M_c^2}{M_* M} \gg 1, \qquad 5.29$$

since $M_c \gg 10^3 M_\odot$. Moreover, $\Delta t_{col}/t_d$ increases outwards, so this inequality is valid throughout the star-star collision region. Thus, the anisotropic portion of f associated with collision-induced migration is much smaller than the isotropic part, and one may neglect the anisotropy in computing the collision rates and related quantities that affect the equilibrium distribution.

The effect of collisions on f is so slight that it may be taken to be constant along the trajectory of a particle in phase space, and since the coordinates $(\mathbf{x}_1, \mathbf{v}_1)$ and $(\mathbf{x}_2, \mathbf{v}_2)$ at two locations on such a trajectory correspond to the same energy, f must be just a function of ε and t.

We are now in a position to see why the supermassive black hole is expected to produce a "cusp" profile for the stellar distribution within its influence. Although one can demonstrate this result accurately by numerically solving the Fokker-Planck equation for the phase-space density $f(\mathbf{x}, \mathbf{v}, t)$, we will here follow the much simpler (heuristic) approach of Binney and Tremaine,[7] who developed their argument using dimensional analysis.

A star at radius r has energy of order $E(r) = -GMM_*/r$, and since f is a function only of E (and possibly t), let us suppose that the equilibrium density near the black hole is a power law

$$\rho(r) \sim \rho_0 \left(\frac{r_S}{r}\right)^\mu.$$

5.30

At any given radius r, we also have that

$$\langle v^2 \rangle \sim \frac{GM}{r}.$$

5.31

Thus, the relaxation time (equation 5.11) is

$$t_r \sim \frac{\langle v^2 \rangle^{3/2}}{8\pi (GM_*)^2 \rho(r) \ln \Lambda} \propto r^{\mu - 3/2}.$$

5.32

The key step in the evaluation of μ is to suppose that stars reaching the tidal radius r_t are destroyed and swallowed by the black hole and that the energy $(\sim GMM_*/r_t)$ thereby gained by the cluster is ported outwards via relaxation among the $N(r)$ $(\propto r^{3-\mu})$ cusp stars interior to r. More specifically, the energy flux $F_E(r)$ carried outwards by the cluster stars

---

[7]See Binney and Tremaine (1987).

through radius r is

$$F_E(r) \sim \frac{N(r)E(r)}{t_r} \propto r^{7/2-\mu}.$$ 5.33

In steady state, the flow of energy must be independent of r, and one therefore predicts a stellar cusp with a power-law index $\mu = 7/4$.

In reality, the expected radial slope of the power-law distribution ranges between ~ -0.5 and ~ -2.5, depending on how the black hole formed and how the stellar cusp was assembled, and on the importance of inelastic stellar collisions. The solution we have just derived describes a relaxed single-mass profile, which, however, is not applicable when, for example, the black hole's growth rate is larger than the inverse relaxation timescale. If the black hole grows adiabatically in such a situation,[8] then the surrounding stellar cusp takes on a power law with index $3/2$ instead of $7/4$. Other initial conditions[9] will result in a variety of density distributions, sometimes as steep as $r^{-5/2}$. Also, in a realistic stellar population with a spectrum of masses, the more massive stars will concentrate closer to the center than the less massive ones. For example, a simple extension[10] of the model we have described here, incorporating two mass components ($M_{*1} < M_{*2}$), shows that the two species settle into different power-law distributions with power-law indices given by

$$\mu_i = \frac{M_{*i}}{4 M_{*2}} + \frac{3}{2}.$$ 5.34

Let us now examine the actual data pertaining to the stellar distribution surrounding Sagittarius A*. The stellar counts versus distance shown in figure 5.1 were obtained using an adaptive optics system/near-infrared camera on the 8.2-meter VLT Yepun telescope at the European Southern Observatory in Paranal, Chile. These data have been extracted from the H-band (1.65 μm), K_s-band (2.16 μm), and L$'$-band images obtained during the observing run of August 2002.

The flattened isothermal sphere model (equation 5.11) accounts quite well for the observed surface density at projected distances $\geq 5''$–$10''$ (i.e., ≥ 0.2–0.4 pc). But clearly an excess of faint stars is seen above the flat core within several arcseconds of Sagittarius A*. Overall,

[8] See Young (1980).
[9] See, e.g., Lee and Goodman (1989) and Quinlan et al. (1995).
[10] This is based on a study by Bahcall and Wolf (1977).

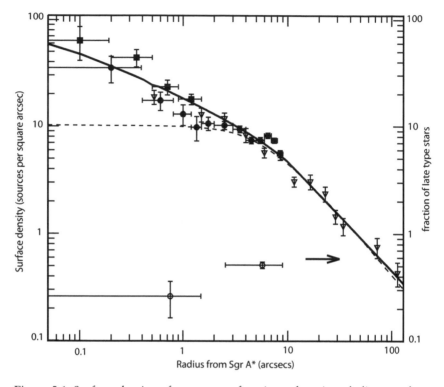

Figure 5.1 Surface density of stars as a function of projected distance from Sagittarius A*. Filled circles are counts of all sources present on both H and $K_s \leq 17$ maps. Squares with crosses denote direct H-band counts by eye to $H \leq 19.8$ near Sagittarius A*. Downward-pointing triangles denote $K \leq 15$ counts in the overlap region beyond a few arcseconds from Sagittarius A*. Open circles at the bottom of the figure denote the fraction of late-type stars in the total $K \leq 15$ sample with proper motions (see Genzel et al. 2003b for a more detailed description). The dashed curve represents the model of a flattened isothermal sphere with core radius $r_c = 0.34$ pc (see equation 5.22). The solid curve is a broken power law, going as r^{-2} beyond $10''$ (≈ 0.4 pc) and $r^{-1.4}$ within $10''$. All vertical error bars are $\pm 1\sigma$ and denote the total uncertainty due to Poisson statistics and, where appropriate, due to incompleteness. (Image from Genzel et al. 2003b)

the surface density increases with decreasing separation from the radio source. The existence of this *cusp* is confirmed visually by the smoothed two-dimensional distribution of faint stars in the H+K map shown in figure 5.2, demonstrating a central peak at RA $= 0.09''$ and Dec $= -0.15''$ relative to Sagittarius A*, with an uncertainty of $\pm 0.2''$. Thus, it appears that the observed stellar density distribution

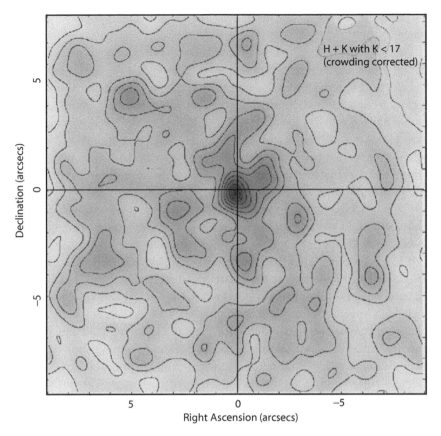

Figure 5.2 Map of the smoothed (with a 1″ Gaussian) surface density in H+K for stars with K≤ 17. Contours are 10%, 20%, ..., 90%, 95%, 99%, and 99.9% of the peak surface density. The maximum of the stellar density lies at RA = 0.09″ and Dec = −0.15″ relative to Sagittarius A*, with an uncertainty of ±0.2″. (Image from Genzel et al. 2003b)

is consistent, within the uncertainties, with the theoretical predictions of a centrally condensed star cluster surrounding the massive black hole.

Before closing this section, we should mention yet another intriguing dynamical phenomenon predicted for the central star cluster almost two decades ago, though tentatively confirmed observationally only in the past year or two. The relative velocity of two stars engaged in a close gravitational encounter while subject to the influence of the black hole is so great that even a slight perturbation may be sufficient to eject one of them out of the Galaxy at very high speed.[11]

The main argument behind this process is fairly easy to understand. A star's specific energy at radius r is $E = v^2/2 + \Phi$, where v is its velocity and Φ is the gravitational potential of the black hole. Should this star approach sufficiently close to the central object that $|\Phi(r)| \gg E$, it would have a velocity $v = [2(E - \Phi)]^{1/2} \approx (2GM/r)^{1/2} = 3 \times 10^3 \text{ km s}^{-1}$ (for $M \sim 3.5 \times 10^6 \, M_\odot$ and $r \sim 3$ light-days). If this star were then to suffer a perturbation in its velocity ($\delta v \ll v$) due to interactions with other stars or due to the tidal disruption of a close binary, its specific energy would increase by an amount $\delta E \approx (v + \delta v)^2/2 - v^2/2 \approx v \, \delta v$, which is much greater than $|E|$. In fact, the predicted velocity for such an object would be $\sim (2v \, \delta v)^{1/2} \approx 1.5 \times 10^3 \text{ km s}^{-1}$, greater than the escape speed of the Galaxy as a whole.

It has been postulated that the discovery of hypervelocity stars would provide strong evidence for the existence of a massive black hole in our Galaxy, since no other process is known to accelerate individual stars to such high speeds. By now, at least five such objects have been found, ranging in velocity from $\sim 548 \text{ km s}^{-1}$ to $\sim 717 \text{ km s}^{-1}$, all apparently moving radially away from the galactic center. Assuming that they are all late-B-type stars, they lie at distances ~ 55–75 kpc and have travel times (~ 100 Myr) to their current location consistent with their lifetime. No doubt, many more such objects will be discovered in the near future, and a more detailed study of their distribution in velocity and space may lead to a better understanding of how stars in the central cluster interact with each other and with the central black hole.

In the next section, we will look at the constituents of the central star cluster and investigate more closely their global dynamical properties, which may hint at their formative history. Several stars are now known from proper motion studies to reside within the central 25 light-days. Six of them have had their orbits traced with sufficient precision for us to know that they are bound to the central object with periods between fifteen and a few hundred years (see §5.3).

---

[11] This suggestion was made by Hills (1988), who postulated that the tidal breakup of a binary by the black hole's strong gravitational field would lead to the capture of one star and the ejection of the second away from the center of the Galaxy. Yu and Tremaine (2003) developed this idea further and identified several other three-body processes by which such an ejection might occur, but it was not until several years later that the first solid evidence of a hypervelocity star was obtained by Brown et al. (2005).

5.2 STELLAR CONSTITUENTS AND DYNAMICS

The K-band spectra of the IRS 16/IRS 13 stars (figures 5.3 and 5.4) contain strong HeI lines, together with products (nitrogen and carbon) in their outer atmospheres of significant nucleosynthesis. These "HeI" stars therefore appear to be members of the blue supergiant variety, characterized by an initial mass $> 40\ M_\odot$, that have evolved off the main sequence. They are probably on their way to becoming Wolf-Rayet stars and then supernovae.

Wolf-Rayet stars are hot ($\sim$25,000–50,000 K), massive ($>20\ M_\odot$) objects with a high rate of mass loss. Strong, broad emission lines, consistent with wind speeds in excess of 750–1,000 km s^{-1}, are apparently produced in the material blown off their surface. These stars can be subdivided further into three broad classes based on their spectrum and implied abundance: the WN stars, whose envelopes are dominated by nitrogen and some carbon; the WC stars, in which carbon is dominant and no nitrogen is present; and the rarer WO stars, in which oxygen is strong compared to carbon.

How the IRS 16/IRS 13 HeI stars were assembled in the first place is not entirely clear, though they may represent the most massive members of a burst of star formation that occurred between 2 and 9 million years ago; presumably several hundred OB stars and thousands of others all formed at about the same time.[12] Given their mass, there would have been sufficient time by now for the HeI stars to have evolved off the main sequence. Additional albeit indirect evidence in favor of this genesis is provided by the presence, within the central parsec, of some thirty very cool ($<$3,000 K) and very bright red giants (AGB stars) with luminosities 10^3–$10^4\ L_\odot$. These may be the relics of other such starburst episodes that occurred much earlier, somewhere between 100 and 1,000 million years ago.

In contrast, the integrated spectrum of the Sagittarius A* cluster, concentrated within 1″ of the radio source, is blue and featureless,[13] suggesting stars hotter than K-type giants. With K-band magnitudes of $\sim$14–16, these cluster members therefore appear to be early B or late O stars. But if the Sagittarius A* cluster members are indeed early-type

---

[12]Proposals along these lines have been made by Rieke and Lebofsky (1982), Lacy, Townes, and Hollenbach (1982), Allen and Sanders (1986), and Krabbe et al. (1995).

[13]See Eckart, Ott, and Genzel (1999) and Gezari et al. (2002).

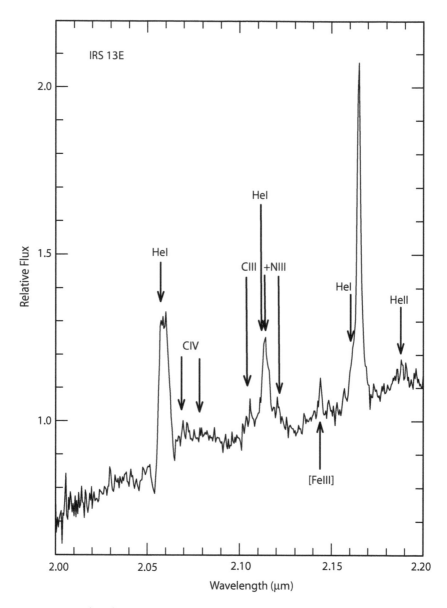

Figure 5.3 K-band spectrum of IRS 13E obtained with the MPE three-dimensional spectrometer on the ESO-MPG 2.2-meter telescope on LaSilla. The arrows mark the wavelengths of prominent transitions/species. IRS 13E has a spectrum characteristic of a late WN or WC star. (Image from Genzel 2001)

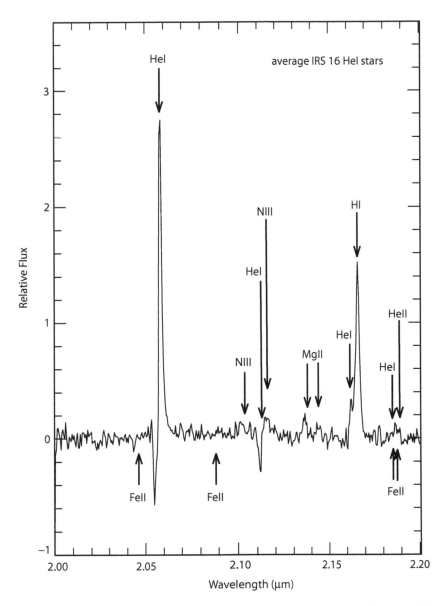

Figure 5.4 K-band spectrum of an average of IRS 16NE, C, NW, and SW obtained with the MPE three-dimensional spectrometer on the ESO-MPG 2.2-meter telescope on LaSilla. The IRS 16 stars have spectra similar to luminous blue variables (LBVs, like AG Car, P-Cyg, or η Car) or ON stars, and they appear to be on their way to becoming hot Wolf-Rayet stars. (Image from Genzel 2001)

103

main sequence stars, they must also be relatively young (<20 Myr), posing a significant difficulty in accounting for how they could have formed so close to the supermassive black hole, where the extreme conditions would presumably have inhibited star formation.[14]

We can get a better picture of how the stars within the Sagittarius A* cluster are related to those farther out by examining the dynamical profiles of these two aggregates. As a group, the starburst (early-type) stars exhibit a well-defined overall angular momentum unlike that expected on the basis of the general galactic rotation. The line-of-sight velocities of twenty-nine emission-line sources, for which adequate data have been acquired, follow a rotation pattern with blueshifted radial velocities north and redshifted velocities south of the dynamical center (see figure 5.5). The corresponding rotation axis of this sample is aligned, to an accuracy of ±20°, in the east-west direction. It therefore appears that the early-type stars are rotating clockwise (on the sky), counter to the overall galactic rotation, which shows blueshifted motion south and redshifted motion north of the galactic center. figure 5.5 also points to a relatively high average rotation rate ($\sim$150 km s^{-1}), consistent with an enclosed central concentration of two to three million solar masses. By comparison, the late-type stars exhibit a slow rotation rate of only a few tens of km s^{-1}.

An even closer inspection of these velocities reveals that there are actually two dominant groups of counterrotating stars within several arcseconds of Sagittarius A*. Within the central 3″, about half of the early stars follow clockwise (and 20% follow counterclockwise) tangential orbits. Between 3″ and 10″, the early stars divide equally into clockwise and counterclockwise rotation, though still none lie on radial trajectories. Most of the early-type stars within 10″ ($\approx$ 0.4 pc) of Sagittarius A* are therefore unrelaxed and belong to one of two rotating and geometrically thin disks with opposite line-of-sight angular momentum vectors.[15] The clockwise disk is more compact, with a characteristic radius of 2″–4″ ($\approx$0.08–0.16 pc), while the counterclockwise disk is larger, with a characteristic radius of 4″–7″ ($\approx$0.16–0.28 pc).

---

[14]Morris (1993) has concluded that gas near the galactic center would have to be compressed to densities five orders of magnitude higher than is typically found in the interstellar medium in order to overcome the strong magnetic fields ($\sim$mG), large turbulent velocities ($\sim$10 km s^{-1}), high temperatures, and strong tidal forces induced by the black hole, all of which conspire to prevent condensations of gas from forming protostars.

[15]This analysis was carried out by Genzel et al. (2003b).

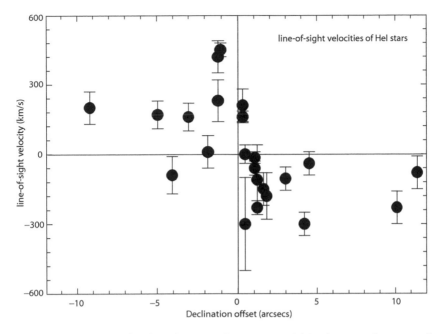

Figure 5.5 The line-of-sight velocities of HeI stars within the central parsec of the galaxy, as a function of declination offset from the position of Sagittarius A*. The error bars are $\pm 1\sigma$. (Image from Genzel et al. 2000)

Though they counterrotate, they still share a common projection of the rotation vector that is opposite to that of the Galaxy, as we noted earlier.

One should note that the stellar contents of the two disks are essentially identical. Of the 14 clockwise members with spectroscopic identifications, 4 are Of/LBV, 5 are WNL, 1 is a WNE, and 4 are WCL stars. Of the 12 spectroscopic counterclockwise members, 2 are Of/LBV, 3 are WNL, 6 are WCL, and 1 is a WCE star. Since the lifetime of the luminous blue variable and Wolf-Rayet stars is no more than several hundred thousand years, the two disks must have formed at about the same time, less than one million years ago.

But what about the inner 1″? By now, the projected and radial motions of more than 100 stars within the central 0.1 pc have been measured. As we alluded to earlier, the acceleration of several sources within only a few light-days of Sagittarius A* has also been determined (see §5.3). To uncover any possible anisotropy in their velocity distribution, we can

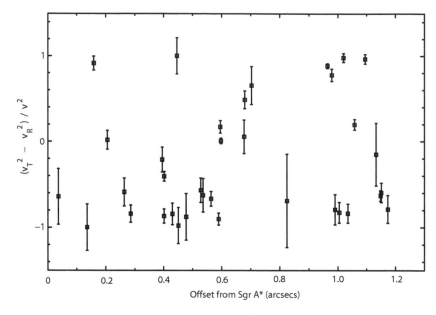

Figure 5.6 The anisotropy parameter $(v_T^2 - v_R^2)/v^2$ plotted against the projected distance from Sagittarius A*. Here, v is the proper motion velocity, and v_T and v_R are, respectively, the projected tangential and radial components. Since most of the stars fall within the negative half-plane, the majority of this sample appears to be on radial orbits. (Image from Schödel et al. 2003)

employ the anisotropy estimator[16]

$$\gamma_{TR} \equiv \frac{v_T^2 - v_R^2}{v^2},\qquad 5.35$$

in which v is the proper motion velocity and v_T and v_R are, respectively, its projected tangential and radial components. At the two extremes, a value of $+1$ signifies projected tangential motion, whereas -1 is the signature for purely radial motion.

Approximately half of these stars have measured proper motions with errors small enough to be used in this analysis. Their anisotropy estimator γ_{TR} is plotted against projected distance from Sagittarius A* in figure 5.6. The number of stars on radial orbits is evidently 2σ–3σ (assuming Poisson errors) above the number on projected tangential

[16]The stellar dynamics in the central arcsecond of our Galaxy has been explored in detail by Schödel et al. (2003).

orbits. Though the statistical significance of the implied predominance of $\gamma_{TR} < 0$ is only moderate for this small sample, the main result is reinforced by the observation that the radial anisotropy seems to increase with decreasing distance from the black hole. In addition, the overall rotation in the Sagittarius A* cluster, though in the same sense (clockwise as seen in the plane of the sky) as that of the HeI stars, is not as pronounced in the former as it is in the latter. So unlike the stellar disk components farther out, the Sagittarius A* cluster sources brighter than $K \sim 16$ evidently possess a *radial* anisotropy, but whether this also holds for the fainter stars is yet to be determined.

The fact that any tangential anisotropy in the Sagittarius A* cluster may be ruled out also renders unlikely the possibility that the galactic center may be harboring more than one supermassive object. Stars on highly eccentric orbits would be ejected or destroyed preferentially by a system whose gravitational potential was dominated by binary black holes,[17] producing instead a predominance of $\gamma_{TR} > 0$, counter to what is actually observed.

It is natural to wonder how these massive (and apparently) young stars within the two disks and the Sagittarius A* cluster came to reside in this chaotic environment. Did they form in situ, or did they migrate toward the black hole from somewhere else? Given their short lifespan, it is difficult to see how they could have simply diffused to the center via two-body interactions.[18]

But whereas a single star would have insufficient time to sink gravitationally to the center via dynamical friction, the timescale for this process goes as the inverse of the object's mass.[19] Thus, a compact young cluster could in principle migrate to the central parsec before getting tidally disrupted.[20] Detailed simulations of this process show that the cluster must be very massive ($\gg 10^4 M_\odot$) and very compact (<0.2–0.4 pc) in order to spiral into the center within the lifetime of its O-stars (a few million years).

---

[17] See Milosavljević and Merritt (2001).

[18] Alexander (2003) has estimated the relaxation time of stars within the cusp to be greater than 10^8 years. By comparison, the lifetime of $\sim 10\,M_\odot$ stars is about an order of magnitude shorter.

[19] For an excellent treatment of all aspects of gravitational dynamics discussed in this chapter, see Binney and Tremaine (1987).

[20] This suggestion was first made by Gerhard (2001), and subsequent numerical simulations were carried out by Portegies Zwart, McMillan, and Gerhard (2003).

According to these calculations, such a cluster could indeed make it to the central $\sim 10''$, but only from within an initial radius of ~ 4–5 pc. However, the problem with this picture is that after disruption the remnant aggregate has a radius of ~ 1–2 pc, much larger than the characteristic size (~ 0.1–0.4 pc) of either of the two disks. In addition, there doesn't appear to be any observational evidence for the presence of such compact, dense clusters within a few parsecs of the nucleus. Finally, the most serious difficulty with this scenario is that the existence of two disks requires the highly improbable migration—at about the same time—of two different clusters into the relatively small cusp volume.

It is also difficult to argue that we are merely seeing stars that form outside the central parsec and then rapidly move through the central region on highly elliptical orbits. Were this the case, we then ought to see around 100 times as many massive stars outside the cusp as in the interior, and this does not appear to be supported by current observations.

Other than the infusion of new massive stars via gravitational friction—whose timescale appears to be inadequate—the most likely scenario for the introduction of early-type stars into the central cusp appears to be in situ condensation from the collapse of infalling molecular clouds. But as we noted earlier, the tidal shear and physical conditions at the galactic center greatly inhibit star formation unless the cloud density is ~ 5–10×10^9 hydrogen atoms cm^{-3}, some five orders of magnitude greater than is generally encountered there.

Thus, in order for this process to work, the infalling clouds must be compressed substantially before they can spawn 10^{3-4} $M_\odot$ of stars during the 2–7 million year starburst. Perhaps the collision between two infalling clouds may have provided the necessary conditions for this to occur. Future detailed numerical simulations will reveal whether such a scenario is consistent with what we now know about the molecular gas and stellar content in the inner 5–10 pc of the Galaxy. In addition, it remains to be seen whether this population could then also account for the young stars plunging into the Sagittarius A* cluster on highly elliptical orbits.

So, to summarize, the stellar cusp centered on the massive black hole consists not only of old, low-mass stars with properties similar to those found in the galactic bulge, but it also contains a surprising number of unrelaxed, apparently young, massive objects, including very short-lived HeI emission line stars with masses 30–100 $M_\odot$. Many of these are members of two counterrotating (0.1–0.4 pc) disks. The nucleus is evidently a region undergoing episodic star formation, producing an influx of early-type unrelaxed stars into the general mix. The situation changes yet again as we look even closer to the black hole. In the next section, we shall examine the orbits of six individual members of the Sagittarius A* cluster. We will see that the angular momentum distribution for them is unlike that of any of the other cusp stars we have discussed so far.

5.3 Stellar Orbits and the Enclosed Mass

The orbits of six early-type stars within 0.″5 of Sagittarius A* are plotted in figure 5.7. As we discussed in the previous section, these sources (and half of the other Sagittarius A* cluster members) show an excess of *radial* orbits, unlike the cusp stars at distances greater than $\sim 2''$. Since they spend most of their time near the apoapse of their trajectory, it is likely— given the higher probability of observing them here rather than elsewhere on their orbit—that they are being viewed at radii roughly twice their semimajor axis ($\sim$25 light-days). These radial stars are evidently very tightly bound to the black hole.

Among the fastest (560–1,350 km s^{-1}) moving objects at the galactic center, they are also the closest (0.004–0.013 pc) individually distinguished stars to the nominal position of Sagittarius A*. And their accelerations of 2–5 mas yr^{-1}, or equivalently $(3-6) \times 10^{-6}$ km s^{-2}, are comparable to that of Earth in its orbit about the Sun. Acceleration vectors are particularly excellent tracers of the central mass distribution, since each of them should point toward the source of gravity even if projected onto the plane of the sky (see figure 5.8). Of course, the combination of accelerations and velocities provides the optimum result, though one must still contend with some fairly sophisticated orbital fitting. The reason for this is that one must deal with at least nine unknowns: the projected center of attraction (two coordinates), the period, the semimajor axis, the eccentricity, the time of periapse passage, the angle of nodes to periapse, the angle of the line of nodes, and the inclination.

With the acceleration uncertainty cones thus determined (see figure 5.8), the overlap region reveals the location of the dark mass. For the three stars shown here, the 1σ intersection point lies 0.05 ± 0.03 arcsec east and 0.01 ± 0.03 arcsec south of Sagittarius A* (situated at the origin), fully consistent with the latter's identification as the radiative

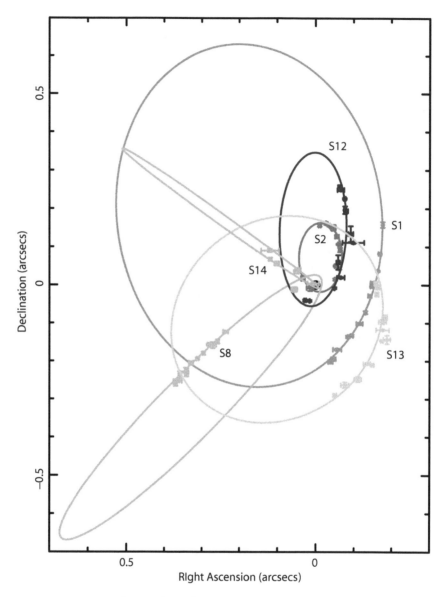

Figure 5.7 Measured time-dependent positions (with errors) of six stars near Sagittarius A*. These are labeled S1, S2, S8, S12, S13, and S14 (in the Eckart and Genzel 1997 naming scheme). The best-fit projected orbits all have moderate to high eccentricities. (Image from Schödel et al. 2003)

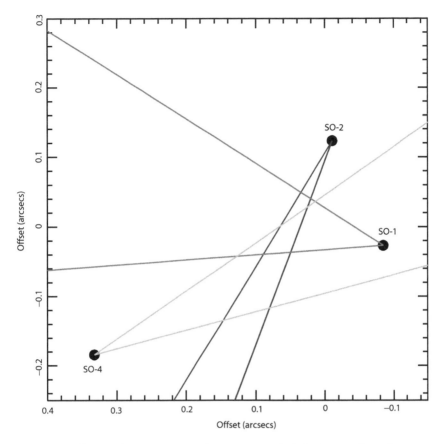

Figure 5.8 The acceleration uncertainty cones and their intersections for stars S0-1, S0-2, and S0-4 in the Ghez et al. (1998) naming convention, also known, respectively, as stars S1, S2, and S8 in the Eckart and Genzel (1997) designation (see figure 5.7). The edges of these cones correspond to directions for which the accelerations deviate by 1σ from their best-fit values. The 1σ intersection point lies 0.05 ± 0.03 arcsec east and 0.02 ± 0.03 arcsec south of Sagittarius A*. (Image from Ghez et al. 2000)

manifestation of the nonluminous mass concentration at the galactic center.

But these orbits do a lot more than simply locate the source of gravity—they also help to constrain the *volume* within which the dark matter is concentrated. The most dramatic orbit seen to date has been associated with the newly identified star S0-16,[21] which, at periapse,

[21] See Ghez et al. (2003a).

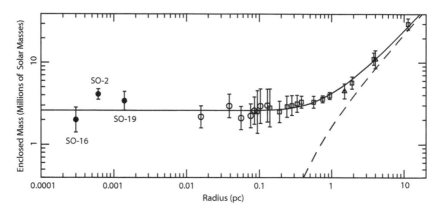

Figure 5.9 Enclosed mass as a function of radius. Solid circles represent the mass determination from individual stars, here S0-2 (S2 in the Eckart and Genzel 1997 scheme), S0-16, and S0-19 (S12 in Eckart and Genzel 1997). The other symbols indicate the enclosed mass inferred from velocity dispersions in the central cluster. The solid curve is the best-fit black hole plus luminous cluster model. The inferred black hole mass is $3.6 \pm 0.4 \times 10^6 \, M_\odot$. The dashed curve represents the enclosed mass due to the stars alone (see figure 5.1). (Image from Ghez et al. 2003a)

passed a mere 90 AU from the center at a velocity of 9,000 km s^{-1}. This event alone makes a compelling argument that the nonluminous matter must be confined within a radius of 0.0004 pc (12 light-hours) or roughly 1,000 Schwarzschild radii for a black hole mass of $(3.6 \pm 0.4) \times 10^6 \, M_\odot$ (see figure 5.9 and equation 6.117). Not quite as restrictive as the size limitations placed on Sagittarius A* by its X-ray and IR variability (see chapter 4), this result is nonetheless just as impressive because it is based on an entirely different technique. Together, these two diverse lines of evidence leave no doubt that we are dealing with $\sim 3 \times 10^6 \, M_\odot$ of mass in a region of solar-system proportions—or smaller.

Before we leave this topic and move on to consider other influences that the black hole can exert on its surrounding environment, we should point out that one should not overinterpret the fact that all six stars shown in figure 5.7 have eccentric orbits. In a spherical system of test particles orbiting a point mass, the number of eccentricities in the range $(e, e + de)$ is proportional to $e \, de$.[22] The cumulative distribution function (number versus eccentricity) is shown as a dashed curve in figure 5.10. Note that 75% of stars have $e > 0.5$. To be sure, the data displayed

[22] See problem 4-22 in Binney and Tremaine (1987).

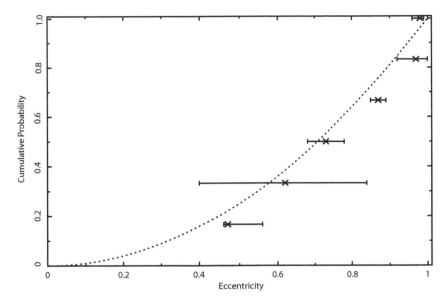

Figure 5.10 The crosses mark the eccentricities inferred for the six stars S1, S2, S8, S12, S13, and S14, whose orbits are plotted in figure 5.7. By comparison, the dotted curve shows the cumulative probability distribution function of the eccentricities expected of test particles orbiting a central potential in a spherical isotropic system. (Image from Schödel et al. 2003)

on this plot all fall to the right and below this theoretical curve, in line with the general anisotropy evident in figure 5.6. So these orbits are more eccentric than they would otherwise be in the isotropic case. But it is the degree of eccentricity that matters, not simply the fact that the orbits are eccentric, which they would be anyway, even in a spherical system.

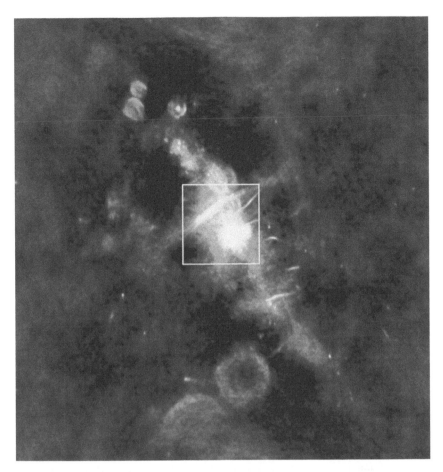

Plate 1. Radiation with a wavelength significantly larger than the size ($\sim 10^{-4}$ cm) of dust particles can pass through the interstellar medium without any noticeable attenuation. Thus, at 90 cm, the galactic center reveals itself as one of the brightest and most intricate regions of the sky. This VLA radio image, spanning an area of about 1,000 light-years on each side, shows a rich morphology produced by supernova remnants (the circular features), wispy, snakelike synchrotron filaments, and highly ionized hydrogen gas. The galactic plane in this image runs from the lower right to the upper left. A schematic diagram of the extended sources seen here is shown in figure 1.1. The central box indicates the bright region magnified in color plate 2. (Produced at the U.S. Naval Research Laboratory by Dr. N. E. Kassim and collaborators from data obtained with the National Radio Astronomy's Very Large Array Telescope, a facility of the National Science Foundation operated under cooperative agreement with Associated Universities, Inc. This image originally appeared in LaRosa, Kassim, and Lazio 2000.)

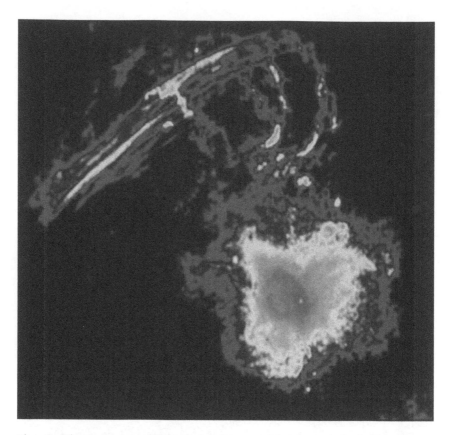

Plate 2. This is the magnified view of the central bright region of color plate 1. This VLA radio image shows the intensity of radiation at 20 cm, produced mostly by magnetized, hot gas between the stars. The size of this magnified region is about one-fifth of that shown in color plate 1, spanning a couple of hundred light-years in either direction. The galactic plane runs from the top left in this image to the bottom right, as in color plate 1. One of the most interesting features appearing here is the system of narrow filaments (some wrapped around each other) with a width of about 3 light-years. These radio filaments are oriented perpendicular to the galactic plane. The bright rosettelike region surrounds the center of our Galaxy, which is magnified further in color plate 3. The spot in the middle of the red spiral identifies the radio source known as Sagittarius A*, believed to be the supermassive black hole at the center of our Galaxy. (Image courtesy of F. Yusef-Zadeh, and the National Radio Astronomy Observatory/Associated Universities, Inc.)

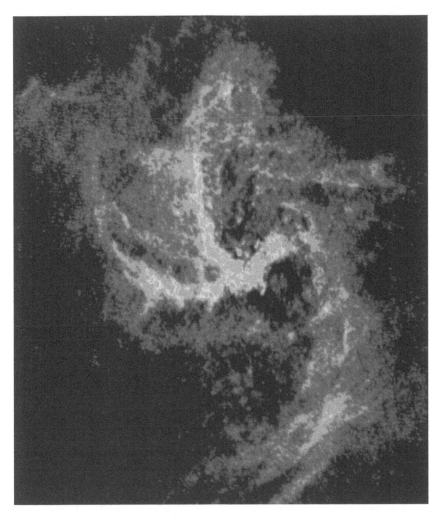

Plate 3. This image shows a magnified view, though now at 6 cm, of the (red) spiral structure in color plate 2. Each of the "arms" is about 3 light-years in length, but it is not clear whether we are witnessing a real spiral pattern or merely a superposition of independent gas flows into the center. It is now known that this gas is moving about the nucleus with a velocity as high as 1,000 km s^{-1}. The central region is magnified further in color plate 4. (Image courtesy of F. Yusef-Zadeh at Northwestern University, and the National Radio Astronomy Observatory)

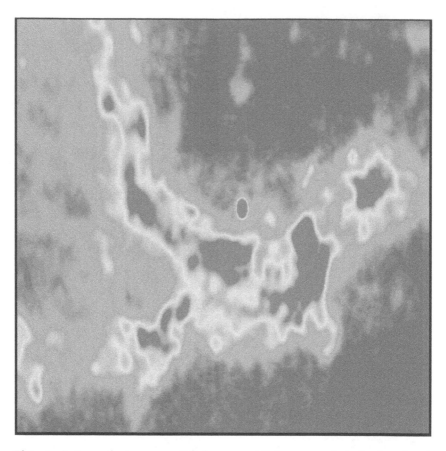

Plate 4. At 2 cm, the innermost 2 light-year × 2 light-year region of color plate 3 is dominated by the central portion of the spiral pattern of Sagittarius A West and a bright pointlike source known as Sagittarius A*, near the middle of the image. This object, which appeared as a spot of emission near the middle of the rosette in color plate 2, is associated with a mass of ≈ 2.6–$3.6 \times 10^6\ M_\odot$. To the north of Sagittarius A*, the cometary-like feature (in light blue against the dark blue background) is associated with the luminous red giant star IRS 7. The gas blown upward from its envelope provides evidence of a strong wind emanating from the region near the supermassive black hole. The distance between Sagittarius A* and the red giant is ∼3/4 light-year. (Image courtesy of F. Yusef-Zadeh at Northwestern University, and the National Radio Astronomy Observatory)

Plate 5. This very sharp 2 light-year × 2 light-year view of the stars surrounding the supermassive black hole at the heart of the Milky Way was created with the European Southern Observatory's 8.2-meter telescope atop Paranal, Chile. The colorization was produced by blending three images between 1.6 and 3.5 microns, using a color scheme in which blue is hot and red is cool. The diffuse emission is produced by interstellar dust. The location of the black hole itself, which coincides with the center of the Galaxy, is indicated by the two yellow arrows in the middle of the image. Sagittarius A* does not radiate perceptibly at infrared wavelengths, so it is not visible in this image. (Photograph courtesy of R. Genzel et al. at the Max-Planck-Institut für Extraterrestrische Physik, and the European Southern Observatory)

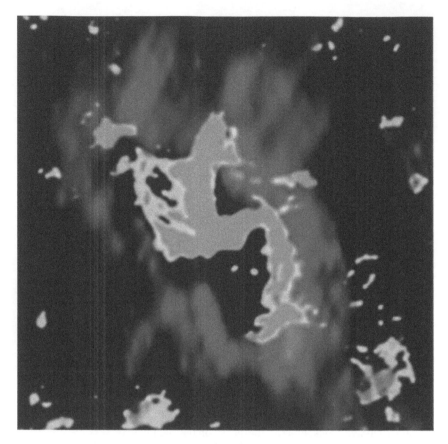

Plate 6. The obscuration toward the galactic center is much less severe in the infrared than at optical wavelengths. This is a radio image of ionized gas (Sgr A West) at $\lambda = 1.2$ cm, with its three-arm appearance (orange) superimposed on the distribution of HCN emission (violet), providing evidence for the presence of a torus of dusty gas in orbit about the central source of gravity, Sagittarius A*. The dust in this ring shines by converting ultraviolet light into an infrared glow. Most of the ionized gas is distributed within the molecular cavity. At the distance to the galactic center, this image corresponds to a size of approximately 4 pc on each side. (Image courtesy of F. Yusef-Zadeh at Northwestern University, M. Wright at the Radio Astronomy Laboratory, University of California at Berkeley, and the National Radio Astronomy Observatory)

Plate 7. This image is a mosaic covering an $\sim 2° \times 0.8°$ band in galactic coordinates centered at $(l^{II}, b^{II}) = (-0.1°, 0°)$. Based on thirty separate observations made in July 2001, this image is color coded to show intensity in three different energy bands: 1–3 keV (red), 3–5 keV (green), and 5–8 keV (blue). The galactic plane runs vertically down the center of this map. (Image courtesy of Q. Daniel Wang at the University of Massachusetts, Amherst, and NASA)

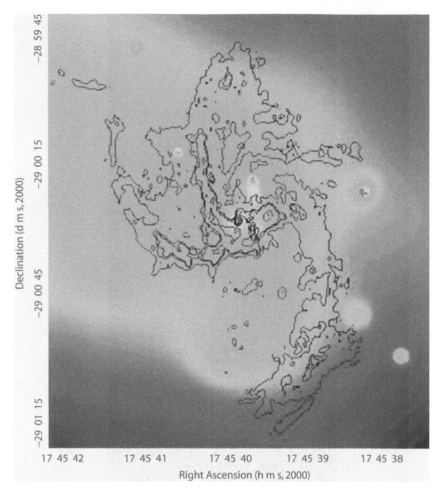

Plate 8. This is a smoothed, colorized version of the *Chandra* 0.5–7 keV image shown in figure 3.4, though with a slightly larger field of view (showing the central 1.′3 × 1.′5 of the Galaxy). Overlaid on the X-ray image are VLA 6 cm contours corresponding to the intensity map shown in color plate 3. The X-ray emission from Sagittarius A* itself appears as a red dot at $17^h45^m40.0^s$, $-29°00'28''$. Bright diffuse emission from hot gas is visible throughout the region and appears to be produced primarily via wind-wind collisions (see chapter 7). (From Baganoff et al. 2003)

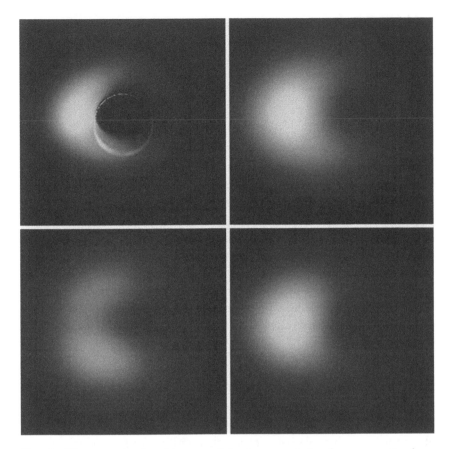

Plate 9. These are simulated images of the 1-mm emission from magnetized gas orbiting about the black hole in a counterclockwise direction. At this wavelength, we are viewing a polarization map near the peak of the mm–to–sub-mm portion of Sagittarius A*'s spectrum. The panel in the top left-hand corner shows the raw ray-tracing output, colorized so that red corresponds to a net vertical polarization, whereas cyan shows the brightness map of radiation polarized in the horizontal direction. The ringlike feature is produced by radiation that orbits once around the black hole before escaping. The corresponding image in the top right-hand corner is blurred to account for the finite VLBI resolution and interstellar scattering. At this wavelength, the obscuring effects of scattering by the interstellar gas and dust are present, though not overwhelming, and we can begin to see the shadow of the black hole cast against the radiant plasma. The clarity continues to improve as the wavelength of the radiation decreases further, though this effect is somewhat mitigated by the fact that the source also gets fainter. The lower two panels show the vertical (red) and horizontal (cyan) components of the polarized emission. The pixel brightness in all images scales linearly with flux. (From Bromley, Melia, and Liu 2001)

CHAPTER 6

The Four-Dimensional Spacetime

Throughout the remainder of this book, we will consider the behavior of gas and stars interacting dynamically with the central concentration of dark matter in regions where the effects of strong gravity are sometimes manifested. We will begin by establishing the basis for the description we shall use, starting with the language of four-dimensional spacetime. A generalization of the simplest theory—that which is relevant only for systems moving at constant speed—to a framework in which the various observers accelerate with respect to each other may be made seamlessly with an appropriate consideration of inertial forces, an exercise that will occupy us toward the end of this chapter.

6.1 THE FLAT SPACETIME METRIC

In Galilean relativity, the space ($\mathbf{x}$) and time (t) coordinates of two frames of reference moving relative to each other with a *constant* velocity $\mathbf{v}_o$ are related to each other by

$$\mathbf{x}' = \mathbf{x} - \mathbf{v}_o t \qquad\qquad 6.1$$

and

$$t' = t. \qquad\qquad 6.2$$

Although the physical laws of *classical* mechanics are invariant under such a Galilean coordinate transformation, it is well known that the Maxwell equations of electrodynamics—themselves forged before the advent of special relativity—are not invariant.[1] It was specifically

---

[1] This can be seen most readily via the application of equations (6.1) and (6.2) to Faraday's law, which changes form from one reference frame to the next. See, for example, Melia (2001b).

the desire to make both classical mechanics and electrodynamics consistent with the same transformation between inertial frames of reference that provided the impetus for a modification of the Galilean relativity theory. And in the process of bringing about this unification, the early workers in relativity theory found that it was the rest of classical mechanics that had to be modified to the point where it could no longer be Galilean invariant, while the laws of electrodynamics remained unchanged when written in the frame-dependent language of a given observer. Nevertheless, even the concept of an electromagnetic field had to be reshaped in order to develop an invariant description of the pertinent physical laws using the vocabulary of four-dimensional spacetime—the structure of the new relativity theory.

The method devised to handle the transformation of one set of coordinates into another had to incorporate two essential postulates: (1) that only *relative* motion is observable, and (2) that the velocity of light in vacuum is a *constant c*, independent of the source and/or observer speed.[2]

Postulate (2) immediately negates the possibility that observers in different frames may agree on the space and time (*spacetime*) coordinates of an event. Suppose that a pulse of light is emitted at time $t = 0$ when the origins O and O' of the two coordinate systems occupy a common point. Then the observer at O will report that the pulse arrives at another location $\mathbf{x}$ in this space after a time

$$t = r/c \tag{6.3}$$

has elapsed, where

$$r = (x^2 + y^2 + z^2)^{1/2}. \tag{6.4}$$

The observer at O', on the other hand, reports that the elapsed time is instead

$$t' = r'/c, \tag{6.5}$$

where now

$$r' = [(x')^2 + (y')^2 + (z')^2]^{1/2}. \tag{6.6}$$

---

[2]For an early critical review of the experimental basis for the second postulate, see Fox (1965, 1967).

The speed of light c is the same in both frames, but because the two origins have separated during the propagation of the pulse, then clearly $r \neq r'$ and $t \neq t'$.

It would seem, however, that the constancy of c should be reflected in the invariance of certain functions of the coordinates. One of the best-known such relationships—and ultimately the most useful—is the "spacetime interval"

$$(\Delta s)^2 \equiv (c\,\Delta t)^2 - |\Delta \mathbf{x}|^2, \qquad\qquad 6.7$$

which is *always* zero along a light path. In terms of the spherical coordinates we have just described, with $t = t' = 0$ at the origin,

$$(\Delta s)^2 = (c\,t)^2 - r^2 = 0 = (c\,t')^2 - (r')^2 = (\Delta s')^2. \qquad 6.8$$

The identification of the transformation laws for the new relativity theory amounts to simply finding a set of relations among the coordinates that guarantee the invariance of $(\Delta s)^2$. With guesses and trials, it is not difficult to convince oneself that in order for the spacetime interval to be preserved the coordinates in the two frames must be related via a so-called Lorentz transformation, in which

$$x' = x,$$

$$y' = y,$$

$$z' = \gamma(z - vt),$$

$$t' = \gamma \left(t - \frac{vz}{c^2} \right),$$

$$\gamma \equiv \left[1 - \left(\frac{v}{c} \right)^2 \right]^{-1/2}. \qquad\qquad 6.9$$

(These equations assume that the primed frame is moving at a constant velocity $\mathbf{v} = v\,\hat{z}$ relative to the unprimed coordinate system.) Other transformations preserving the speed of light—for instance, a rigid rotation—are also permissible; they are all members of the homogeneous Lorentz group. However, only the relations in equation (6.9) can describe the transformation of the coordinates from one frame moving relative to another with a constant uniform velocity. The symmetry between the two observers is preserved by a reciprocal transformation, which, as one would expect, is represented by

$$x = x',$$

$$y = y',$$

$$z = \gamma(z' + vt'),$$

$$t = \gamma\left(t' + \frac{vz'}{c^2}\right). \qquad 6.10$$

Notice that in this instance the observer at O' sees her counterpart at O moving with a velocity $-v\hat{z}$.

Let us now broaden our analysis of this new relativity and consider additional consequences of a Lorentz transformation. Measurements of length and time are essential for the determination of many physical quantities, certainly those that involve velocities and accelerations. Even inertia, for example, is characterized in terms of how much acceleration is induced by a particular force, and because the intervals of length and time change from observer to observer, inertia's inferred value must therefore be dependent on the frame of reference. But though spatial and time intervals may differ in frames moving relative to each other, *dimensionless* quantities, such as the number of events, must be preserved. (Remember, the physical laws must be invariant, so if a π° decays into two photons in one frame, then it must also be seen to decay into two photons in every other frame.)

An important example of this invariance is the phase of a wave, which has nothing to do with either distance or time separately but rather is measured as a fraction of a complete oscillation:

$$\Omega \equiv \omega t - \mathbf{k} \cdot \mathbf{x} = \omega' t' - \mathbf{k}' \cdot \mathbf{x}'. \qquad 6.11$$

Notice what happens if we introduce the Lorentz transformation equations for the coordinates into this expression:

$$\omega t - kz = \omega'\left(\gamma t - \gamma\frac{vz}{c^2}\right) - k'\left(\gamma z - \gamma vt\right) \qquad 6.12$$

or

$$\omega t - kz = \left(\omega'\gamma + \gamma k'v\right)t - \left(\gamma\frac{\omega'v}{c^2} + \gamma k'\right)z, \qquad 6.13$$

again assuming that the wave is propagating in the z-direction. This must be true for all values of t and z, and we are therefore compelled to conclude that

$$\omega = \gamma(\omega' + k'v) \qquad\qquad 6.14$$

and

$$k = \gamma\left(k' + \frac{v\omega'}{c^2}\right). \qquad\qquad 6.15$$

Something quite remarkable happens when we apply these expressions to light, for which the frequency and wavenumber are related as $k = \omega/c$ and $k' = \omega'/c$, respectively, in the two frames. Evidently,

$$\omega = \gamma\omega'(1+\beta) \qquad\qquad 6.16$$

or, in the reciprocal sense,

$$\omega' = \gamma\omega(1-\beta), \qquad\qquad 6.17$$

where $\beta \equiv v/c$. Generalizing this to situations in which the wavevector $\mathbf{k}$ is not directed along $\hat{z}$ is straightforward, since only the component parallel to $\mathbf{v}$ is affected. In that case, we obtain the *Doppler shift formula*,[3] written as

$$\omega' = \gamma\omega(1 - \beta\cos\theta), \qquad\qquad 6.18$$

where $\mathbf{k}\cdot\mathbf{v} = \cos\theta|\mathbf{k}||\mathbf{v}|$.

A striking feature of equations (6.15) and (6.16) is that the quantities $(\omega/c, \mathbf{k})$ have the same (generalized) transformation properties as $(ct, \mathbf{x})$—clearly an inkling of a more elaborate substructure in special relativity.[4] The four-dimensional groupings

$$x^\mu \equiv (x^0, \mathbf{x}) \quad (\mu = 0, 1, 2, 3) \qquad\qquad 6.19$$

---

[3]This expression also predicts a *transverse* Doppler shift, which was verified by several early experiments, including one using the Mössbauer effect (Hay et al. 1960).

[4]Two highly recommended books on this subject are Sard (1970) and the more detailed text by Weinberg (1972).

and

$$k^\mu \equiv (k^0, \mathbf{k}) \quad (\mu = 0, 1, 2, 3), \qquad\qquad 6.20$$

where $x^0 \equiv ct$ and $k^0 \equiv \omega/c$, are examples of four-vectors in the four-dimensional spacetime. More precisely, x^μ and k^μ *transform* as four-vectors under a Lorentz boost from one frame to the next. Any quantity that transforms in the same fashion as x^μ is called a four-vector in this space.

Well then, physical quantities (such as x^μ and k^μ) that involve both a magnitude and a direction must necessarily involve four components that transmute into each other upon a transformation of frames. But not all measurable elements depend on direction—for example, the phase of a wave or the number of particles that we introduced above. Some of them simply require denumerability. For these physical entities, a single function is sufficient. There also exist certain interactions where one must consider the transport of a vector along different directions. A practical example in this category is the stress applied to a surface by an electromagnetic field, for which the components of force may vary across the enclosure. We shall see that for physical quantities such as this the description in four-dimensional spacetime requires sixteen transmutable components.

The spacetime interval Δs is itself a special quantity because it is *invariant*, which we can show via the application of the transformation relations in equation (6.9):

$$\begin{aligned}
(\Delta s)^2 &= (x^0)^2 - (x^1)^2 - (x^2)^2 - (x^3)^2 \\
&= \gamma^2 \left(x'^0 + \beta \frac{x'^3}{c} \right)^2 - (x^1)^2 \\
&\quad -(x^2)^2 - \gamma^2 \left(x'^3 + \frac{vx'^0}{c} \right)^2 \\
&= (x'^0)^2 - (x'^1)^2 - (x'^2)^2 - (x'^3)^2 \\
&= (\Delta s')^2.
\end{aligned} \qquad\qquad 6.21$$

The interval s is therefore a *scalar* in four-dimensional spacetime. This property has practical value because it allows one observer to compare his intervals of space and time directly with those measured in another frame. For example, in the rest frame of an observer for whom $|\mathbf{x}_2 - \mathbf{x}_1| = 0$, the interval Δs is due entirely to a segment of time τ, known as the *proper* time, that is, the time measured in a frame where the events that define it occur at the *same* spatial location:

$$s^2 = c^2 \tau^2. \tag{6.22}$$

In a different frame,

$$s^2 = (x^0)^2 \left(1 - \frac{|\mathbf{x}|^2}{(x^0)^2} \right)$$

$$= c^2 t^2 (1 - \beta^2)$$

$$= c^2 t^2 / \gamma^2, \tag{6.23}$$

which produces a result known as *time dilatation* (or stretching) from one frame to the next: $t = \gamma \tau$.

By the way, an extension of this theory to situations in which the relative velocity of two frames is changing requires that we consider a Lorentz transformation to be valid only for an infinitesimal interval of time, during which the change in velocity is minimized. In other words, by considering the *differential* elements of physical quantities, we retain the validity of a Lorentz transformation as long as it is understood that it applies solely to that given instant. The interval s, for example, would then be considered in its infinitesimal limit

$$(ds)^2 = (c\, dt)^2 - (dx^1)^2 - (dx^2)^2 - (dx^3)^2 \tag{6.24}$$

and

$$dt = \gamma\, d\tau. \tag{6.25}$$

In the infinitesimal limit, ds becomes the Lorentz invariant.

The Lorentz group encompasses all the transformations that leave s^2 (or $[ds]^2$) invariant. Another way of expressing our first postulate is that the laws of nature must be invariant in *form* under the transformations of this group, which may however intermix the four components of the coordinate four-vector:

$$x'^{\alpha} = x'^{\alpha}\,(x^0, x^1, x^2, x^3) \qquad (\alpha = 0, 1, 2, 3). \qquad 6.26$$

Four-dimensional spacetime, however, encompasses other groupings of components, so x^{α} is but one member of the general category of so-called tensors in this set. Their properties depend on a rank k, determined by how they transform under the translation

$$x \to x'. \qquad 6.27$$

(In special relativity, the four-vector $[x^0, x^1, x^2, x^3]$ is sometimes written using the shorthand notation x.) The three quantities x^{μ}, k^{μ}, and s that we have encountered thus far all belong to the tensor class. A *scalar* (i.e., a tensor of rank zero) is a single function of x whose value is *not* changed by the transformation. The Lorentz interval $ds(x^0, x^1, x^2, x^3)$ is a scalar.

A *vector* (i.e., a tensor of rank 1) is a set of ordered numbers that transform according to the rule

$$V'^{\alpha} = \frac{\partial x'^{\alpha}}{\partial x^{\beta}}\, V^{\beta} \qquad 6.28$$

defining a *contravariant* vector and

$$V'_{\alpha} = \frac{\partial x^{\beta}}{\partial x'^{\alpha}}\, V_{\beta} \qquad 6.29$$

for a *covariant* vector. These two versions of a four-vector contain the same physics, but they incorporate a sign change in their spatial coordinates that ultimately simplifies the mathematical formalism. In addition, they allow us to easily handle a differentiation with respect to a contravariant coordinate x^{μ}, which evidently produces a covariant four-vector

$$\frac{\partial}{\partial x'^{\alpha}} = \frac{\partial x^{\beta}}{\partial x'^{\alpha}}\frac{\partial}{\partial x^{\beta}}. \qquad 6.30$$

Note that in this expression and throughout this book unless otherwise excepted a repeated index means that the term in which the index appears is to be summed over all its possible values. So, for example, $x^{\alpha}x_{\alpha} \equiv x^0 x_0 + x^1 x_1 + x^2 x_2 + x^3 x_3$.

The transformation of a vector is often written in the more compact form

$$V'^{\alpha} = a^{\alpha}{}_{\beta} V^{\beta},$$

6.31

where

$$a^{\alpha}{}_{\beta} \equiv \frac{\partial x'^{\alpha}}{\partial x^{\beta}}.$$

6.32

And for the Lorentz transformation represented in equation (6.9), the matrix of coefficients can immediately be written down in the form

$$[a^{\alpha}{}_{\beta}] = \begin{pmatrix} \gamma & 0 & 0 & -v\gamma/c \\ 0 & 1 & 0 & 0 \\ 0 & 0 & 1 & 0 \\ -v\gamma/c & 0 & 0 & \gamma \end{pmatrix}.$$

6.33

For a covariant four-vector, the coefficients are the same as these, though with a sign change that ultimately arises because of the reversed sense of velocity inferred by one observer relative to the other—compare equations (6.9) and (6.10):

$$[a_{\alpha}{}^{\beta}] = \begin{pmatrix} \gamma & 0 & 0 & v\gamma/c \\ 0 & 1 & 0 & 0 \\ 0 & 0 & 1 & 0 \\ v\gamma/c & 0 & 0 & \gamma \end{pmatrix}.$$

6.34

Notationally, it is important to remember that the indices in these coefficients may be either subscripts or superscripts and that their horizontal ordering is not arbitrary. A superscript means that its corresponding coordinate is being differentiated, and the first horizontal position is reserved for the "primed" coordinate. So, for example, $a_{\mu}{}^{\nu} = \partial x^{\nu}/\partial x'^{\mu}$.

A tensor of rank $k = 2$ is a four-vector of four-vectors and is therefore a grouping of sixteen functions of x that transform according to an obvious generalization of the rules for scalars and vectors. The components of a *contravariant* tensor $T^{\alpha\beta}$ satisfy the transformation relations

$$T'^{\alpha\beta} = a^{\alpha}{}_{\gamma} a^{\beta}{}_{\delta} T^{\gamma\delta},$$

6.35

whereas a *covariant* tensor of rank 2 transforms according to

$$T'_{\alpha\beta} = a_\alpha{}^\gamma a_\beta{}^\delta T_{\gamma\delta}. \qquad 6.36$$

Next, we consider carrying out algebra with these tensors, a process that is is greatly facilitated with certain simple rules obeyed by the coefficients $a^\alpha{}_\beta$ and $a_\alpha{}^\beta$. One of the most important is

$$a^\alpha{}_\beta a_\alpha{}^\gamma = \delta^\gamma_\beta, \qquad 6.37$$

where

$$\delta^\alpha_\beta = \begin{cases} 1, & \text{if } \alpha = \beta, \\ 0, & \text{if } \alpha \neq \beta, \end{cases} \qquad 6.38$$

is the Kronecker delta. Equation (6.37) may be verified by direct application of the chain rule of differentiation, which gives

$$\frac{\partial x'^\alpha}{\partial x^\beta} \frac{\partial x^\gamma}{\partial x'^\alpha} = \frac{\partial x^\gamma}{\partial x^\beta} = \delta^\gamma_\beta. \qquad 6.39$$

An illustration of the useful relations that follow from this property is the decomposition of a Lorentz scalar into covariant and contravariant components. It is said that a scalar is the *contraction* of a covariant four-vector with its contravariant counterpart, which follows from the fact that $V^\alpha W_\alpha$ is invariant:

$$\begin{aligned} V'^\alpha W'_\alpha &= a^\alpha{}_\beta V^\beta a_\alpha{}^\gamma W_\gamma \\ &= \delta^\gamma_\beta V^\beta W_\gamma \\ &= V^\beta W_\beta. \end{aligned} \qquad 6.40$$

In contrast, the product $S_{\alpha\beta} \equiv V_\alpha W_\beta$ is a tensor of rank 2, since

$$\begin{aligned} S'_{\alpha\beta} &= V'_\alpha W'_\beta \\ &= a_\alpha{}^\gamma V_\gamma a_\beta{}^\delta W_\delta \\ &= a_\alpha{}^\gamma a_\beta{}^\delta S_{\gamma\delta}. \end{aligned} \qquad 6.41$$

By the same token, $T_\alpha{}^\alpha$ is a scalar, and $W_\alpha \equiv T_\alpha{}^\beta V_\beta$ is a covariant four-vector.

So in order for ds to be a scalar, it must be the product of contravariant and covariant vectors (following the decomposition in equation 6.40). Indeed, we may write

$$(ds)^2 = (dx^0)^2 - (dx^1)^2 - (dx^2)^2 - (dx^3)^2 \qquad 6.42$$

as a special case of the differential element

$$(ds)^2 = \eta_{\alpha\beta}\, dx^\alpha dx^\beta, \qquad 6.43$$

where $\eta_{\alpha\beta} = \eta_{\beta\alpha}$ is known as the *metric tensor*. In special relativity (as opposed to the general theory that we will consider next), $\eta_{\alpha\beta}$ is diagonal for certain coordinate systems, such as Cartesian, for which

$$[\eta_{\alpha\beta}] = \begin{pmatrix} 1 & 0 & 0 & 0 \\ 0 & -1 & 0 & 0 \\ 0 & 0 & -1 & 0 \\ 0 & 0 & 0 & -1 \end{pmatrix}. \qquad 6.44$$

With the definition of ds in equation (6.42) and $\eta_{\alpha\beta}$ in equation (6.44), it is also clear that

$$dx_\beta = \eta_{\beta\alpha}\, dx^\alpha. \qquad 6.45$$

Contraction with $\eta_{\alpha\beta}$ is in fact the general procedure for converting a contravariant index on any tensor into a covariant one. Thus, another way to write equation (6.42) is

$$(ds)^2 = dx_\beta\, dx^\beta. \qquad 6.46$$

To complete this description, we need a method for converting a covariant index into a contravariant index, and not surprisingly, this may be effected with the use of the contravariant metric tensor $\eta^{\alpha\beta}$, the inverse of $\eta_{\alpha\beta}$, defined via the requirement that

$$x^\alpha = \eta^{\alpha\beta}(\eta_{\beta\gamma} x^\gamma). \qquad 6.47$$

This is satisfied as long as

$$\eta^{\alpha\beta}\eta_{\beta\gamma} = \delta^\alpha_\gamma. \qquad 6.48$$

Thus, in special relativity, the coefficients $\eta^{\alpha\beta}$ and $\eta_{\beta\gamma}$ are the same when they are written in terms of Cartesian coordinates:

$$\eta^{\alpha\beta} = \eta_{\alpha\beta}. \qquad\qquad 6.49$$

In the coming sections, we shall see that all physical quantities can be described compactly and completely using the language of tensors in four-dimensional spacetime. In particular, we will find that only a description of nature written in this way can be made invariant under a Lorentz transformation, which provides us with the means of preserving the essence and relevance of all physical laws. Thus far, we have encountered the interval ds, the coordinate four-vector x^{μ}, and the four-wavevector k^{μ}. The latter two are groupings of components that may mix with each other under a Lorentz transformation but only in such a way that $x^{\mu}x_{\mu}$ and $k^{\mu}k_{\mu}$ are scalar invariants. In three-space, this would be analogous to the condition that the magnitude of a three-vector must remain unchanged under a rotation or translation, though its individual components may change.

Before broaching the subject of how physical laws may be cast in this new language, however, we shall introduce one more four-vector by bootstrapping our way up from basic principles—a technique that frequently uncovers groupings of functions that together form a tensor. In time-dependent electrodynamics, equations are often written in the Lorenz gauge, defined by the condition

$$\partial\Phi/\partial x^0 + \vec{\nabla}\cdot\mathbf{A} = 0, \qquad\qquad 6.50$$

for which the scalar (Φ) and vector ($\mathbf{A}$) potentials satisfy the wave equation. One can show that the quantity $\partial\Phi/\partial x^0 + \vec{\nabla}\cdot\mathbf{A}$ is a Lorentz scalar, thus satisfying the first postulate of special relativity, by supposing that Φ and $\mathbf{A}$ are elements of a new four-vector potential $A^{\alpha} \equiv (\Phi, \mathbf{A})$. In Cartesian coordinates, this can be done as

$$\frac{\partial\Phi'}{\partial x'^0} + \vec{\nabla}'\cdot\mathbf{A}' \equiv \frac{\partial A'^{\sigma}(x')}{\partial x'^{\sigma}}$$

$$= \frac{\partial}{\partial x'^{\sigma}}\left[a^{\sigma}{}_{\beta}A^{\beta}(x)\right]$$

$$= a^{\sigma}{}_{\beta}\frac{\partial A^{\beta}(x)}{\partial x'^{\sigma}}$$

125

$$= a^{\sigma}{}_{\beta} \frac{\partial A^{\beta}(x)}{\partial x^{\alpha}} \frac{\partial x^{\alpha}}{\partial x'^{\sigma}}$$

$$= a^{\sigma}{}_{\beta} a_{\sigma}{}^{\alpha} \frac{\partial A^{\beta}(x)}{\partial x^{\alpha}}$$

$$= \delta^{\alpha}_{\beta} \frac{\partial A^{\beta}(x)}{\partial x^{\alpha}}$$

$$= \frac{\partial \Phi}{\partial x^0} + \vec{\nabla} \cdot \mathbf{A}. \qquad 6.51$$

In other words, the left-hand side of this equation must be a Lorentz scalar, and since $\partial/\partial x^{\alpha}$ transforms as a covariant four-vector, we conclude from equation (6.40) that the potential A^{α} is therefore a contravariant four-vector in this spacetime.

6.2 Relativistic Transformation of Physical Laws

To develop a description of physical laws using the tensor language of four-dimensional spacetime, we must contend with the fact that inertia is not an invariant quantity, which presents an obvious difficulty when we attempt to quantify the effects on particles due to the influence of others. Without knowing a priori how to calculate the force on a moving object, we have no direct means of translating a physical law known in terms of our frame-dependent language into its invariant form mandated by special relativity. The most reliable approach, it turns out, is to carry out our analysis of the interactions in carefully selected frames first. As one can imagine, the most special frame is that in which the particle is at *rest*. Correspondingly, the *rest mass m*, representing the particle's inertia in its own frame, is a quantity upon which all observers can agree, and we shall therefore always mean this value when we refer to the "mass." We shall see that the variation in the particle's inertia from frame to frame may be represented by something other than *m*.

Newton's laws are valid when the particle's velocity v is very small compared to c. In the particle's rest frame, therefore, where the velocity is in fact zero, the relativistic equations of motion must reduce to their classically derived form. The idea is to calculate the applied force in this frame using the techniques of classical mechanics and to guess from it an invariant version using a combination of four-dimensional tensors. According to the postulates of special relativity, this new representation of the physical law should then be valid for all observers.

Let's look at a specific example to see how this works in practice. We will first use our intuition to guess the form of Newton's law, which ought to show the particle's acceleration in proportion to the applied force. Although the rest mass represents inertia just in the particle's rest frame, it is nonetheless the only such quantity upon which all observers can agree. So it makes sense to *define* the following four-vector as a possible relativistic representation of Newton's force law $\mathbf{F} = m\mathbf{a}$ and then to test its limiting form at low velocity:

$$f^\alpha \equiv m\frac{d^2 x^\alpha}{d\tau^2}. \qquad 6.52$$

In the particle's rest frame, where $d\tau = dt$, we have

$$f^i = F^i \quad (i = 1, 2, 3),$$
$$f^0 = 0, \qquad 6.53$$

where F^i are the Cartesian components of $\mathbf{F}$. This four-vector f^α looks like it might provide what we need, because it apparently reduces to the correct form in the classical limit, and it is written in terms of a four-vector x^α, whose transformation properties are known (see equation 6.31), and two scalars—the rest mass m and the proper time τ. But what happens to it under a transformation? We know that

$$dx'^\alpha = \alpha^\alpha{}_\beta \, dx^\beta \qquad 6.54$$

and that $d\tau$ is invariant. Thus,

$$f'^\alpha = \alpha^\alpha{}_\beta \, f^\beta, \qquad 6.55$$

as expected of a four-vector. Substituting for f^α from equation (6.53), we immediately get

$$f_L^0 = \frac{\gamma}{c}\mathbf{v} \cdot \mathbf{F},$$
$$\mathbf{f}_L = \mathbf{F} + (\gamma - 1)\mathbf{v}\frac{\mathbf{v} \cdot \mathbf{F}}{v^2}, \qquad 6.56$$

where f_L^α is the "four-force" in the frame moving with velocity $\mathbf{v}$ relative to the particle (i.e., in the lab frame). Evidently, f_L^0 is proportional to the power exerted by the force on the particle.

127

Newton also defined a linear momentum **p**, whose time rate of change was considered to be an equivalent representation of the force. It would seem, therefore, that the relativistic force f^α might also be written in terms of a relativistic version of **p**:

$$\frac{dp^\alpha}{d\tau} = f^\alpha,$$

6.57

where evidently

$$p^\alpha \equiv m\frac{dx^\alpha}{d\tau}.$$

6.58

Since $d\tau = dt/\gamma$, we see that

$$p^0 = \gamma mc,$$

$$\mathbf{p} = \gamma m\mathbf{v},$$

6.59

so the vector component **p** of p^α has the correct limiting form when $\gamma \to 1$. We shall see shortly that p^0 is proportional to the particle's energy. Together, the four components $(p^0, \mathbf{p})$ form what is known as the energy-momentum four-vector.

The equation $\mathbf{p} = \gamma m\mathbf{v}$ is particularly intriguing, because it suggests that the modification introduced to the momentum by special relativity is due entirely to the inertia, which varies with particle speed as hypothesized earlier. The mass of a particle moving at velocity v relative to an observer is apparently γm, where $\gamma = [1 - (v/c)^2]^{-1/2}$. But what does one make of p^0? Since p^α is a product of a four-vector and a scalar, it too is a four-vector, and the contraction $p^\alpha p_\alpha$ must be invariant. Indeed,

$$p^\alpha p_\alpha = (p^0)^2 - |\mathbf{p}|^2$$

$$= \gamma^2 m^2 c^2 - m^2 \gamma^2 |\mathbf{v}|^2$$

$$= m^2 c^2 = \text{constant},$$

6.60

which directly couples p^0 to **p** in all frames. In other words, every observer agrees that p^0 is related to the particle's momentum. We can get an even better idea of its meaning by considering what happens in the limit $v \to 0$, where

$$\gamma mc^2 = \frac{mc^2}{\sqrt{1-(v/c)^2}}$$

$$\approx mc^2 \left[1 + \frac{1}{2}\left(\frac{v}{c}\right)^2 + \frac{3}{8}\left(\frac{v}{c}\right)^4 + \cdots \right]$$

$$\approx mc^2 + \frac{1}{2}mv^2. \qquad 6.61$$

It looks like p^0/c is just the sum of the classical kinetic energy and another form of energy that does not vanish even when $v = 0$. For obvious reasons, we refer to the latter as the particle's *rest energy* and to the sum

$$E \equiv \gamma mc^2 \qquad 6.62$$

as its *relativistic energy*. So clearly

$$p^0 = E/c. \qquad 6.63$$

Using equations (6.59) and (6.62), we can also write the particle's velocity in terms of $\mathbf{p}$ and E,

$$\mathbf{v} = \frac{\mathbf{p}}{\gamma m} = \frac{c^2 \mathbf{p}}{E}, \qquad 6.64$$

and combining (6.60) and (6.63), we obtain one of the most celebrated equations in special relativity:

$$E^2 = |\mathbf{p}|^2 c^2 + m^2 c^4. \qquad 6.65$$

Let us now put this formalism to work and see how a theory can be made consistent with special relativity using the structure of our four-dimensional spacetime. Since it was the set of Maxwell equations that brought the issue of invariance under transformations to light in the first place, we might as well start with the equations describing the electric $\mathbf{E}$ and magnetic $\mathbf{B}$ fields. In terms of the four-potential $A^\alpha = (\Phi, \mathbf{A})$, these fields are given as

$$\mathbf{E} = -\frac{1}{c}\frac{\partial \mathbf{A}}{\partial t} - \vec{\nabla}\Phi \qquad\qquad 6.66$$

and

$$\mathbf{B} = \vec{\nabla} \times \mathbf{A}. \qquad\qquad 6.67$$

The goal is to reformulate these definitions into invariant expressions that involve only scalars, four-vectors, and four-tensors. So looking at each individual component of $\mathbf{E}$, say, we find that

$$
\begin{aligned}
E^i &= -\frac{1}{c}\frac{\partial A^i}{\partial t} - \frac{\partial \Phi}{\partial(-x_i)} \\
&= -\frac{\partial A^i}{\partial x^0} - \frac{\partial A^0}{\partial(-x_i)} \\
&= -\frac{\partial A^i}{\partial x_0} + \frac{\partial A^0}{\partial x_i} \\
&= -\left(\partial^0 A^i - \partial^i A^0\right).
\end{aligned}
\qquad 6.68
$$

And with a similar derivation, we find that

$$
\begin{aligned}
B_i &= -\varepsilon_{ijk}\frac{\partial A^j}{\partial(-x_k)} \\
&= \varepsilon_{ijk}\,\partial^k A^j,
\end{aligned}
\qquad 6.69
$$

where $\partial^k \equiv \partial/\partial x_k$ and ε_{ijk} is defined by

$$
\varepsilon = \begin{cases}
+1, & \text{if } ijk \text{ forms an even permutation of 123,} \\
-1, & \text{if } ijk \text{ forms an odd permutation of 123,} \\
0, & \text{otherwise.}
\end{cases}
\qquad 6.70
$$

The form of E^i and B_i suggests that the components of $\mathbf{E}$ and $\mathbf{B}$ are elements of a second-rank, antisymmetric *field-strength tensor*

$$F^{\alpha\beta} = \partial^\alpha A^\beta - \partial^\beta A^\alpha, \qquad\qquad 6.71$$

which written out explicitly is

$$F^{\alpha\beta} = \begin{pmatrix} 0 & -E^x & -E^y & -E^z \\ E^x & 0 & -B^z & B^y \\ E^y & B^z & 0 & -B^x \\ E^z & -B^y & B^x & 0 \end{pmatrix}. \qquad 6.72$$

Special relativity therefore knits the two three-vector fields **E** and **B** into a single entity—a field-tensor—resolvable on the four dimensions of spacetime in ways that convert electric into magnetic fields and vice versa. Which of the fields an observer sees depends merely on his frame of reference. To be convinced of this notion, we need simply to use $F^{\alpha\beta}$ for our reformulation of the Maxwell equations into an explicitly covariant form. First, the inhomogeneous equations are

$$\vec{\nabla} \cdot \mathbf{E} = 4\pi\rho \qquad 6.73$$

or

$$\partial_i F^{i0} = \frac{4\pi}{c} J^0, \qquad 6.74$$

and

$$\vec{\nabla} \times \mathbf{B} - \frac{1}{c}\frac{\partial \mathbf{E}}{\partial t} = \frac{4\pi}{c}\mathbf{J}, \qquad 6.75$$

where $J^\alpha \equiv (\rho c, \mathbf{J})$ is the four-dimensional current density. These can be written together in covariant form as

$$\partial_\alpha F^{\alpha\beta} = 4\pi J^\beta / c, \qquad 6.76$$

as one can verify with the definitions of $F^{\alpha\beta}$ and J^β. The decomposition of equation (6.76) into its different components may be made with equal validity in any frame, consistent with the relativity principle.

Second, the homogeneous equations are

$$\vec{\nabla} \cdot \mathbf{B} = 0 \qquad 6.77$$

or

$$\partial^1 F^{32} + \partial^2 F^{13} + \partial^3 F^{21} = 0, \qquad 6.78$$

and

$$\vec{\nabla} \times \mathbf{E} + \frac{1}{c} \frac{\partial \mathbf{B}}{\partial t} = 0. \qquad 6.79$$

Both of these equations may be combined into the single form

$$\partial^\alpha F^{\beta\gamma} + \partial^\beta F^{\gamma\alpha} + \partial^\gamma F^{\alpha\beta} = 0. \qquad 6.80$$

The point in claiming that there is only one *electromagnetic* field is that, whereas the frame-dependent field projections $\mathbf{E}$ and $\mathbf{B}$ require four Maxwell equations for a complete description, in covariant form there are only two—the first showing how a source J^α produces a change in $F^{\alpha\beta}$, and the second accounting for the propagation of the field once it is produced.

6.3 Accelerated Frames

To generalize the validity of the transformation equations described in the previous section, it will be essential to introduce next the possibility of acceleration between frames of reference. Two years after the advent of special relativity, Einstein (1879–1955) was writing a review on the new physics, which at the time he called *invariance theory*, when he began to wonder how Newtonian gravitation could be modified to make it consistent with the new relativity theory. The idea that took form in his mind, described later by him as the "happiest thought of my life," was that an observer who is falling from the roof of a house experiences no gravitational field. The unfortunate homeowner is in *free fall* within Earth's gravity. But every loose object in his vicinity falls with him at exactly the same rate, since the acceleration they experience due to Earth's pull is completely independent of their inertial mass. Be it a hammer, a nail, or the human body, they all exist within a region where no one can tell if something else is being pulled or if they themselves are being pulled, since every object is accelerated in tandem with everything else.

Thus, regardless of what actually causes the pull of gravity, its effect is entirely equivalent to a uniform acceleration throughout a given volume of space. Einstein called this the *principle of equivalence*. He further hypothesized that since all gravitational fields vanish inside a free-falling frame, special relativity ought to apply to all measurements of distances and times within that frame. And in a leap of faith (or inspiration, or both), he argued that the two postulates of the special theory should apply even in cases where we compare the measurements of an observer in this frame with those of a distant observer, for whom the effects of the gravitational field are negligible. Thus, in an elegant and all-encompassing way, the theory of gravity may be merged with the framework of special relativity. Gravity is described by its equivalence to an accelerated frame, and its effects are thereby fully incorporated into the laws of physics via the properties of special relativity. This is the essence of the *general* theory of relativity.

Let us begin the process of merging gravity with special relativity by considering what happens when we express the interval ds in terms of an arbitrary coordinate system. Writing the generalized coordinates as $X^\alpha \equiv (X^0, X^1, X^2, X^3)$, the chain rule of differentiation allows us to transform equation (6.43) as follows:

$$
\begin{aligned}
(ds)^2 &= \eta_{\alpha\beta} \, dx^\alpha dx^\beta \\
&= \eta_{\alpha\beta} \left(\frac{\partial x^\alpha}{\partial X^\gamma} dX^\gamma \right) \left(\frac{\partial x^\beta}{\partial X^\delta} dX^\delta \right) \\
&= \left(\eta_{\alpha\beta} \frac{\partial x^\alpha}{\partial X^\gamma} \frac{\partial x^\beta}{\partial X^\delta} \right) dX^\gamma dX^\delta,
\end{aligned}
\tag{6.81}
$$

so that

$$
(ds)^2 = g_{\alpha\beta} \, dX^\alpha dX^\beta,
\tag{6.82}
$$

where

$$
g_{\alpha\beta} \equiv \eta_{\mu\nu} \frac{\partial x^\mu}{\partial X^\alpha} \frac{\partial x^\nu}{\partial X^\beta},
\tag{6.83}
$$

in terms of the inertial Cartesian coordinates x^α.

In the first simple application of this formalism, we will convert the interval ds into spherical polar coordinates,

$$
X^\alpha \equiv (cT, r, \theta, \phi),
\tag{6.84}
$$

133

where the coordinate transformation is given as

$$t = T,$$
$$x = r \sin \theta \cos \phi,$$
$$y = r \sin \theta \sin \phi,$$
$$z = r \cos \theta. \qquad \qquad 6.85$$

Looking at the definition of $g_{\alpha\beta}$ in equation (6.83), it is clear that

$$g_{00} = +1,$$
$$g_{0j} = g_{j0} = 0 \qquad (j = 1, 2, 3),$$
$$g_{11} \equiv g_{rr} = (\sin^2 \theta \cos^2 \phi + \sin^2 \theta \sin^2 \phi + \cos^2 \theta)(-1)$$
$$= -1,$$
$$g_{22} \equiv g_{\theta\theta} = r^2(\cos^2 \theta \cos^2 \phi + \cos^2 \theta \sin^2 \phi + \sin^2 \theta)(-1)$$
$$= -r^2,$$
$$g_{33} \equiv g_{\phi\phi} = r^2(\sin^2 \theta \sin^2 \phi + \sin^2 \theta \cos^2 \phi)(-1)$$
$$= -r^2 \sin^2 \theta. \qquad \qquad 6.86$$

Therefore,

$$(ds)^2 = c^2(dT)^2 - (dr)^2 - r^2(d\theta)^2 - r^2 \sin^2 \theta (d\phi)^2, \qquad 6.87$$

and the metric tensor

$$[g_{\alpha\beta}] = \begin{pmatrix} 1 & 0 & 0 & 0 \\ 0 & -1 & 0 & 0 \\ 0 & 0 & -r^2 & 0 \\ 0 & 0 & 0 & -r^2 \sin^2 \theta \end{pmatrix}, \qquad 6.88$$

though still diagonal in this case, now has coefficients that depend on the coordinates.

Next, let us see what happens when we mix spatial and time coordinates, as would happen when one of the reference frames is accelerated at a rate a relative to the other. Adopting a Cartesian system $X^\alpha \equiv (cT, X, Y, Z)$, for which

$$T = t,$$
$$X = x - (1/2)at^2,$$
$$Y = y,$$
$$Z = z, \qquad\qquad 6.89$$

we can now determine that the metric coefficients must be

$$g_{00} = 1 - (1/2)(at/c)^2 - (1/2)(at/c)^2$$
$$= 1 - (at/c)^2,$$
$$g_{11} = -1,$$
$$g_{22} = -1,$$
$$g_{33} = -1,$$
$$g_{01} = g_{10} = -(at/c). \qquad\qquad 6.90$$

All others are zero. So in this case

$$(ds)^2 = (1 - a^2 t^2/c^2)c^2(dT)^2 - (dX)^2$$
$$- (dY)^2 - (dZ)^2 - 2at\, dX\, dT \qquad\qquad 6.91$$

and

$$[g_{\alpha\beta}] = \begin{pmatrix} (1 - a^2 t^2/c^2) & -at/c & 0 & 0 \\ -at/c & -1 & 0 & 0 \\ 0 & 0 & -1 & 0 \\ 0 & 0 & 0 & -1 \end{pmatrix}. \qquad\qquad 6.92$$

Not only are the coefficients $g_{\alpha\beta}$ now functions of the coordinates, but the metric tensor $[g_{\alpha\beta}]$ is no longer diagonal.

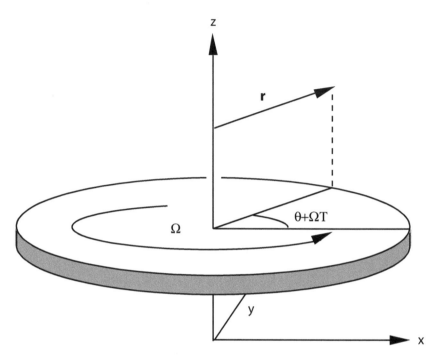

Figure 6.1 Transformation of coordinates from a rotating system (cT, r, θ, Z) to the laboratory Cartesian frame (ct, x, y, z). The accelerated frame is rotating about the z-axis with an angular velocity Ω. The time-dependent angle $\theta + \Omega T$ is measured relative to the x-axis.

An even more interesting situation is that in which a transformation is made from Cartesian coordinates x^α into a rotating system $X^\alpha \equiv (cT, r, \theta, Z)$, where

$$t = T,$$

$$x = r\cos(\theta + \Omega T),$$

$$y = r\sin(\theta + \Omega T),$$

$$z = Z. \qquad\qquad 6.93$$

The angular velocity Ω of the accelerated frame is measured relative to the inertial system x^α (see figure 6.1). Leaving the evaluation of the individual coefficients $g_{\alpha\beta}$ as a straightforward exercise for the reader, we note that the corresponding interval here is

$$(ds)^2 = (1 - \Omega^2 r^2/c^2)c^2(dT)^2 - (dr)^2 - r^2(d\theta)^2$$
$$- (dZ)^2 - 2\Omega r^2 \, d\theta \, dT, \qquad 6.94$$

and the metric tensor may be written as

$$[g_{\alpha\beta}] = \begin{pmatrix} (1 - \Omega^2 r^2/c^2) & 0 & -\Omega r^2 & 0 \\ 0 & -1 & 0 & 0 \\ -\Omega r^2 & 0 & -r^2 & 0 \\ 0 & 0 & 0 & -1 \end{pmatrix}. \qquad 6.95$$

6.4 GENERAL RELATIVITY

These coordinate translations are interesting and apropos because the essence of general relativity is the transformation from an inertial frame (usually in free fall) x^α into an appropriately accelerated (typically the laboratory) frame X^α. The principal of equivalence is based on the observed equivalence of the gravitational and inertial masses, and it states that at every spacetime point in an arbitrary gravitational field one can choose a coordinate system—the local inertial frame—such that within a small region around that point the laws of nature take the same form as in a Cartesian inertial frame (usually called the CIF) *without* gravitation. The local effects of gravitation may then be determined by transforming to the accelerated coordinate system.

In the latter, matter is seen to produce a gravitational field that is described by the metric $g_{\alpha\beta}(x)$, different from the "flat-space" metric $\eta_{\alpha\beta}(x)$. Because this altered $g_{\alpha\beta}(x)$ has an effect on the particle *trajectories*, it is often said that the gravitational field changes the geometry of spacetime. In an inertial frame x^α, the particle's equation of motion (see equation 6.52) is

$$m\frac{du^\alpha}{d\tau} = f^\alpha, \qquad 6.96$$

where

$$u^\alpha \equiv \frac{dx^\alpha}{d\tau} \qquad 6.97$$

137

is the four-velocity and f^α is the four-force. A transformation to another inertial frame obviously leaves equation (6.96) unaffected because that is how special relativity is structured. However, we expect modifications due to "inertial" forces when transforming to an accelerated frame, and these alterations are expressible in terms of the $g_{\alpha\beta}(x)$ coefficients.

To see how this comes about, let us consider the derivative of a four-vector. Suppose we begin with the tensor $\partial A_\alpha / \partial x^\beta$ in a CIF and transform it to an accelerated frame X^α, calling the four-vector A_α in this new frame $\bar{A}_\alpha$. Then

$$
\begin{aligned}
\frac{\partial \bar{A}_\alpha}{\partial X^\beta} &= \frac{\partial}{\partial X^\beta}\left(\frac{\partial x^\lambda}{\partial X^\alpha} A_\lambda\right) \\
&= \frac{\partial^2 x^\lambda}{\partial X^\beta \partial X^\alpha} A_\lambda + \frac{\partial x^\lambda}{\partial X^\alpha}\frac{\partial A_\lambda}{\partial X^\beta} \\
&= \frac{\partial^2 x^\lambda}{\partial X^\beta \partial X^\alpha} A_\lambda + \frac{\partial x^\lambda}{\partial X^\alpha}\frac{\partial x^\mu}{\partial X^\beta}\frac{\partial A_\lambda}{\partial x^\mu}.
\end{aligned}
\qquad 6.98
$$

In special relativity (i.e., a transformation from a CIF to another CIF), the first term is zero because $\partial x^\lambda / \partial X^\alpha$ are constant coefficients. In general relativity, however, where $g_{\alpha\beta} \neq \eta_{\alpha\beta}$, the first term introduces inertial forces.

In the accelerated (i.e., the laboratory) frame, equation (6.96) therefore takes on additional terms, which we can evaluate as follows:

$$
\begin{aligned}
\bar{f}^\delta &= \frac{\partial X^\delta}{\partial x^\alpha} f^\alpha \\
&= m\frac{\partial X^\delta}{\partial x^\alpha}\left(\frac{\partial x^\alpha}{\partial X^\beta}\frac{d^2 X^\beta}{d\tau^2} + \frac{\partial^2 x^\alpha}{\partial X^\beta \partial X^\gamma}\frac{dX^\gamma}{d\tau}\frac{dX^\beta}{d\tau}\right) \\
&= m\frac{d^2 X^\delta}{d\tau^2} + m\frac{\partial^2 x^\alpha}{\partial X^\beta \partial X^\gamma}\frac{\partial X^\delta}{\partial x^\alpha}\frac{dX^\gamma}{d\tau}\frac{dX^\beta}{d\tau}.
\end{aligned}
\qquad 6.99
$$

In other words, the generalized Newton's law in the accelerated frame of reference is

$$
m\frac{d^2 X^\alpha}{d\tau^2} = \bar{f}^\alpha - m\Gamma^\alpha{}_{\beta\gamma}\frac{dX^\beta}{d\tau}\frac{dX^\gamma}{d\tau},
\qquad 6.100
$$

where

$$\Gamma^\alpha{}_{\beta\gamma} = \frac{\partial^2 x^\lambda}{\partial X^\beta \partial X^\gamma} \frac{\partial X^\alpha}{\partial x^\lambda},$$

6.101

known as the Christoffel symbols. A convenient way of calculating these quantities is to determine the transformation rule from the local inertial (i.e., free-fall) frame x^α to the accelerated (i.e., laboratory) frame X^α. To explicitly bring out the dependence of the equation of motion on the "gravitational" field, these Christoffel symbols are sometimes written directly in terms of the metric coefficients $g_{\alpha\beta}$:

$$\Gamma^\alpha{}_{\beta\gamma} = \frac{1}{2} g^{\alpha\lambda} \left(\frac{\partial g_{\lambda\beta}}{\partial X^\gamma} + \frac{\partial g_{\lambda\gamma}}{\partial X^\beta} - \frac{\partial g_{\gamma\beta}}{\partial X^\lambda} \right),$$

6.102

where

$$g^{\alpha\beta} g_{\alpha\gamma} = \delta^\beta_\gamma$$

6.103

and $g_{\alpha\beta} = g_{\beta\alpha}$. That is,

$$[g^{\alpha\beta}] = [g_{\alpha\beta}]^{-1}$$

6.104

(the matrix inverse). Needless to say, $\Gamma^\alpha{}_{\beta\gamma} = 0$ when there is no gravitational field, since in that case $g^{\alpha\beta} = $ constant.

Thinking back to the transformation between a CIF and a second frame accelerating steadily relative to it (see equation 6.89), we should be able to demonstrate the correct behavior of a particle moving near Earth's surface using equation (6.100) and the metric of equation (6.92). The equivalence principle espouses the view that a uniform gravitational field, such as we encounter locally, is indistinguishable from a constant acceleration of the frame in the opposite direction. In the example we considered earlier, the X^α-frame is accelerated in the positive x-direction, which should therefore represent a gravitational field increasing toward negative values of x. Well, it is straightforward to show from equation (6.102) that

$$\Gamma^1{}_{00} = \frac{a}{c^2},$$

6.105

and so

$$m\frac{d^2 X^1}{d\tau^2} \approx m\frac{d^2 X^1}{dT^2}$$

$$= \bar{f}^1 - m\Gamma^1{}_{00}\frac{dX^0}{d\tau}\frac{dX^0}{d\tau}$$

$$= F^1 - ma. \tag{6.106}$$

In this derivation, we have assumed nonrelativistic motion to argue that $\tau \approx T$ and that $\bar{f}^1$ is just the x-component F^1 of the Newtonian force. As expected, the "inertial" effect is here equivalent to a gravitational force $-ma$ pointing in the negative x-direction.

6.5 Particle Orbits and Trajectories

The geometry of a spacetime is dictated by the Christoffel symbols $\Gamma^\alpha{}_{\beta\gamma}$, which directly impact the curvature of the extremal lengths followed by particles moving freely, except subject to a gravitational field. These paths, known as geodesics, are described by the equation

$$\frac{d^2 X^\alpha}{d\tau^2} + \Gamma^\alpha{}_{\beta\gamma}\frac{dX^\beta}{d\tau}\frac{dX^\gamma}{d\tau} = 0, \tag{6.107}$$

which follows directly from equation (6.100) with $\bar{f}^\alpha = 0$. In flat spacetime, where $g_{\alpha\beta} = \eta_{\alpha\beta}$, $\Gamma^\alpha{}_{\beta\gamma} = 0$, so the particle path is always a straight line, as we would expect in special relativity, where the velocity of the particle's rest frame is constant relative to a CIF.

The presence of a source of gravity—mass and/or energy—creates gradients in $g_{\alpha\beta}$ as functions of the coordinates that translate into nonzero values of $\Gamma^\alpha{}_{\beta\gamma}$. The differential field equations of general relativity "simply" express these gradients in terms of the local gravitational source, in the same way that the Poisson equation in electrodynamics accounts for the gradients in the electrostatic potential due to local charges. And other than the flat spacetime metric, in which the gradients of $g_{\alpha\beta}$ are zero because the source is absent, the simplest spherically symmetric vacuum solution to these equations has the following metric, derived by Karl Schwarzschild (1873–1916) in 1916, soon after general relativity was developed:

$$(ds)^2 = B(r)c^2(dT)^2 - B^{-1}(r)(dr)^2$$
$$-r^2(d\theta)^2 - r^2 \sin^2 \theta (d\phi)^2, \qquad \text{6.108}$$

where

$$B(r) \equiv \left(1 - \frac{2GM}{c^2 r}\right) \qquad \text{6.109}$$

and the gravitational effect due to a central mass M is now represented through its inertial terms via the metric coefficients

$$[g_{\alpha\beta}] = \begin{pmatrix} B(r) & 0 & 0 & 0 \\ 0 & -B^{-1}(r) & 0 & 0 \\ 0 & 0 & -r^2 & 0 \\ 0 & 0 & 0 & -r^2 \sin^2 \theta \end{pmatrix}. \qquad \text{6.110}$$

This solution and metric describe the relativistic gravitational field everywhere outside a gravitating, nonrotating body, such as a star or a black hole.

Incidentally, although it is outside the scope of the present text to extend the analysis of this solution to situations in which the configuration of the central mass is changing in time, we nonetheless mention an important theorem developed several years later that bears on the nature of spacetime surrounding a collapsing object. This idea actually traces its roots to the work of Sir Isaac Newton (1642–1727), who, in order to describe the Moon's motion around the Earth, used the newly invented calculus to derive a very important property of his universal law of gravitation. He showed that the gravitational field outside a spherically symmetric body behaves as if the whole mass were concentrated at its center. In other words, the Moon feels exactly the same gravitational influence from the Earth as it would from an object with the same mass, though only the size of a grain, situated at the center of where Earth now stands.

In 1923, not long after general relativity was established, George Birkhoff (1984–1944) made the surprising discovery that Newton's theorem was valid even for this more comprehensive description of gravity, though with some appropriate corrections. He demonstrated that even if a spherically symmetric body were collapsing or expanding radially, the Schwarzschild metric describing its gravitational field in empty space would not change in time. In other words, the effect of gravity outside a spherically symmetric body does not depend on how big that object is—it is based solely on how much mass is enclosed within its surface.

The Birkhoff theorem seems peculiar because in general relativity a nonstatic body generally radiates gravitational waves. We now know that in fact no gravitational radiation can escape into empty space from an isotropic configuration of mass, and so the fact that the Schwarzschild metric does not depend on the size of the central object is fully consistent with the idea that even a collapsing clump of matter retains all of its gravitational energy.

Let us next consider several simple situations to acquire an intuitive understanding of the effects associated with the Schwarzschild metric. To begin with, suppose we measure a time interval dT at some fixed radius r and angles θ and ϕ. How does this compare with the proper time $d\tau$? Well, from equation (6.108), we see that

$$ds = c\,d\tau = \left(1 - \frac{2GM}{c^2 r}\right)^{1/2} c\,dT, \qquad 6.111$$

which means that

$$d\tau = \left(1 - \frac{2GM}{c^2 r}\right)^{1/2} dT. \qquad 6.112$$

Clocks evidently run slow in gravitational fields, an effect that produces both a time dilation and a gravitational redshift in the presence of a gravitating body.

The fact that an acceleration, in this case due to gravity, produces such an effect is rather easy to understand. The point is that in special relativity the distortions to intervals of time and distance are entirely dependent upon the relative velocity between two different observers (see equation 6.25). General relativity, however, introduces additional distortions that constitute more than a mere subtlety. To be sure, the special relativistic time dilation must still apply to any pair of reference frames, whether or not they are imbued with a gravitational field, but the acceleration does something new—it causes clocks to run more slowly in the presence of gravity. The reason for this effect can be attributed to the meaning of time and its dependence on change.

Time passes when something is changing. To measure a second on a clock, the hand must turn one-sixtieth of a complete cycle across the clock's face. Regardless of how fast an accelerated frame is moving at a given instant, during the next second on the clock, its speed will have increased. So as long as the frame is accelerating, there is no way to avoid the fact that its velocity relative to a CIF is different at the end of that elapsed second compared to its value at the beginning. And no matter how much we shrink the interval of time, the starting and ending velocities are always different—that's the nature of acceleration.

An observer falling freely under the influence of a black hole looks at his clock and measures an interval of time. But in the act of allowing time to pass, he has crossed into a frame of reference moving even faster relative to a distant surveyor than the previous one. According to special relativity, there should be an additional time dilation associated with this increase in speed. The effect is greatly magnified when the acceleration is so great that even a minute interval can bring the magnitude of the falling frame's velocity close to the speed of light. The ensuing time dilation can then appear to freeze the action completely. Thus, as long as we accept the fact that time intervals are altered during a transformation from one frame to another (but always in such a way as to preserve the constancy of the speed of light), we must also accept the conclusion that the acceleration of one frame relative to another itself incurs an additional time dilation. Gravitational fields, therefore, slow down the passage of time as viewed from distant vantage points, and the retardation effect is greater the stronger the field.

The time dilation implied by equation (6.112) also means that the wavelength of radiation emitted within a gravitational field is stretched when viewed by an observer using the accelerated coordinates (cT, r, θ, ϕ), producing what is commonly known as a *gravitational redshift*. Suppose a source at rest at radius r_e is emitting radiation with frequency v_e as measured in the proper frame (one that is free of any gravitational effects). Then the emitted radiation frequency measured in the accelerated (i.e., the laboratory) frame is

$$\bar{v} = v_e \left. \frac{d\tau}{dT} \right|_{r_e} = v_e \left(1 - \frac{2GM}{c^2 r_e} \right)^{1/2}.$$ 6.113

Correspondingly, an observer in the proper frame at r_o will measure a frequency for this radiation of

$$v_o = \bar{v} \left. \frac{dT}{d\tau} \right|_{r_o},$$ 6.114

so that the overall change in frequency between the points of emission and observation is given by

$$\left(\frac{v_o}{v_e}\right) = \left(1 - \frac{2GM}{c^2 r_e}\right)^{1/2} \left(1 - \frac{2GM}{c^2 r_o}\right)^{-1/2}.$$ 6.115

The observer is usually at a great distance from the source, which means that to very high accuracy (with $v_o \to v_\infty$)

$$\left(\frac{v_\infty}{v_e}\right) = \left(1 - \frac{2GM}{c^2 r_e}\right)^{1/2}.$$ 6.116

An analogous effect occurs when the interval ds is due entirely to a length, meaning that the starting and end points of the interval are determined at the same time, for which $d\tau = dT = 0$. In that case, the proper distance between two radii r_1 and r_2 is not simply $r_2 - r_1$, as a laboratory observer would conclude. This is not surprising given the inertial distortions to ds caused by the gravitational field. In fact, a particle moving from r_1 to r_2 would measure a proper distance

$$\int_{r_1}^{r_2} -ds \approx (r_2 - r_1)\left(1 - \frac{2GM}{c^2 r}\right)^{-1/2},$$ 6.117

where we have assumed that $r_2 - r_1 \ll r_2$ and $r \equiv (r_1 + r_2)/2$.

The third aspect we will consider arising from the Schwarzschild metric is the consequence of selecting only particle trajectories that correspond to speeds smaller than c. This means that displacements in space and time must occur in such a way that $(ds)^2 > 0$ or, since $(ds)^2 = c^2 (d\tau)^2$, that $(d\tau)^2 > 0$. This is the condition that defines *timelike* world lines. Now consider a particle situated at fixed spatial coordinates (r, θ, ϕ). According to equation (6.108), this condition is satisfied only when $r > 2GM/c^2$. At smaller radii, no force can maintain a constant value of r, and the particle is pulled inexorably toward the singularity at $r = 0$. The critical radius

$$r_S \equiv \frac{2GM}{c^2}$$ 6.118

is named in honor of Karl Schwarzschild; it specifies the location of the so-called event horizon—effectively the "size" of the black hole.

The particle trajectories in a Schwarzschild geometry may be found for motion in the r-ϕ plane using equation (6.107), which yields three separate expressions for r, ϕ, and T, respectively. The angle θ is of course constant in this plane. We won't go through all the algebra here, which includes the evaluation of the Christoffel symbols $\Gamma^{\alpha}{}_{\beta\gamma}$. The ϕ equation gives

$$r^2 \frac{d\phi}{d\tau} = \text{constant} \equiv \tilde{l}, \qquad\qquad 6.119$$

where $\tilde{l}$ is the conserved angular momentum per unit mass. The T equation yields

$$\left(1 - \frac{2GM}{c^2 r}\right) \frac{dT}{d\tau} = \text{constant} \equiv \tilde{E}/c^2. \qquad\qquad 6.120$$

This constant $\tilde{E}$ is just the conserved "energy at infinity," including the rest mass energy of the particle, per unit mass. Finally, the r equation can also be written in terms of the above two constants of the motion:

$$c^2 \left(\frac{dr}{d\tau}\right)^2 = \tilde{E}^2 - V_{\text{eff}}(r), \qquad\qquad 6.121$$

where

$$V_{\text{eff}}(r) = \left(1 - \frac{2GM}{c^2 r}\right)\left(1 + \frac{\tilde{l}^2}{c^2 r^2}\right)(c^2)^2 \qquad\qquad 6.122$$

is the effective potential.

Far away from the source, where $r \gg r_S$ and $2Gm/c^2 r \ll 1$, the proper time and the accelerated (i.e., laboratory) frame time T approach each other, and for any finite specific angular momentum $\tilde{l}$, we can also put $\tilde{l}/cr \ll 1$. If we further define the Newtonian energy per unit mass $\tilde{E}_N$ via

$$\tilde{E} = c^2 + \tilde{E}_N, \qquad\qquad 6.123$$

then the equations of motion reduce to the following limiting forms:

$$\tilde{E}_N = \text{constant} \qquad\qquad 6.124$$

and

$$c^2 \left(\frac{dr}{dT} \right)^2 \approx 2c^2 \tilde{E}_N + \left(\frac{2GM}{c^2 r} - \frac{\tilde{l}^2}{c^2 r^2} \right) c^4. \qquad 6.125$$

That is,

$$\frac{1}{2} \left(\frac{dr}{dT} \right)^2 \approx \tilde{E}_N - V_N(r), \qquad 6.126$$

where

$$V_N(r) = -\frac{GM}{r} + \frac{\tilde{l}^2}{2r^2}. \qquad 6.127$$

This should look familiar from classical mechanics, where we would simply have calculated the velocity v using

$$\frac{1}{2} v^2 = E - V, \qquad 6.128$$

in terms of the total (non-rest-mass) energy E and the potential energy V. The solution to equations (6.127) and (6.128) leads to the standard elliptical or hyperbolic orbits, in which knowledge of $\tilde{E}_N$ and $\tilde{l}$ allows us to use the condition $dr/dT = 0$ to determine the minimum and (if it exists) maximum radii of the trajectory.

Returning now to the more general problem (equations 6.121 and 6.122), we see that the potential in a Schwarzschild metric has an additional term $-2GM\tilde{l}^2/r^3$, which can overwhelm even the angular momentum barrier term $c^2 \tilde{l}^2/r^2$ at a finite radius r causing $V_{\text{eff}}(r)$ to approach $-\infty$ toward the origin. Figure 6.2 shows a plot of $V_{\text{eff}}(r)^{1/2}/c^2$ as a function of $c^2 r/2GM$ for various values of the specific angular momentum parameter $\tilde{l}c/2GM$, ranging from zero (i.e., no angular momentum) to 2.165 (representing a high angular momentum barrier).

It is clear now that in the Schwarzschild metric there exist three different types of orbit. First, since the potential is not strictly $1/r$, as in Newtonian gravity, no closed orbits (such as ellipses) are permitted. Instead, orbits that would otherwise have been closed now *precess*, an effect that has received elegant confirmation with the advancing perihelion of Mercury. Second, a particle approaching the central object with a trajectory beginning at infinity may have sufficient energy to

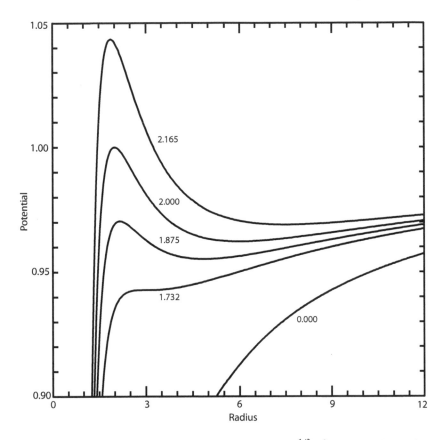

Figure 6.2 A plot of the effective potential $V_{\text{eff}}(r)^{1/2}/c^2$ versus the radius $c^2 r/2GM$ for various values of the specific angular momentum $\bar{l}c/2GM$ ranging from zero (i.e., no angular momentum) to 2.165 (representing a high angular momentum barrier).

recede to infinity after a single passage. This type of particle path is analogous to a Newtonian hyperbolic orbit. Third, particles may be bound but still possess sufficient energy to overcome the potential hump at small radii (see figure 6.2) and fall into the black hole. There is no analog for these orbits in Newtonian mechanics.

However, circular orbits are still possible, because precession for them is not evident. In order to achieve such a trajectory, a particle must have

$$\frac{\partial V_{\text{eff}}}{\partial r} = 0 \qquad\qquad 6.129$$

and

$$\frac{dr}{d\tau} = 0, \qquad\qquad 6.130$$

which result in two very useful conditions on $\tilde{l}$ and $\tilde{E}$:

$$\tilde{l}^2 = \frac{GMr^2}{r - 3GM/c^2} \qquad\qquad 6.131$$

and

$$\tilde{E} = \frac{c^2(r - 2GM/c^2)^2}{r(r - 3GM/c^2)}. \qquad\qquad 6.132$$

Note that circular orbits are not permitted for radii $r < 3GM/c^2$. In fact, as $r \to 3GM/c^2$, both $\tilde{l}$ and $\tilde{E}$ become excessively large. A circular trajectory at this radius must therefore correspond to a photon orbit— remember that $\tilde{l}$ and $\tilde{E}$ are calculated per unit mass, so they become infinite for massless particles. This orbit is also unstable, given that a slight perturbation on the photon will induce it to either fall into the black hole or escape to infinity.

But look at what also happens to $\tilde{l}^2$ when $r \to \infty$. Evidently, there exists a minimum value of $\tilde{l}^2$ for which circular orbits are realized, and this occurs at precisely $r = 6GM/c^2 = 3r_S$, when $\tilde{l} = 2\sqrt{3}\, GM/c$. Looking at figure 6.2, we see that this is the *last* (or marginally) stable circular orbit. A potential with $\tilde{l}$ smaller than this has no minimum, while one with greater angular momentum has a minimum at larger radii. A particle falling below this radius (due to the dissipative loss of angular momentum, say, in an accretion disk) finds itself in progressively less stable orbits and spirals toward the black hole. It no longer has sufficient angular momentum to maintain a circular orbit below $3r_S$.

Incidentally, the last stable circular orbit corresponds to an energy squared $\tilde{E}^2 = (8/9)c^4$ at $r = 3r_S$, so the amount of energy extracted from the infalling particle that reaches this point is $c^2 - \tilde{E} \approx 0.057\, c^2$ per unit mass. Evidently, there exists a maximum fraction of the rest-mass energy that can be extracted by disk accretion onto a Schwarzschild black hole, and one often uses 6% as a reasonable approximation for this quantity.

Finally, at $r = 2r_S$, a circular orbit requires a specific angular momentum $\tilde{l} = 4GM/c$ and a specific energy $\tilde{E} = c^2$. Since this is the same as the rest energy (per unit mass) at infinity, one often refers to this radius as the *marginally bound* orbit. However, a particle spiraling in due to the viscous action of friction will not actually achieve such an orbit, since it would need to acquire more angular momentum (increasing the value of $\tilde{l}$) and energy (which is not available locally) as it moves inward.

6.6 THE KERR METRIC

Schwarzschild's solution to the gravitational field equations would have retained its pivotal significance even to this day had it not been for another breakthrough solution that eventually eclipsed it. Fifty years after the first formulation of spacetime in a gravitational field, Roy Kerr produced a description of intervals and time surrounding a *rotating* black hole.[5]

Most objects in the universe spin at least a little because it is virtually impossible to assemble an aggregate of matter without any angular momentum, which is so prevalent in nature that it often functions as a powerful diagnostic of the forces that shape the various concentrations of mass. The conservation of angular momentum can produce furiously spinning entities when their progenitors collapse into ultracompact volumes. Should our Sun approach old age as a white dwarf, shrinking by a factor of 100 in size, its current rotational cycle of twenty-six days would shorten into a period of only four minutes. And if it were to condense into a neutron star, it would rotate several hundred times per second.

Black holes are born spinning and grow with the additional acquisition of matter throughout their existence. Even a black hole with a mass of $10^7 \, M_\odot$ and a circumference stretching more than 100 million miles could be rotating with a period of only 1.75 hours. Schwarzschild's metric cannot handle this because a rotating black hole impacts the spacetime around it in unexpected and challenging ways. The interval ds, for example, is no longer given by equation (6.108) for the same reason that an observer watching someone jumping onto a merry-go-round cannot measure the total distance he covers by simply counting his steps. Though she may still be able to monitor the passage of time T by tracking the ticks on her watch, the actual distance traveled by the jumper is now augmented by the merry-go-round's sideways motion, which carries him along for the ride.

---

[5]Kerr first reported the discovery of this solution in *Physical Review Letters* in 1963.

As we shall see, Kerr's solution to the equations of general relativity shows that the spacetime itself, like water in a whirlpool, swirls around the black hole with a speed wedded to the latter's spin, though decreasing with distance from the center. This phenomenon is known as *frame dragging*, meaning that the spacetime itself and all its contents are forced into co-rotation with the source of gravity, even if objects in that frame are completely stationary relative to the space itself. This is quite a bizarre concept and difficult to accept at face value, for it seems to imply that, even if we could somehow place a particle with zero angular momentum in the vicinity of a spinning black hole, the fact that its underlying spacetime is rotating means that the particle would still appear to be moving sideways from the perspective of a distant observer.

Having said this, the exterior field solution is still rather manageable because of something known as the *no hair theorem*. Regardless of the constituents that originally went into making the collapsed object or the sequence in which the black hole was assembled, once it has settled down, it is completely described by its mass M, its angular momentum J, and its charge Q. It is difficult to maintain nonzero values of Q over large distances, so the most general astrophysical black holes we can have are described completely by assigning them just a mass M and an angular momentum J. These are the two primary parameters that characterize the Kerr family of solutions.

The spacetime interval for a rotating black hole is given by the expression

$$(ds)^2 = A(r, \theta)c^2\,(dT)^2 + C(r, \theta)c\,dT\,d\phi$$
$$- (\Sigma/\Delta)(dr)^2 - \Sigma(d\theta)^2$$
$$- D(r, \theta)\sin^2\theta(d\phi)^2, \qquad\qquad 6.133$$

where

$$A(r, \theta) \equiv \left(1 - \frac{2GMr}{c^2\Sigma}\right), \qquad\qquad 6.134$$

$$C(r, \theta) \equiv \frac{4aGMr\sin^2\theta}{c^2\Sigma}, \qquad\qquad 6.135$$

$$D(r, \theta) \equiv \left(r^2 + a^2 + \frac{2GMra^2\sin^2\theta}{c^2\Sigma}\right), \qquad\qquad 6.136$$

with

$$a \equiv \frac{J}{cM},$$ 6.137

$$\Delta \equiv r^2 - \frac{2GM}{c^2}r + a^2,$$ 6.138

and

$$\Sigma \equiv r^2 + a^2 \cos^2 \theta.$$ 6.139

The metric coefficients are therefore

$$[g_{\alpha\beta}] = \begin{pmatrix} A(r) & 0 & 0 & C(r,\theta)/2 \\ 0 & -\Sigma/\Delta & 0 & 0 \\ 0 & 0 & -\Sigma & 0 \\ C(r,\theta)/2 & 0 & 0 & -D(r,\theta)\sin^2\theta \end{pmatrix}.$$ 6.140

With this matrix, it is straightforward to show that the Kerr metric reduces to the Schwarzschild metric when the spin angular momentum parameter a is zero. However, compared to the Schwarzschild metric, the full Kerr solution has two new effects. First, the metric coefficients $g_{\alpha\beta}$ depend on θ, so rotation has broken the spherical symmetry, though we still have azimuthal symmetry. Second, there is now a nonzero off-diagonal term $g_{T\phi}$, whose origin is analogous to that of the off-diagonal terms in equation (6.95). Transforming to a rotating frame, we inevitably mix the angle of rotation (here ϕ) with time.

The polar axis ($\theta = 0$) is parallel to the angular momentum vector **J**. Let us first consider what happens along this direction, where the metric simplifies to

$$(ds)^2 = \left(1 - \frac{2GMr}{c^2(r^2 + a^2)}\right) c^2 (dT)^2 - \frac{r^2 + a^2}{\Delta}(dr)^2$$
$$- (r^2 + a^2)(d\theta)^2.$$ 6.141

The corresponding metric tensor is diagonal, meaning that the swirling action of the spinning body has no impact on the spacetime since the interval ds does not depend on the azimuthal angle ϕ. As with the Schwarzschild solution, *static* observers are permitted only when $g_{TT} > 0$, that is, only for speeds less than c. The limiting radius for this timelike region is therefore given by the condition

$$2GMr < c^2(r^2 + a^2) \tag{6.142}$$

or

$$r > r_+ \equiv \frac{GM}{c^2} + \left(\left[\frac{GM}{c^2} \right]^2 - a^2 \right)^{1/2}, \tag{6.143}$$

the negative root corresponding instead to a region inside the limiting radius, which is beyond the derived static limit. The radius r_+ represents the static limit along $\theta = 0$. Note that

$$r_+ \to \frac{2GM}{c^2} = r_S, \tag{6.144}$$

when $a \to 0$, as expected.

In the equatorial plane, where $\theta = \pi/2$, we have instead

$$(ds)^2 = \left(1 - \frac{2GM}{c^2 r}\right) c^2 (dT)^2 + \frac{4aGM}{c^2 r} c \, dT \, d\phi$$

$$- \frac{r^2}{r^2 - 2GMr/c^2 + a^2} (dr)^2 - r^2 (d\theta)^2$$

$$- \left(r^2 + a^2 + \frac{2GMa^2}{c^2 r}\right) (d\phi)^2. \tag{6.145}$$

The static region with timelike intervals (i.e., $g_{TT} > 0$) now corresponds to

$$r > r_0 = r_S \equiv \frac{2GM}{c^2}. \tag{6.146}$$

And for an arbitrary angle θ, the static limit (from equation 6.133) is evidently

$$r_0(\theta) = \frac{GM}{c^2} + \left(\left[\frac{GM}{c^2} \right]^2 - a^2 \cos^2 \theta \right)^{1/2}. \tag{6.147}$$

Note, however, that the static limit in the case of a Kerr black hole is *not* the event horizon, since the interval ds can still be greater than zero (i.e., timelike) even when $r < r_0(\theta)$ due to the presence of the off-diagonal term $dT\,d\phi$ in equation (6.133). Observers are permitted to exist at a fixed radius r inside $r_0(\theta)$ but only if the positive contribution to $(ds)^2$ from this term is large enough to offset the negative value of $g_{TT}\,c^2\,(dT)^2$. Since

$$2g_{T\phi}c\,dT\,d\phi = 2g_{T\phi}c\,(dT)^2\,\frac{d\phi}{dT}, \qquad\qquad 6.148$$

observers may exist at fixed r inside $r_0(\theta)$ only if they are rotating along with the black hole at a rate $d\phi/dT > 0$. This phenomenon, as we have seen, is known as *frame dragging* and gives rise to another critical radius called the *stationary limit*.

To see exactly how this comes about, let us rewrite equation (6.145) in the following form:

$$(ds)^2 = c^2\,(dT)^2 \left[-\left\{ \left(\frac{r^3 + a^2 r + 2GMa^2/c^2}{r} \right)^{1/2} \frac{d\phi}{c\,dT} \right. \right.$$

$$\left. - \frac{2a\,GM/c^2}{r^{1/2}\left(r^3 + a^2 r + 2GMa^2/c^2\right)^{1/2}} \right\}^2$$

$$+ \frac{4a^2(GM/c^2)^2}{r\left(r^3 + a^2 r + 2GMa^2/c^2\right)} + \left(1 - \frac{2GM}{c^2 r}\right) \right]$$

$$- \frac{r^2}{r^2 - 2GMr/c^2 + a^2}\,(dr)^2$$

$$+ r^2\,(d\theta)^2. \qquad\qquad 6.149$$

Inside the square brackets, the first term gives the $(d\phi)^2$ piece in equation (6.145), the second term squared cancels with the third term, the first term plus the second gives the $dT\,d\phi$ piece, and the fourth term reproduces the $(dT)^2$ contribution. The advantage of writing $(ds)^2$ in this way is that one can directly extract the conditions required to make the interval timelike. Evidently, $(ds)^2 > 0$ (with $dr = d\theta = 0$) below $r_0(\theta)$ for special values of $d\phi/dT$. The factor inside the curly brackets of equation (6.149) has a minimum at

$$\left(\frac{d\phi}{dT}\right)_0 = \frac{2a\,GM/c^2}{r^3 + a^2 r + 2GMa^2/c^2}. \qquad\qquad 6.150$$

(Note that this is essentially just $\sim G|J|/c^2r^3$.) The stationary limit is reached when the factor multiplying $(dT)^2$ in this equation goes to zero, and with the value of $(d\phi/dT)_0$ given in equation (6.150), this is attained when

$$r = r_+ \equiv \frac{GM}{c^2} + \left(\left[\frac{GM}{c^2}\right]^2 - a^2\right)^{1/2} \quad \text{for all } \theta. \qquad 6.151$$

We interpret $(d\phi/dT)_0$ as being the angular velocity forced on a zero angular momentum object by the spin of the black hole, a phenomenon that is also called the *dragging of inertial frames*.

Thus, r_+ is the true event horizon for a Kerr black hole. Inside r_+, no stationary observers are allowed, regardless of the value of $d\phi/dT$. The region between r_+ and $r_0(\theta)$ is the so-called ergosphere, where stationary observers are permitted (as long as they have the requisite $d\phi/dT$) but not the static ones. Outside $r_0(\theta)$, all observers are allowed to exist with a fixed value of r. In other words, the static horizon is a flattened sphere, with the semiminor axis parallel to J, whereas the stationary—or event— horizon is a sphere with radius r_+, which is equal to the Schwarzschild radius r_S when the spin parameter a is zero but is otherwise always smaller than r_S. In addition, we gather from equation (6.151) that these relations make sense only so long as

$$a \leq r_G \equiv \frac{GM}{c^2}, \qquad 6.152$$

where we have now defined the so-called gravitational radius $r_G \equiv r_S/2$. Thus, for a *maximally* spinning Kerr black hole, the true event horizon has a radius $r_+ = r_G$.

Qualitatively, we expect that the same range of orbits should be accessible in this metric as in the Schwarzschild case, though the frame-dragging effect can introduce quantitative differences. In particular, it is straightforward to show from equation (6.145)—for motion in the equatorial plane—that the expressions in (6.131) and (6.132) generalize to the form[6]

$$l^2 = \frac{\pm GM(r^2 \mp 2ar_G^{1/2}r^{1/2} + a^2)^2}{r^{3/2}(r^{3/2} - 3r_Gr^{1/2} \pm 2ar_G^{1/2})} \qquad 6.153$$

---

[6]The full details of this derivation may be found in Bardeen, Press, and Teukolsky (1972).

and

$$\tilde{E} = \frac{c^2(r^{3/2} - 2r_G r^{1/2} \pm ar_G^{1/2})^2}{r^{3/2}(r^{3/2} - 3r_G r^{1/2} \pm 2ar_G^{1/2})}, \qquad 6.154$$

when $a \neq 0$. As usual, the upper sign in these equations refers to direct orbits (i.e., co-rotating with $\tilde{l} > 0$), while the lower sign refers to retrograde orbits (counterrotating with $\tilde{l} < 0$). The coordinate angular velocity of a circular orbit is evidently

$$\Omega \equiv \frac{d\phi}{dT} = \pm \frac{cr_G^{1/2}}{r^{3/2} \pm ar_G^{1/2}}. \qquad 6.155$$

We shall see in chapter 8 how the latest timing studies of Sagittarius A* use this expression in order to estimate[7] the black hole spin parameter a.

A particularly interesting phenomenon arises with orbits off the equatorial plane, where the θ-dependent gravity induces precession of the particle trajectory around the angular momentum axis of the spinning black hole. In active galactic nuclei, this precession may be the reason why jets are so stable over millions of years. As we indicated toward the end of chapter 4 and as we will revist later in chapter 9, some evidence for this precession, with a period of about 100 days, may already have been found in the radio emission of Sagittarius A*.

[7]See also Melia et al. (2001).

CHAPTER 7

Mass Accretion and Expulsion

In chapter 3, we learned how the deployment of the *Chandra* X-ray Observatory has allowed us to study the X-ray emission from Sagittarius A* and its surroundings with sub-arcsecond resolution and a broad X-ray band imaging detector (see figure 3.4). These capabilities provide the means to discriminate point sources from the diffuse emission due to hot plasma in the ambient medium. Indeed, the central $17' \times 17'$ region of figure 1.8 contains remarkable structure with sufficient detail to allow a comparison with features seen in the radio and IR wavebands (see, e.g., color plate 8). The X-ray emission from Sagittarius A East appears to be concentrated within the central 2–3 pc of the 6 pc $\times$ 9 pc radio shell and is offset by about 2 pc from Sagittarius A* itself. The thermal plasma at the center of Sagittarius A East has a temperature $kT \approx 2\,\mathrm{keV}$ and appears to have strongly enhanced metal abundances with elemental stratification. We saw in chapter 1 that these X-ray properties identify it as a rare type of metal-rich "mixed-morphology" SNR, produced by the Type II explosion of a 13–20 $M_\odot$ progenitor.

However, this is not the sole component of hot gas surrounding Sagittarius A*. *Chandra* also detected X-ray emission extending perpendicular to the galactic plane in both directions through the position of the black hole, which would seem to indicate the presence of a hot, "bipolar" outflow from the nucleus. And an additional component of X-ray emitting gas appears to have been found within the molecular ring (see figures 1.8 and 3.4), much closer to Sagittarius A*. The western boundary of the brightest diffuse X-ray emission within 2 pc of the galactic center in fact coincides very tightly with the shape of the Western Arc of the thermal radio source Sagittarius A West (see color plate 8). Given that the Western Arc is believed to be the ionized inner edge of the molecular ring, this coincident morphology suggests that the brightest X-ray-emitting plasma is being confined by the western side of the gaseous torus. In contrast, the emission continues rather smoothly into the heart of Sagittarius A East toward the eastern side.

When one fits the emission from the hot gas within $10''$ of Sagittarius A*, it is clear that this is not a simple extension of the X-ray-emitting plasma at the center of Sagittarius A East.[1] The inferred 2–10 keV flux and luminosity of the local diffuse emission are $(1.9 \pm 0.1) \times 10^{-15}$ ergs cm^{-2} s^{-1} arcsec^{-2} and $(7.6^{+2.6}_{-1.9}) \times 10^{31}$ ergs s^{-1} arcsec^{-2}, respectively. Based on the parameters of the best-fit emissivity model, it is estimated that the local hot diffuse plasma has an RMS electron density $\langle n_e^2 \rangle^{1/2} \approx 26$ cm^{-3}. Assuming twice solar abundances (typical for this part of the Galaxy), the total inferred mass of this plasma is $M_{\text{local}} > 0.1 \, M_\odot$. In addition, the local plasma around Sagittarius A* appears to be distinctly cooler than the Sagittarius A East gas, with a temperature $kT = 1.3^{+0.2}_{-0.1}$ keV. By comparison, the best-fit model for Sagittarius A East indicates that the plasma there has about four times solar abundances and twice this temperature (see above).

At least some of the local hot plasma within a few parsecs of Sagittarius A* must be injected into the interstellar medium via stellar winds, and the diffuse X-ray emission therefore constitutes an excellent probe of the gas dynamics near the black hole. There is ample observational evidence in this region for the existence of rather strong outflows in and around the nucleus. Measurements of high velocities associated with IR sources in Sagittarius A West[2] and in IRS 16;[3] the H_2 emission in the molecular ring and from molecular gas being shocked by a nuclear mass outflow;[4] broad Brα, Brγ, and He I emission lines from the vicinity of IRS 16 (whose Doppler shifts and equivalent widths yield crucial information on the gas velocity and column depth, respectively); and radio continuum observations of IRS 7 (which measure the radiative signature of wind-wind collisions)[5] provide clear evidence of a hypersonic wind, with a velocity $v_w \sim 500$–$1,000$ km s^{-1}, a number density $n_w \sim 10^{3-4}$ cm^{-3} near the mass-ejecting stars, and a total mass loss rate $\dot{M}_w \sim 3$–$4 \times 10^{-3} \, \dot{M}_\odot$, pervading the inner parsec of the Galaxy.

---

[1] See Baganoff et al. (2003).

[2] These have been measured and reported by Krabbe et al. (1991).

[3] See §6.2. The original data were first published by Geballe et al. (1991).

[4] See Gatley et al. (1986), but note Jackson et al. (1993) for the potential importance of UV photodissociation in promoting this H_2 emission.

[5] See, e.g., Yusef-Zadeh and Melia (1991).

7.1 SAGITTARIUS A*'S GASEOUS ENVIRONMENT

The possible link between—and probative value of—the diffuse X-ray emission and stellar wind collisions has been under consideration since the early 1990s.[6] Strong outflows from OB stars and WR stars (see chapter 5) can produce shocks that heat up the gas to temperatures as high as $\sim 10^7$ K.

It is now possible to carry out comprehensive high-resolution numerical simulations of the wind-wind interactions using a detailed suite of stellar wind sources (see figure 7.1) and their inferred wind velocities and outflow rates. Unfortunately, there is still no way of knowing the radial coordinate (along the z-axis in this figure), but it seems from these simulations that the overall X-ray emissivity is insensitive to the choice of this "unknown" parameter. At the time of this writing, the state-of-the-art three-dimensional hydrodynamics simulations of this process have been carried out using a smooth particle hydrodynamics (SPH) code developed from a basic algorithm written and tested over a ten-year span at Los Alamos's Theory Division.[7] This code has been tested on many of the largest computer systems in the world and has been shown to scale well out to large processor numbers. The SPH formalism has been tested on a number of problems, from structure formation in cosmology to core collapse supernovas, and is ideal for a problem such as that described here, in which new particles are continuously added to the system. In these simulations, the number of particles grows rapidly at first but reaches a steady state (at about 7 million) when the addition of particles from wind sources is compensated by the particle loss out of the computational domain or onto the central black hole. The particle masses here vary from 9.4×10^{-7} $M_\odot$ to 1.0×10^{-5} $M_\odot$.

The wind-wind collisions create a complex configuration of shocks that efficiently convert the kinetic energy of the outflows into internal

---

[6]Chevalier (1992) discussed the properties of stellar winds in the galactic center and their ability to produce X-rays, under the assumption that the colliding winds create a uniform medium with a correspondingly uniform mass and power input within the central 0.8 pc of the Galaxy. A preliminary, low-resolution three-dimensional hydrodynamical simulation by Melia and Coker (1999) revealed that, in actuality, a tessellated pattern of wind-wind collision shocks arises and that these can shift on a local dynamical timescale. See also Ozernoy, Genzel, and Usov (1997).

[7]This work was carried out jointly at Los Alamos and the University of Arizona by Rockefeller et al. (2004) using the tree-based algorithm described in Warren and Salmon (1993, 1995) with the intent to calculate the multidimensional effects of gravity and to find SPH neighbors.

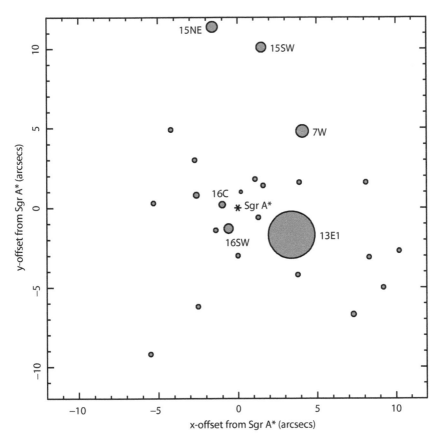

Figure 7.1 Location of the twenty-five strongest wind-producing stars relative to the position of Sagittarius A*, which is here indicated by the asterisk in the middle. The radius of each circle corresponds (on a linear scale) to that star's mass loss rate. Setting the scale is 13E1, with $\dot{M} = 7.9 \times 10^{-4}\ M_\odot\ \mathrm{yr}^{-1}$. (Image from Rockefeller et al. 2004)

energy of the gas. Figure 7.2 shows isosurfaces of specific internal energy in the central cubic parsec around 10,000 years after the beginning of the calculation. The dark surfaces indicate regions of gas with low specific internal energy; these tend to lie near the wind sources themselves. The gray surfaces mark regions of high specific internal energy, where gas has passed through multiple shocks. From simulations such as this, it is evident that about a quarter of the total energy in the central parsec is converted to internal energy via multiple shocks. The total kinetic energy of material there is $\sim 8 \times 10^{48}$ ergs, while the total internal energy is $\sim 3 \times 10^{48}$ ergs.

159

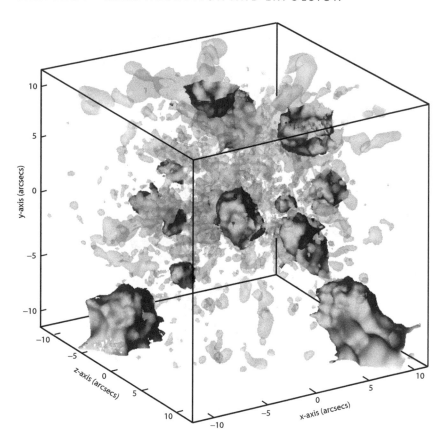

Figure 7.2 About 10,000 years after the winds are "turned on" in a three-dimensional hydrodynamics simulation using the stellar sources shown in figure 7.1, the 2–10 keV luminosity from the central 3 pc of the Galaxy reaches steady state (see figure 7.3). Shown here are the isosurfaces of specific internal energy of the shocked gas within the inner 20″ cube of the Galaxy. The line of sight is along the z-axis (so that the stars in figure 7.1 are seen projected onto the x-y plane). The darkest surfaces correspond to a specific internal energy of 2.5×10^{12} ergs g^{-1}; the gray surfaces correspond to 3.8×10^{15} ergs g^{-1}. (Image from Rockefeller et al. 2004)

The observed continuum spectrum from the perspective of an observer positioned along the negative z-axis in figure 7.2 is shown in figure 7.3. In the inner 3 pc of the Galaxy, scattering is negligible and the optical depth is always less than unity, so the dominant components of emissivity are optically thin electron-ion and electron-electron bremsstrahlung. The winds fill the volume of solution after only about 4,000 years, which coincides with the point at which the luminosity stops rising rapidly and

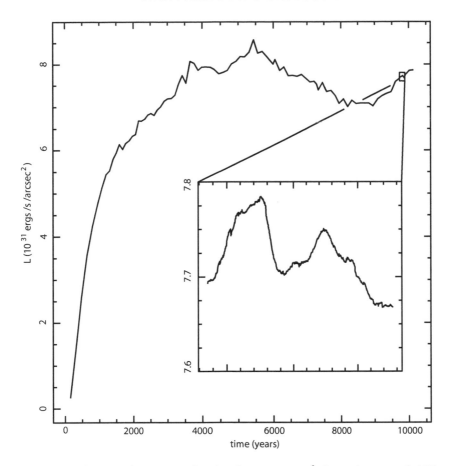

Figure 7.3 The 2–10 keV X-ray luminosity per arcsec^2 from the central 10″ versus time for the three-dimensional hydrodynamics simulation pertaining to figures 7.1 and 7.2. The large plot shows the variation in luminosity over the entire calculation; the inset shows a magnified view over a timescale of about ten years. The winds fill the volume after only around 4,000 years. By comparison, the luminosity measured by *Chandra* is $7.6^{+2.6}_{-1.9} \times 10^{31}$ ergs s^{-1} arcsec^{-2}. The implication is that the entire diffuse X-ray emission from the inner $\sim$3 pc of the Galaxy is due to stellar wind-wind collisions, providing a reliable constraint on the gas environment surrounding Sagittarius A*. (Image from Rockefeller et al. 2004)

begins varying by only $\sim$5% around a steady value. The inset plot shows the variation in luminosity on a timescale of around ten years. These fluctuations reflect noticeable changes in the temperature and density of the gas in the central 10″.

The conclusion from all this is that the diffuse X-ray luminosity and temperature within $10''$ of Sagittarius A* are explained entirely by the shocked winds of stars in the central cluster. This is rather significant in view of the fact that, were the diffuse X-rays being produced by another mechanism, the overall mass-loss rate or wind velocity would need to be scaled down by a factor of two or more in order to comply with the observed limits. In other words, the census of wind-producing stars in the central 3 pc seems to provide a fairly accurate accounting of the gas content and dynamics in this region. It is important also to emphasize the fact that these outflows bring the environment near the black hole back to a steady state within only around 4,000 years. It is unlikely, therefore, that the supernova that spawned Sagittarius A East less than 10,000 years ago could have created a current environment with an anomalously low density by sweeping out most of the material from the central parsecs.

7.2 Bondi-Hoyle Capture from Distributed Sources

In the classical Hoyle-Lyttleton theory,[8] matter in the interstellar medium is captured by the central accretor once its total specific energy (gravitational potential energy per unit mass plus kinetic energy per unit mass) is negative. This happens when the gas with velocity v_w approaches the black hole to within a distance shorter than the so-called accretion radius $r_{\rm acc}$, where

$$\frac{1}{2}v_w^2 = \frac{GM}{r_{\rm acc}}.$$

7.1

Hoyle and Lyttleton argued that matter flowing within a cross section $\pi r_{\rm acc}^2$ will be captured by the accretor, and since the mass flux carried by the gas breezing past the black hole is $\rho_w v_w$, in terms of the mass density ρ_w and wind velocity v_w, the effective accretion rate must be

$$\dot{M} \approx \dot{M}_{HL} \equiv \pi r_{\rm acc}^2 \rho_w v_w = \left(\frac{c}{v_w}\right)^4 \pi r_S^2 \rho_w v_w,$$

7.2

---

[8] See Hoyle and Lyttleton (1941) for an early application of this theory.

where r_S is the Schwarzschild radius defined in equation (6.118). At the galactic center, where $M \approx 3 \times 10^6 \, M_\odot$, the Hoyle-Lyttleton accretion rate depends sensitively on the (as yet imprecisely known) value of v_w and ρ_w. The former may be as high as $1{,}000 \, \mathrm{km \, s^{-1}}$ if the dominant flow past the black hole is unshocked; it may be as low as one-fourth of this value otherwise. The density may be as high as $\sim 10^4 \, m_p \, \mathrm{cm^{-3}}$ upstream of a shock (in terms of the proton mass m_p), or it could be as low as the value $26 \, m_p \, \mathrm{cm^{-3}}$, derived empirically from the *Chandra* observations of the diffuse X-ray emission (see §7.1). In either case, taking the most conservative view, the accretion rate $\dot{M}$ onto Sagittarius A* derived from this simple theory appears to be $> 10^{20} \, \mathrm{g \, s^{-1}}$. The problem with this number, as we shall see shortly, is that it is several orders of magnitude larger than what the observations seem to be telling us.

The earliest three-dimensional hydrodynamic simulations of accretion onto Sagittarius A* from the ambient medium[9] did indeed confirm that the rate at which matter is captured by the black hole is typically within a factor of two of the Hoyle-Lyttleton value. But does this capture rate really represent what happens as the plasma descends below r_{acc}, which for Sagittarius A* is $\sim 8 \times 10^{16} \, \mathrm{cm}$ ($\approx 0.03 \, \mathrm{pc}$ or $\sim 1''$)? These calculations were highly simplified versions of the necessarily complicated processes associated with the accretion. For example, the captured plasma is known to be magnetized, and as we shall see below, the dissipation of the magnetic field entrained by the highly ionized flow is expected to be a dominant heating agent toward smaller radii.[10] This effect was not included in the thermodynamic state of the plasma, nor on the flow dynamics. Magnetic bremsstrahlung (also known as cyclosynchrotron) emission can provide substantial cooling. In addition, these early simulations assumed a uniform flow past the accretor; in reality, the medium surrounding Sagittarius A* is fed by the winds of many stars.

Nonetheless, several important properties of Bondi-Hoyle accretion (as the better-developed hydrodynamic process is known[11]) onto Sagittarius A* emerged from these calculations and survived to the modern version of the theory. Chief among these are, first, that most of the specific

---

[9] See Ruffert and Melia (1994).
[10] These ideas were first introduced by the Russian astrophysicist Shvartsman (1971) and elaborated upon for Sagittarius A* by Melia (1992a, 1994).
[11] See the classic paper by Bondi and Hoyle (1944).

angular momentum $l \equiv \lambda r_S c$ (here written in terms of the dimensionless quantity λ) is canceled in the postshock region, though local fluctuations in the accretion rate $\dot{M}$ can be as large as 10%–20% on a timescale of several years. These are associated with transient excesses in l (or λ) that can lead to alternately prograde and retrograde disk structures. As the name suggests, the former is a state in which the disk's angular momentum vector is roughly aligned with the black hole's spin axis; in the latter, the two vectors are counteraligned. A second result is that the clumps of matter reaching Sagittarius A* tend to be preferentially those with small λ. These circularize (i.e., settle into a Keplerian orbit) at a radius

$$r_{\mathrm{circ}} = 2\lambda^2 r_S. \qquad\qquad 7.3$$

Since λ can never be exactly zero, a disk must be present in Sagittarius A* but is probably smaller than 10–100 r_S, since most of the infalling plasma has $\lambda \sim 3$–20.

When one includes multiple wind sources in the simulation, many shocks form due to wind-wind collisions even before the gas reaches the accretion radius r_{acc}. With the consequent conversion of bulk kinetic energy into heat, the black hole accretes at an even larger rate (typically twice the Hoyle-Lyttleton value) than in the uniform case.[12] Not surprising, a substantial difference between the uniform and nonuniform flows past the black hole is the presence (or absence) of a large-scale bow shock, and this affects primarily how much specific angular momentum is carried inwards with the infalling plasma. In figure 7.4, the captured λ is seen to vary by 50% over $\sim$200 years with an average equilibrium value of 37 ± 10. This is larger than in the uniform case, because the wind sources are not distributed isotropically about the black hole. Even so, clumps of gas with a high specific angular momentum do not penetrate within r_{acc}. The variability in the sign of the components of λ suggests that if an accretion disk forms at all, it dissolves and reforms (perhaps) with an opposite sense of spin on a timescale of $\sim$100 years.

But even though the three-dimensional hydrodynamic simulations (for a uniform flow or multiple wind sources) confirm the basic expectation of mass capture gleaned from the simple Hoyle-Lyttleton approach,

---

[12]Some of the early hydrodynamic simulations of multiple wind sources feeding the accretion onto Sagittarius A* were carried out by Coker and Melia (1997).

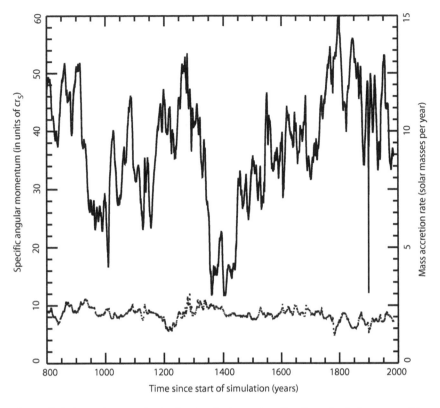

Figure 7.4 Accretion results in the simulation using distributed wind sources. The upper solid curve is accreted specific angular momentum λ (in units of $c r_S$) shown on the left-hand vertical axis. The lower dotted curve is the mass accretion rate $\dot{M}$ (10^{-4} $M_\odot$ yr^{-1}) versus time. The scale for $\dot{M}$ is shown on the right side. (Image from Coker and Melia 1997)

the empirical evidence for Sagittarius A* suggests that the calculated accretion rate is grossly overestimated.

7.3 ACCRETION CLOSE TO SAGITTARIUS A*

In §3.2, we examined the degree of Faraday depolarization arising between us and the galactic center, and we concluded that the rotation measure RM_{depol} produced by the intervening medium is too small to significantly affect our measurements of Sagittarius A*'s polarized emission at radio and mm/sub-mm wavelengths. However, a similar

165

argument applied to the infalling gas within $r_{\rm acc}$ produces the opposite result and may be used to constrain the actual accretion rate close to the black hole.[13]

To proceed with this analysis, our expression (equation 3.18) for the rotation measure must be modified here, since the gas in the emitting region close to Sagittarius A* is relativistic, even if thermal (see §3.1). The dominant correction to the nonrelativistic expression is the relativistic inertia, in which the mass of the electron $m_e \rightarrow \gamma m_e$. One must solve the Vlasov equation—the collisionless form of the Boltzmann equation describing the evolution of a particle distribution function f in phase space—in order to calculate the plasma permittivity and hence the dispersion relation for electromagnetic waves.[14] When the magnetic field and wavenumber are aligned along the z-axis, the first-order permittivity is given by

$$\epsilon_{xy} = \frac{-i}{2\omega} \frac{4\pi e^2}{m_e} \frac{eB}{m_e c} \int d^3p \, \frac{1}{(\omega - kv_z)^2} \frac{p_\perp^2}{p} \frac{df}{dp} \frac{m_e^2 c^2}{p^2 + m_e^2 c^2}, \qquad 7.4$$

where the unperturbed distribution function is normalized by the condition $\int d^3p \, f(p) = n_e$ (the number density), $p_\perp^2 \equiv p_x^2 + p_y^2$, and $\omega = ck$ is the frequency of the radio wave.

When the plasma is ultrarelativistic, equation (7.4) integrates out to

$$\epsilon_{xy} = \frac{i \, \omega_p^2 \omega_B}{\omega^3} \frac{\log \gamma}{2\gamma^2}, \qquad 7.5$$

where $\gamma \equiv kT_e/m_e c^2$ (in terms of T_e, the temperature of the relativistic electron Maxwell-Boltzmann distribution), $\omega_B = eB/m_e c$ is the cyclotron frequency, and $\omega_p = (4\pi n_e e^2/m_e)^{1/2}$ is the plasma frequency. Using the same arguments as in the nonrelativistic case,[15] it is straightforward to see that in the ultrarelativistic case equation (3.18) transforms to

$$\mathrm{RM}_\gamma \equiv 2.63 \times 10^{-13} \, n_e \, B_\parallel \, L \, \frac{\log \gamma}{2\gamma^2} \quad \mathrm{rad\,m^{-2}} \qquad 7.6$$

(with B now in G and L in cm).

[13] This was cleverly done by Quataert and Gruzinov (2000).

[14] A detailed treatment of this derivation may be found in the textbook by Melrose (1991).

[15] As before, a straightforward derivation of this result may be found in Rybicki and Lightman (1979).

So taking $\gamma \sim 10$ (see §3.1) and the most conservative lower limit on n_e, which arises from mass conservation $\dot{M} = 4\pi r^2 n_e m_p v_r$ with hydrogen abundance and the most extreme radial velocity $v_r = c$, together with an equipartition magnetic field $B_\parallel \sim B_{eq} \equiv (8\pi n_e \gamma m_e c^2)^{1/2}$ and $L \sim r$, one finds that

$$\mathrm{RM}_\gamma \approx 90.5 \, \dot{M}^{3/2} \, r^{-2} \, \mathrm{rad} \, \mathrm{m}^{-2}. \qquad 7.7$$

With the limit $\mathrm{RM}_\gamma < 3 \times 10^6 \, \mathrm{rad} \, \mathrm{m}^{-2}$ derived in chapter 3 and a characteristic radius $r = 3r_S \approx 3 \times 10^{12} \, \mathrm{cm}$ (for a $\sim 3 \times 10^6 \, M_\odot$ black hole), the implied upper limit to $\dot{M}$ is therefore $\sim 4 \times 10^{19} \, \mathrm{g} \, \mathrm{s}^{-1}$—and this is really an upper limit, since we assumed spherical accretion with $v_r = c$, which tends to dilute n_e unrealistically.[16]

By comparison, the three-dimensional hydrodynamic simulation shown in figure 7.4 suggests that the black hole is capturing matter at a rate closer to $\sim 10^{22} \, \mathrm{g} \, \mathrm{s}^{-1}$. Something must be happening to the infalling plasma somewhere between the accretion radius r_{acc} and the region (within tens of Schwarzschild radii) where most of Sagittarius A*'s spectrum is produced. This problem has yet to find a full resolution, though some indications are now emerging that an ingredient hitherto ignored may in fact play a dominant dynamic role below r_{acc}. The magnetic field entrained by the ionized plasma may be considerably altering the energetics of the accreting gas.

7.4 Magnetic Field Dissipation

The importance of magnetic field dissipation within the accreting gas was recognized from the earliest thinking on this subject.[17] Simply stated, we have a situation in which the highly ionized plasma "freezes" the magnetic field and intensifies it due to flux conservation in the flow converging toward the black hole.

---

[16] However, one should keep in mind that a strong rotation measure does not in principle limit densities and magnetic fields when the field has a very small bias superimposed on it and the dispersion of the projected magnetic field on the sky is not too large; see Ruszkowski and Begelman (2002). There will be additional discussion toward the end of this chapter, when we address the likely origin of Sagittarius A*'s circular polarization.

[17] The very first paper on this topic (Shvartsman 1971) already foresaw the consequence of this process on the energetics, though its influence on the dynamics has become more apparent in recent years.

Let us take the simplest possible approach and estimate the size of this effect.[18] For an idealized spherical infall of pure hydrogen, the electron number density is

$$n_e(r) = \frac{\dot{M}}{4\pi r^2 m_p v_r},$$
7.8

where v_r is the free-fall velocity $(2GM/r)^{1/2}$. Thus,

$$n_e(r) \approx 2 \times 10^{11} \left(\frac{r_S}{r}\right)^{3/2} \text{cm}^{-3}$$
7.9

for $\dot{M} \sim 10^{22}\,\text{g s}^{-1}$ (see figure 7.4).

This plasma is highly ionized even at the capture radius r_{acc}, with a temperature well above $10^4\,\text{K}$ (see color plate 8), and the implied high conductivity therefore prevents any diffusion of magnetic field lines across the gas. As the accreting matter converges toward the black hole, the area element of any given flux tube decreases as $\sim r^2$, so flux conservation implies that the magnetic field intensity must go as

$$B(r) \propto r^{-2},$$
7.10

and hence the energy density goes as r^{-4}. By comparison, the specific kinetic energy density $v_r^2 n_e/2$ is proportional to $r^{-5/2}$. Evidently, the magnetic field intensity becomes divergent very quickly as the radius shrinks. But this cannot go on indefinitely since eventually the magnetic pressure would overwhelm that of the gas and any further constriction would cease. The crossover point occurs at equipartition, where all the specific energies are equal, including that of the magnetic field:

$$\frac{B_{\text{eq}}^2}{8\pi} \equiv \frac{1}{2} v_r^2 n_e(r) m_p.$$
7.11

Numerically, this equation may be written as

$$B_{\text{eq}}^2 \approx 3.4 \times 10^9 \left(\frac{r_S}{r}\right)^{5/2} \text{G}^2.$$
7.12

[18] An early application of these ideas to Sagittarius A* may be found in Melia (1992a, 1994).

As the dominant agent, any tangled super-equipartition field would realign itself, possibly even annihilate adjacent sheared field components until the intensity again falls below the equipartition value. The caveat here is that the physics of magnetic reconnection is poorly known. Some attempts have been made to incorporate the magnetic tearing mode instability into a model of super-equipartition field annihilation in Sagittarius A*, but there is still precious little agreement between our understanding of plasma processes such as this and what is actually observed in real systems, such as the Sun. We shall return to this issue below, when we consider the magnetized accretion flow in more detail.[19]

Notwithstanding the uncertainty of how the magnetic field actually reconnects to maintain its equipartition value, let us nonetheless assume that at any given radius the rate at which magnetic field energy is dissipated into heat is determined by the excess of the actual field intensity above its equipartition value. That is, we will write the dissipation rate as

$$\Gamma_{\mathrm{mag}} = \frac{d}{dt}\left(\frac{B^2}{8\pi n_e}\right) - \frac{d}{dt}\left(\frac{B_{\mathrm{eq}}^2}{8\pi n_e}\right), \qquad 7.13$$

where B is the magnetic field produced at any given radius and time by the conservation of flux within the ionized plasma.

The rate Γ_{mag} represents the energy per unit volume per unit time transferred from the magnetic field to the gas. In this simple model,

$$\Gamma_{\mathrm{mag}} = \frac{3}{2}\frac{v_r}{r}\frac{B^2}{8\pi}, \qquad 7.14$$

the factor 3/2 arising from the difference between the two time derivatives in equation (7.13). Numerically,

$$\Gamma_{\mathrm{mag}} \approx 10^6 \left(\frac{r_S}{r}\right)^4 \mathrm{ergs\ cm}^{-3}\ \mathrm{s}^{-1}. \qquad 7.15$$

This is actually quite a significant rate of heating for the gas; its importance may be appreciated with greater clarity through a comparison of

[19]The reader interested in learning more about magnetic dissipation in converging flows may want to read Kowalenko and Melia (1999) and Melia and Kowalenko (2001).

the "magnetic dissipation" timescale

$$\tau_{mag} \equiv \frac{B_{eq}^2}{8\pi} \div \Gamma_{mag} \approx 1.4 \times 10^2 \left(\frac{r}{r_S}\right)^{3/2} \ s, \qquad 7.16$$

with the dynamical timescale

$$\tau_{dyn} \equiv \frac{v_r}{r} \approx \frac{r^{3/2}}{c} \left(\frac{1}{r_S}\right)^{1/2}. \qquad 7.17$$

Evidently,

$$\frac{\tau_{mag}}{\tau_{dyn}} \sim 4. \qquad 7.18$$

The heating of the gas due to magnetic dissipation is therefore considerable, and given that the accretion rate at capture (i.e., at r_{acc}) does not match what is seen near Sagittarius A*, the impact of this energy transfer on the gas dynamics must be an effect we cannot ignore.

This supposition was finally put to a numerical test in 2002, with the first magnetohydrodynamic (MHD) simulation of Bondi-Hoyle accretion onto Sagittarius A*.[20] Though several important caveats (described in detail below) must be appended to this work, the results are quite promising in demonstrating how the entrained magnetic field may modify the dynamics of the flow to the point where the accretion rate onto Sagittarius A* is heavily attenuated. Starting with the Bondi-Hoyle solution described above as the outer boundary condition, these simulations solve the MHD equations in the one-fluid approximation:

$$\frac{d\rho}{dt} + \rho \vec{\nabla} \cdot \mathbf{v} = 0, \qquad 7.19$$

$$\rho \frac{d\mathbf{v}}{dt} = -\vec{\nabla}(P + Q) - \rho \vec{\nabla}\Phi + \frac{1}{4\pi}(\vec{\nabla} \times \mathbf{B}) \times \mathbf{B}, \qquad 7.20$$

$$\rho \frac{d\epsilon}{dt} = -(P + Q)\vec{\nabla} \cdot \mathbf{v} + \frac{1}{4\pi} \eta J^2, \qquad 7.21$$

$$\frac{\partial \mathbf{B}}{\partial t} = \vec{\nabla} \times (\mathbf{v} \times \mathbf{B} - c\eta J), \qquad 7.22$$

[20] See Igumenshchev and Narayan (2002).

where ρ is the density, $\mathbf{v}$ is the velocity, P is the pressure, Φ is the gravitational potential, $\mathbf{B}$ is the magnetic induction, ϵ is the specific *internal* energy, $\mathbf{J} = (c/4\pi)\vec{\nabla} \times \mathbf{B}$ is the current density, and η is the resistivity. The appearance of Q in equations (7.20) and (7.21) corresponds to an artificial viscosity introduced to correctly treat shocks. Additional approximations are (1) use of the ideal gas equation of state, for which

$$P = (\gamma - 1)\rho\epsilon, \qquad\qquad 7.23$$

with an adiabatic index $\gamma = 5/3$; (2) the exclusion of any radiative cooling; and (3) a representation of the compact mass M at the center via a pseudo-Newtonian gravitational potential[21]

$$\Phi = -\frac{GM}{r - r_S} \qquad\qquad 7.24$$

that mimics the effect of general relativity near the Schwarzschild radius.

This set of simultaneous equations includes several conservation laws and the condition for resistive MHD: equation (7.19) is the conservation of mass, and (7.20) expresses the conservation of momentum, in which the momentum density of the gas changes in response to a force density from gradients in pressure and the gravitational potential. The last term in this equation is the Lorentz force density with $\mathbf{J}$ replaced with Ampère's law (ignoring the displacement current); (7.21) is the conservation of energy equation, in which the internal energy density of the gas changes through PdV work (first term) or resistive dissipation (second term); and (7.22) is the equation of resistive MHD.

To understand the origin of this expression, consider that in the rest (primed) frame of the fluid, Ohm's law says

$$\mathbf{J}' = \frac{1}{\eta}\mathbf{E}', \qquad\qquad 7.25$$

where $1/\eta$ is the conductivity. From the Lorentz transformations derived in §5.2, we know that

$$\mathbf{E}' = \gamma\left(\mathbf{E} + \frac{\mathbf{v}}{c} \times \mathbf{B}\right), \qquad\qquad 7.26$$

[21] This approximation was introduced for numerical simplicity by Paczyński and Wiita (1980).

so that for nonrelativistic motion

$$\mathbf{J}' \approx \frac{1}{\eta}\left(\mathbf{E} + \frac{\mathbf{v}}{c} \times \mathbf{B}\right).$$

7.27

Thus, if the medium is neutral, for which there are no advected currents,

$$\mathbf{J} \approx \frac{1}{\eta}\left(\mathbf{E} + \frac{\mathbf{v}}{c} \times \mathbf{B}\right).$$

7.28

In *ideal* MHD, the conductivity goes to infinity (or equivalently, η goes to zero), whereupon

$$\mathbf{E} \approx -\frac{\mathbf{v}}{c} \times \mathbf{B}.$$

7.29

In this limit, equation (7.22) is then simply a restatement of Faraday's law with $\mathbf{E}$ replaced with this expression (and $\eta = 0$). In *resistive* MHD, η is finite and cannot be ignored. Equation (7.22) is therefore the more general application of Faraday's law with $\mathbf{E}$ replaced with equation (7.28).

Before we proceed to discuss the results of this simulation, we first acknowledge the two principal caveats in this work. First, along with Bondi-Hoyle outer boundary conditions, this calculation assumes that the initial magnetic field at large radii is uniform, with only one nonzero component B_z. Second, as we noted earlier, the rate at which magnetic reconnection occurs is still poorly understood.

In reality, the magnetic field in the ambient medium, from which Sagittarius A* is accreting, is likely to be nonuniform. At the very least, the seed magnetic field should mirror the tessellated pattern of gas inhomogeneities emerging from the simulations of wind-wind interactions occurring in this environment (see §7.1). For the work described here, the initial strength of $\mathbf{B} = B_z\,\hat{z}$ is specified by a ("sub-equipartition") parameter b_0 defined according to

$$\frac{B_z^2}{8\pi} = b_0^2\,\frac{GM\rho_0}{r_0}.$$

7.30

Rather than starting the calculation at r_{acc}, this simulation adopts Bondi-Hoyle conditions down to the initial radius $r_0 = 256\,r_S$, where the starting density is ρ_0.

The question of magnetic reconnection may take much longer to resolve and, in the end, may rely more on experimental verification than on theoretical discourse. Rapid reconnection of the field occurs wherever any component of the magnetic field changes sign and is not held apart by suitable fluid pressure. In the Petschek mechanism,[22] dissipation of the sheared magnetic field occurs in the form of shock waves surrounding special neutral points in the current sheets, and thus nearly all the dissipated magnetic energy is converted into the magnetic energy carried by the emergent shocks. Rapid reconnection may occur at speeds 10^{-2}–10^{-1} times the Alfvén velocity $v_A = B/\sqrt{4\pi\rho}$ and causes vigorous dissipation of the magnetic field.[23]

In an alternative picture, the reconnection process develops as a result of resistive diffusion of the magnetic field lines through the plasma. It has been pointed out[24] that, aside from the fact that the Petschek mechanism suffers conceptually from the lack of an observable timescale and a predictable energy output, there is also a question as to whether it exists at all, since it has never been observed in either laboratory or astrophysical applications.[25] In the alternative picture, referred to as *resistive magnetic tearing* (or the tearing instability), the instability grows temporally and is driven by the free energy of a sheared magnetic field.

Another complication is that we have little guidance on what to take for the geometry of the magnetic field itself. Since the dissipation presumably occurs only for counteraligned field components, it clearly matters whether the field is completely turbulent or whether it acquires a partial coordination with the flow direction.

In the end, even tearing-mode instabilities may be ineffective at spreading the reconnection layer, since they may be stabilized by shear and grow slowly past the linear stage. But there is no question that rapid reconnection does take place (e.g., in the Sun, where observations point to reconnection at the Alfvén speed, at least when the magnetic field is dynamically important). So the growth rate calculated from either

[22]This subject was first broached in connection with solar flares. See Petschek (1964).

[23]See also Parker (1979).

[24]A leading proponent of an alternative to the Petschek mechanism has been Van Hoven (1976), who also carried out laboratory experiments to measure the speed of reconnection in sheared magnetic fields.

[25]However, more recent *Yohkoh* observations may have provided an indication that Petschek-like reconnection may be taking place in the Sun after all. See Blackman (1997). Theoretically, Petschek reconnection is perceived to be possible as long as some enhanced anomalous resistivity is active at the origin.

of these two mechanisms may simply be an underestimate of the actual value.[26]

The approach taken in the simulation described here is much simpler, and we must await confirmation of its validity from more elaborate future calculations. An explicit artificial resistivity

$$\eta = \eta_0 \frac{|\vec{\nabla} \times \mathbf{B}|}{\sqrt{4\pi \, \rho}} \Delta^2 \qquad 7.31$$

is introduced with a magnitude set to be larger than the effective numerical resistivity one would achieve with numerical reconnection.[27] Here, Δ is the grid spacing, and η_0 is a dimensionless parameter. In essence, this prescription guarantees that the magnetic field reconnects vigorously once it exceeds its equipartition value.

Figure 7.5 shows the magnetic field configuration at time $t = 0.5 \, t_{ff}$, where t_{ff} is the free-fall time from the edge of the computational grid. For this particular calculation, the two critical parameters are set to $\eta_0 = 0.5$ and $b_0 = 0.3$. One sees that the initially parallel field lines are swept in toward the black hole by the converging streamlines of the flow. As the simulation continues, a dense core forms in the inner region where the magnetic field settles into equipartition with the gas, surrounded by a quasi-spherical shock (at roughly $80 \, r_S$) where the infalling supersonic plasma first impacts this turbulent region. At a relatively late time, $t = 7.88 \, t_{ff}$, the magnetic field morphology is that depicted in figure 7.6, where we see a much more tangled configuration.

By this time, the gas distribution is highly inhomogeneous (see figure 7.7), with filaments of higher and lower density sandwiched between each other. This structure is the result of convection in the core, driven by the heat released during episodes of magnetic reconnection. As we suspected, the impact of this heating is considerable, as one may gauge from figure 7.8, which shows the accretion rate (in units of the Hoyle-Lyttleton value) as a function of time. This figure shows the results of three calculations, one strictly hydrodynamic (thick solid line), another with $\eta_0 = 0.5$ and $b_0 = 0.3$ (thin solid curve), and the final one with $\eta_0 = 0.5$ and $b_0 = 0.1$ (dashed).

---

[26] A detailed study of magnetic reconnection in Sagittarius A* was undertaken by Kowalenko and Melia (1999) and Melia and Kowalenko (2001).

[27] This specification was first made by Stone and Pringle (2001) and is designed to mimic the magnetic Reynolds number one would get from the resistivity associated with simple, local numerical fluctuations.

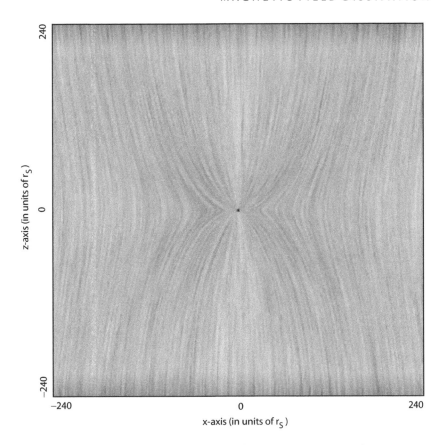

Figure 7.5 Magnetic field lines in the x-z plane in an MHD simulation at time $t = 0.5$, measured in units of the free-fall time from the edge of the grid. The black hole is at the origin. The axes are labeled in units of r_S. At the start of the simulation, the magnetic field is uniform and pointing along $\hat{z}$. Over time, the spherically converging accretion flow deforms the frozen-in field, forming an axisymmetric structure around the z-axis. (Image from Igumenshchev and Narayan 2002)

The most important consequence of the reconnection-driven convection that appears in these simulations is the considerable attenuation—by a factor of $\sim$10 or more—of the mass accretion rate onto the black hole compared with that of a standard Bondi-Hoyle flow. Convection efficiently transports energy from the deep interior of the flow, where the bulk of the gravitational energy is released, to the outer regions of the envelope. The dynamics of the gas is heavily influenced by this displacement in energy, and the otherwise free-fall structure of accretion is greatly inhibited.

175

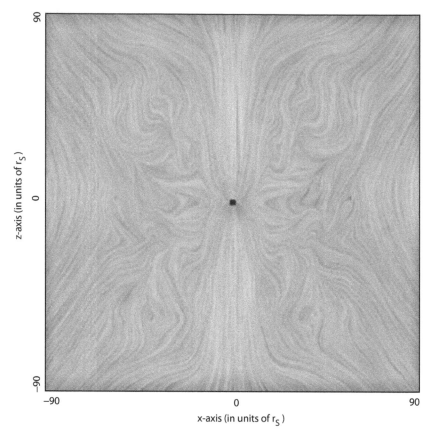

Figure 7.6 Same as figure 7.5, except here at $t = 7.88$. The axes are labeled in units of r_S. Only components of the magnetic field parallel to the x-z plane are shown. By this time, the field has become highly tangled due to the effects of convection. (Image from Igumenshchev and Narayan 2002)

This process is an interesting resolution of the accretion rate puzzle in Sagittarius A*, but as we have emphasized all along, one must be wary of these results until future, more elaborate simulations will have provided a more compelling case for this argument.

7.5 A Compact Magnetized Disk

All the evidence we presented in chapter 4 suggests that the *active* region accounting for Sagittarius A*'s spectrum—certainly when this source is in a variable state—is confined to a tiny region, probably no bigger than

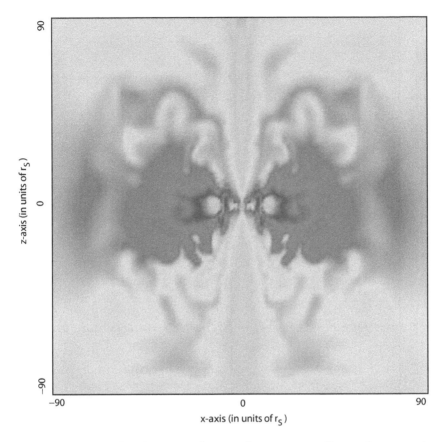

z-axis (in units of r_S)

x-axis (in units of r_S)

Figure 7.7 Density distribution in the x-z plane corresponding to the magnetic field line profile shown in figure 7.6 (at time $t = 7.88$). The darkest regions correspond to a particle density of $10^7\,\mathrm{cm}^{-3}$, whereas the lightest regions (along the polar "funnels") represent a density of $10^6\,\mathrm{cm}^{-3}$. The axes are labeled in units of r_S. Matter is concentrated toward the equatorial plane, with density inhomogeneities produced by the convective motions. There is a shock around $80 r_S$, where the supersonically infalling gas from the outer region impacts the convective core. (Image from Igumenshchev and Narayan 2002)

Earth's orbit about the Sun. The numerical simulations we examined above, first the hydro calculations whose results are summarized in figure 7.4, and then the MHD calculations presented in figure 7.7, lend weight to this notion that the brightest portion of the emitting region appears to reside within only tens of Schwarzschild radii of the black hole.

Having said this, we still have our work cut out for us in discerning the various emission components and their origins. A disk may be present

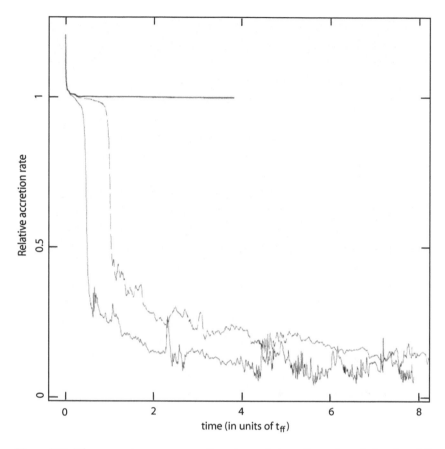

Figure 7.8 The accretion rate as a function of time (in units of the free-fall time), normalized to the Hoyle-Lyttleton rate that one would expect to get in a (nonmagnetic) hydrodynamic calculation. The solid line shows the purely hydrodynamic simulation with no magnetic field. The thin solid line is a model in which the initial magnetic field parameter $b_0 = 0.3$ (see text). The dashed line is for a model in which $b_0 = 0.1$. Note the strong suppression of the accretion rate when the magnetic field is present. (Image from Igumenshchev and Narayan 2002)

in Sagittarius A*, but it is not necessarily the sole contributor to its spectrum. In this section and the next, we will examine the various empirical constraints one infers from the observations and, at the very least, configure them within the context of what we know theoretically about this source. Sagittarius A*, like many other radiating, compact objects, appears to manifest itself through several principal emitting regions, and it will be our task to determine how these all function together to produce the radiative flux and polarization fraction we see originating from the galactic nucleus.

The argument in §7.4—that dissipation of the magnetic field entrained by the converging plasma is the likely explanation for Sagittarius A*'s low accretion rate—at the same time suggests that the infalling matter is radiatively inefficient, at least until it settles into a disk. Of course, this is not obvious, since the captured gas is highly ionized (and magnetized), so it radiates via several mechanisms, including bremsstrahlung, cyclo-synchrotron (i.e., the full range of cyclotron to synchrotron emission in a magnetic field), and inverse Compton scattering. Thus, the overall power depends on several critical physical parameters: the particle density, the temperature (or Lorentz factor, if the dominant emitters are nonthermal), and the magnetic field.

But Sagittarius A* is not a typical AGN. The rate at which it accretes is apparently less than $\sim 10^{19}\,\mathrm{g\,s^{-1}}$, and if one naively integrates the particle number density implied by this value (see equation 7.9) from, say, $r = 40\,r_S$ out to infinity, then one finds that the scattering optical depth $\int n_e(r)\sigma_T\,dr$ (where σ_T is the Thomson scattering cross section) is smaller than 10^{-16}. Clearly, the medium surrounding the black hole is extremely thin, and the integrated emissivity must therefore be correspondingly small.[28]

This can change dramatically once the infalling gas circularizes and forms a disk. The three-dimensional hydrodynamic simulations we discussed above indicate that the accreted specific angular momentum λ (in units of $c r_S$) can vary by 50% over <200 years, with an average equilibrium value of $\sim 40 \pm 10$. However, λ is never zero, meaning that the gas cannot simply flow radially inwards all the way to the event horizon. Nonetheless, even with a possibly large quantity of angular momentum present in the environment surrounding the nucleus, relatively little specific angular momentum is accreted. This is understandable since clumps of gas with a high specific angular momentum cannot penetrate inside the accretion radius r_{acc} (see equation 7.1). The associated variability in the sign of the components of λ suggests that if an accretion disk forms at all—presumably with a size comparable to the circularization radius r_{circ} (see equation 7.3)—it dissolves and reforms (perhaps) with an opposite sense of spin on a timescale of around 100 years or less.

---

[28] For a more detailed discussion of this point, see, e.g., Shapiro (1973), Ipser and Price (1977), and Melia (1992a).

Within the Keplerian structure, the captured plasma falls prey to the action of a magnetic dynamo fed by the differential rotation; this overwhelms the field annihilation and leads to a saturated field intensity (at a fraction of the equipartition value). The discovery of this magnetorotational instability (MRI) in weakly magnetized disks[29] spawned several numerical simulations that confirmed the crucial role played by this process in rotational accretion systems.[30] The magnetic field generated in this fashion produces a significant viscosity via the Maxwell stress, dominating the transfer of angular momentum across the disk.

The Magnetorotational Instability in Sagittarius A*

Before examining the structure of Sagittarius A*'s accretion flow under the influence of the MRI, let us first briefly analyze how this instability develops in a weakly magnetized Keplerian flow. The basic dynamical equations we need to consider are (7.19), (7.20), and (7.22). For simplicity and given that we will find the MRI to be very efficient compared to magnetic field annihilation due to resistive diffusion, we will set $\eta = 0$ in this application. We will adopt standard cylindrical coordinates $\mathbf{r} = (R, \phi, z)$, where R is the perpendicular distance from the z-axis.

Again for simplicity, we will consider perturbations of an initial field $\mathbf{B} = (0, 0, B_z)$. It turns out that the maximal growth rate is reached in the axisymmetric case with a weak B_z, so this is a good initial configuration for pedagogical purposes as well. We will denote the Eulerian perturbations by δv, δB, and so forth, each modulated by the function $e^{i(k_R R + k_z z - \omega t)}$, where k_R and k_z are, respectively, the radial and vertical components of the wavevector.

The numerical simulations show that buoyancy is not a significant factor influencing the instability, nor is the compressibility of the fluid. By neglecting these terms and assuming incompressibility (so that $\delta\rho = 0$ in all the equations other than the equation of motion and the equation of state) and that $\delta P = 0$ in the equation of state, one obtains[31] the

---

[29] See Balbus and Hawley (1991).

[30] Some of these simulations have been reported in Hawley and Balbus (1991, 1992), Hawley, Gammie, and Balbus (1995), and Stone et al. (1996).

[31] These follow simply by putting $v_R \rightarrow v_R + \delta v_R$ (and similarly for the other variables) in equations (7.19), (7.20), and (7.22) and then using these same equations to eliminate terms that depend solely on the unperturbed quantities.

following linearized dynamical equations:

$$k_R \, \delta v_R + k_z \, \delta v_z = 0, \tag{7.32}$$

$$\frac{\partial \delta v_R}{\partial t} - 2\Omega \, \delta v_\phi = i \frac{k_z B_z}{4\pi\rho} \, \delta B_R - i k_R \left(\frac{\delta P}{\rho} + \frac{B_z \, \delta B_z}{4\pi\rho} \right), \tag{7.33}$$

$$\frac{\partial \delta v_z}{\partial t} = -i k_z \frac{\delta P}{\rho}, \tag{7.34}$$

$$\frac{\partial \delta v_\phi}{\partial t} + \frac{\kappa^2}{2\Omega} \, \delta v_R = i \frac{k_z B_z}{4\pi\rho} \, \delta B_\phi, \tag{7.35}$$

$$\frac{\partial \delta B_R}{\partial t} = i k_z B_z \, \delta v_R, \tag{7.36}$$

$$\frac{\partial \delta B_z}{\partial t} = i k_z B_z \, \delta v_z, \tag{7.37}$$

$$\frac{\partial \delta B_\phi}{\partial t} = \frac{R \, d\Omega}{dR} \, \delta B_R + i k_z B_z \, \delta v_\phi, \tag{7.38}$$

where Ω is the angular velocity in the circularized flow and $\kappa^2 = (2\Omega/R) \, d(R^2\Omega)/dR$ is the square of the epicyclic frequency.

Replacing the Lagrangian time derivatives with $-i\omega$ in the linearized equations and eliminating the Eulerian perturbations, one obtains the dispersion relation

$$(\omega^2 - k_z^2 \, v_{A,z}^2)^2 - \frac{k_z^2}{k^2} \kappa^2(\omega^2 - k_z^2 \, v_{A,z}^2) - 4\Omega^2 \frac{k_z^4 v_{A,z}^2}{k^2} = 0, \tag{7.39}$$

where $k^2 = k_z^2 + k_R^2$ and the Alfvén speed in the z-direction is defined as $v_{A,z} = (B_z^2/4\pi\rho)^{1/2}$. For the Keplerian region, $\kappa = \Omega$. This equation can be solved easily for ω, which yields

$$\omega_0^2 = k_{z0}^2 + \frac{k_{z0}^2}{2k_0^2} - 2\sqrt{\frac{k_{z0}^4}{k_0^2} + \frac{k_{z0}^4}{16k_0^4}}. \tag{7.40}$$

Here, $\omega_0 \equiv \omega/\Omega$, and k_0 and k_{z0} are, respectively, k and k_z expressed in units of $\Omega/v_{A,z}$. Note that ω^2 reaches its minimum value of $-(9/16)\Omega^2$ when $k^2 = k_z^2 = (15/16)(\Omega/v_{A,z})^2$. For $k_R = 0$, the modes become stable when $k_z^2 > 3(\Omega/v_{A,z})^2$, and in the long wavelength limit $\omega^2 \simeq -3(v_{A,z}k_z)^2$.

In order to appreciate the physical meaning of this instability, let us examine the fastest growing mode, which occurs when $k^2 = k_z^2 = (15/16)(\Omega/v_{A,z})^2$ and $\omega^2 = -(9/16)\Omega^2$. Solving the linearized equations (7.32)–(7.38) with these values, we get

$$\delta v_R = \delta v_\phi, \qquad\qquad 7.41$$

$$\delta B_R = -\delta B_\phi, \qquad\qquad 7.42$$

$$\delta B_R = i\frac{4}{3}\frac{k_z B_z}{\Omega}\delta v_R, \qquad\qquad 7.43$$

$$\delta B_\phi = -i\frac{5}{4}\frac{4\pi\rho\Omega}{k_z B_z}\delta v_\phi, \qquad\qquad 7.44$$

$$\frac{|\delta B_R|^2}{8\pi} = \frac{5}{3}\frac{\rho|\delta v_R|^2}{2}. \qquad\qquad 7.45$$

With these solutions, we can now return once more to equations (7.32)–(7.38) and see how the instability grows. The perturbation δv_R, which is generated by δv_ϕ through the Coriolis force term $2\Omega\,\delta v_\phi$ in equation (7.33), induces the perturbation δB_R through equation (7.36). The shearing in the disk, affecting the term $(R\,d\Omega/dR)\,\delta B_R$ in equation (7.38), leads to the perturbation δB_ϕ through the term δB_R. This in turn enhances δv_ϕ through the right-hand side of equation (7.35). Thus, a positive feedback loop is established. Some of the other terms in the linearized equations act to stabilize the perturbation, but they are overwhelmed by the positive feedback in the unstable modes. However, for modes with a large wavenumber, the term $ik_z B_z\,\delta v_\phi$ in equation (7.38) and the term $ik_z B_z\,\delta B_R/4\pi\rho$ in equation (7.33) will overwhelm the positive feedback and render the mode stable.

For the unstable modes, equation (7.45) says that the turbulent kinetic energy density is approximately equal to the turbulent magnetic field energy density. Apparently, the final saturated state of the system approaches equipartition. Numerical simulations[32] confirm this basic result and go further in demonstrating that the ratio of final turbulent energy densities is only weakly dependent on the initial and subsequent physical conditions. We shall therefore find it convenient to parametrize the dynamo-generated energies in a rotational flow according to

$$\frac{\langle \delta B^2 \rangle}{8\pi} = C_0 \frac{1}{2} \langle \rho \delta v^2 \rangle, \qquad 7.46$$

where the constant C_0 has a value between 1 and 10, depending on the vertical profile of the Keplerian structure. Note, however, that this formulation does not yet tell us what the actual magnetic energy density is—only that it is comparable to that of the turbulent gas. The final piece of the puzzle that will permit this evaluation will come from the solution of the energy equations themselves (see equations 7.49–7.51).

Another physical effect that may quench the instability and inhibit the growth of a turbulent magnetic field is damping by Ohmic diffusion.[33] The instability is effectively damped when the diffusion length becomes comparable to the wavelength of the most unstable mode within the latter's growth timescale, which is roughly one revolution period. The most unstable wave mode in the Keplerian flow is $k^2 = k_z^2 = (15/16)(\Omega/v_{A,z})^2$. As we shall see, the magnetic field in Sagittarius A* is relatively weak (of order 5–10 G), so $v_{A,z}$ is small, and all the modes with $k_z < 3\Omega/v_{A,z}$ are unstable. But as the instability evolves, $\langle B_z^2 \rangle$ increases, which in turn leads to an increase in the value of $v_{A,z}$. Thus, the wavelength $\lambda_z = 2\pi/k_z$ of the most unstable mode increases with time, and there does not appear to be any physical reason why the magnetic field should stop increasing before λ_z has reached H, the disk scale height. The wavelength of the most unstable mode is therefore expected to be H.

---

[32] See Brandenburg et al. (1995), Hawley, Gammie, and Balbus (1995), and Stone et al. (1996).

[33] Jin (1996) first discussed the possible damping of the shear instability using linear analysis. More recently, this issue was addressed numerically by Fleming, Stone, and Hawley (2000).

In Sagittarius A*, the diffusion length scale over one period is much smaller than this wavelength. To see this, we note that the angular velocity near 10 Schwarzschild radii of a $\sim 3.6 \times 10^6 \, M_\odot$ black hole is $\Omega \approx (GM/10 \, r_S^3)^{1/2} \approx 0.0007 \, \text{s}^{-1}$. We already know from the large-scale simulations of accretion onto this object that the infalling plasma radiates inefficiently and therefore retains most of its heat content. By the time the gas settles into a compact disk, its temperature is $>10^{10} \, \text{K}$, so the Keplerian region is fully ionized, and its resistivity is

$$\eta = 7.26 \times 10^{-9} \frac{\ln \Lambda}{T^{3/2}} \quad \text{(c.g.s.)}, \qquad 7.47$$

where the Coulomb logarithm $\ln \Lambda$ is a slowly varying function of the electron density and temperature; 30 is a reasonable value for the conditions at the galactic center. Thus, for these parameter values, $\eta \simeq 0.22 \times 10^{-21}$ (c.g.s.), and the corresponding diffusion length is

$$L = \sqrt{\frac{\eta c^2 \tau}{4\pi}}, \qquad 7.48$$

where τ is the diffusion time, which may be set equal to the growth time of the most unstable mode, that is, about 1,000 seconds here. The diffusion length is evidently $\sim 4 \, \text{cm}$, clearly much smaller than any reasonable value of H. So the dynamo in Sagittarius A* cannot be damped by Ohmic diffusion.

The remaining question concerns which magnetic field component (or components) dominates. Assuming that the turbulent field generated by the dynamo constitutes the total field in this system, let us examine the equations governing the evolution of the magnetic energy density:

$$\frac{1}{2} \frac{\partial B_\phi^2}{\partial t} = R \frac{d\Omega}{dR} B_\phi B_R + B_\phi \, [\vec{\nabla} \times (\delta \mathbf{v} \times \mathbf{B})]_\phi + \frac{\eta c^2}{4\pi} B_\phi |\nabla^2 \mathbf{B}|_\phi, \quad 7.49$$

$$\frac{1}{2} \frac{\partial B_R^2}{\partial t} = B_R [\vec{\nabla} \times (\delta \mathbf{v} \times \mathbf{B})]_R + \frac{\eta c^2}{4\pi} B_R |\nabla^2 \mathbf{B}|_R, \qquad 7.50$$

$$\frac{1}{2} \frac{\partial B_z^2}{\partial t} = B_z [\vec{\nabla} \times (\delta \mathbf{v} \times \mathbf{B})]_z + \frac{\eta c^2}{4\pi} B_z |\nabla^2 \mathbf{B}|_z, \qquad 7.51$$

where η is again the resistivity of the plasma. These equations all follow from the chain rule of differentiation and the application of equation (7.22) in which **J** is replaced with the curl of **B** from Ampère's law.

We saw in equation (7.42) that in the linear regime the amplitudes of the azimuthal and radial components of the magnetic field are equal when the perturbation is small. However, the final turbulent state is affected by the nonlinear character of the magnetohydrodynamic equations. As the amplitudes of the perturbation increase, nonlinear effects become more important. Due to shearing in the Keplerian flow, the average value of $RB_\phi B_R \, d\Omega/dR$ is positive. The energy equations (7.49)–(7.51) show that this term contributes to a growing anisotropy of the turbulent magnetic field, in the sense that more and more magnetic field energy is channeled into the azimuthal direction. For a Keplerian flow with $R \, d\Omega/dR = -(3/2)\Omega$, the growth rate due to the shearing of this structure is larger than that associated with any other dynamo process. So in the final state, the azimuthal component of the magnetic field dominates the field energy density.

This result has several important consequences, including an explanation for Sagittarius A*'s linear polarization at mm/sub-mm wavelengths (see Figure 3.2 and the discussion following equation 7.83). We shall revist this important result in chapter 9, where we discuss in detail the future polarimetric imaging of the supermassive black hole at the center of our Galaxy.

In summary, then, we have the following situation with regard to Sagittarius A*'s magnetized disk. As the gas flows inwards and spirals into an approximately Keplerian structure at a distance from the black hole corresponding to the circularization radius, a linear instability first stretches the magnetic field lines carried by the gas and produces a radial component of **B**. The magnetic field generated during this step is approximately in equipartition with the turbulent kinetic energy and counts for a small fraction of its final intensity, but it nonetheless provides the seed for the next step. Second, the shearing in the Keplerian flow stretches the radial magnetic field in the azimuthal direction, increasing the magnetic field energy density. The energy comes from the rotation of the gas, and during this process, a significant fraction of angular momentum may be transported outward due to torques arising from the Maxwell stress $B_R B_\phi/4\pi$. Finally, some of the magnetic field energy is converted into kinetic energy, which is eventually converted into thermal energy by viscous dissipation through the Lorentz force term $\mathbf{B} \cdot [\vec{\nabla} \times (\delta\mathbf{v} \times \mathbf{B})]$. In addition, the magnetic field energy may be dissipated through Ohmic resistivity. However, in the case of the galactic center, the actual η is rather small, so that $\eta c^2/4\pi r_S \simeq 10^{-13}$ cm s^{-1}, which is insignificant compared to any velocity within the plasma. The dissipation of the magnetic field in this fashion is therefore negligible.

The final step is to solve the energy equations (7.49)–(7.51) and find the magnetic energy density in the turbulent plasma. From dimensional analysis, we infer that

$$\frac{\langle \rho \delta v^2 \rangle^{1/2}}{\langle \rho \rangle^{1/2} H} \propto R \frac{d\Omega}{dR}, \qquad 7.52$$

where H is the scale height of the flow, and we use $\langle \rho \delta v^2 \rangle^{1/2}/\langle \rho \rangle^{1/2}$ to represent the turbulent velocity. Thus, from equations (7.46) and (7.52), we obtain

$$\langle B^2 \rangle^{1/2} \propto \langle \rho \rangle^{1/2} H R \frac{d\Omega}{dR}, \qquad 7.53$$

which is one of the main results we have been seeking.[34] For a Keplerian flow, $\Omega = (GM/R^3)^{1/2}$, and therefore $R\, d\Omega/dR = -(3/2)\Omega$.

The Anomalous Viscosity

The mathematical formulation of accretion disk theory is derived from the basic equations of continuity, expressing the conservation of mass, momentum, and energy.[35] In a Keplerian disk with column density Σ and angular velocity $\Omega = (GM/R^3)^{1/2}$, the radial velocity v_R at (cylindrical) radius R is given by

$$v_R = -\frac{3}{R^{1/2}\Sigma}\frac{\partial}{\partial R}\left(v\Sigma R^{1/2}\right), \qquad 7.54$$

[34] The validity of this relationship has been confirmed by the numerical simulations of Brandenburg et al. (1995). This work includes an analysis of the magnetohydrodynamic dynamo at two different radii, one of which is five times smaller than the other. More recent calculations by Hawley (2000) have added additional support for this result.

[35] The equations of standard accretion disk theory have been discussed extensively in the literature, and we will not rederive them here. The interested reader may wish to consult the book *Accretion Power in Astrophysics*, by Frank, King, and Raine (2002).

where ν is the kinematic viscosity (in units of velocity times distance)

$$\nu = \frac{2}{3} \frac{W_{R\phi}}{\Sigma \Omega} \qquad 7.55$$

and $W_{R\phi}$ is the vertically integrated sum of the Maxwell and Reynolds stresses. The stress is a pressure, or a force per unit area, which is generally not isotropic—hence its tensor structure, which allows one to write all three force components acting on a surface element with any given unit normal vector. For example, in the case of the electromagnetic field, with electric and magnetic components E_i and B_i (see equation 6.72), the Maxwell stress tensor may be written as[36]

$$T_{ij} = \frac{1}{4\pi} [E_i E_j + B_i B_j - \frac{1}{2}(E^2 + B^2)\delta_{ij}], \qquad 7.56$$

where δ_{ij} is the Kronecker delta, and the indices (i, j) here run over the three cylindrical coordinates (R, ϕ, z). The parcel $T_{\alpha\beta}$ of sixteen quantities (in four-dimensional spacetime) gives the force per unit area applied to any surface by the electromagnetic field.

With the dynamo in full force, $\mathbf{B}$ saturates at near its equipartition value (see above), and T_{ij} easily dominates over the Reynolds stress.[37] Thus, we may write

$$W_{R\phi} \approx \beta_\nu \int dz \frac{\langle B^2 \rangle}{8\pi} \qquad 7.57$$

(the average inside the integral being taken over time). Note that even though this approximation is valid simply on the basis that the Reynolds stress is relatively small, its validity is enhanced by the fact that the turbulent velocity (which accounts for the kinetic, or Reynolds, stress) and $\mathbf{B}$ are generated by the same process, so both should be scalable by $\mathbf{B}$ (see equation 7.46). Numerical simulations show that the proportionality constant β_ν changes very slowly with R and typically falls in the range ~ 0.1–0.2.

However, in order to derive $W_{R\phi}$ from equation (7.53), we need to know the scale height H, whose determination is most easily made for disk gas in hydrostatic equilibrium. Under this condition, we have in the

---

[36] See, e.g., Melia (2001b).
[37] See, e.g., Balbus, Gammie, and Hawley (1994).

vertical direction

$$\frac{1}{\rho}\frac{dP}{dz} = -\frac{GMz}{R^3},$$

7.58

where M is the mass of the accreting black hole and P is the gas pressure. And with the additional simplification that the plasma is approximately isothermal in the z-direction (so that T is a function of R only),

$$P = \frac{R_g}{\mu}\rho T(R),$$

7.59

where R_g is the gas constant and μ is the molecular weight. In that case, equation (7.58) leads to

$$\frac{1}{\rho}\frac{d\rho}{dz} = -\left(\frac{GM}{R^3}\frac{\mu}{R_g T}\right)z,$$

7.60

which integrates easily to yield ρ as a function of R and z:

$$\rho(R, z) = \rho_0(R)\exp\left[-\frac{z^2}{2H^2}\right],$$

7.61

where $\rho_0(R)$ is the density in the midplane of the disk, and we have now defined the vertical scale height

$$H \equiv \left(\frac{R^3 R_g T}{GM\mu}\right)^{1/2}.$$

7.62

In equation (7.53), we have $\langle B^2 \rangle^{1/2} \propto \langle \rho \rangle^{1/2} H\Omega$. Evidently, $(H\Omega)^2 \propto P/\rho$, and we can therefore write

$$\frac{\langle B^2 \rangle}{8\pi} = \beta_P P = \beta_P \frac{R_g \Sigma T}{\mu},$$

7.63

where β_P is another proportionality constant, whose value (indicated from the numerical simulations) is ~ 0.03.

In the α-disk prescription, the viscosity is said to arise from the exchange of small parcels of plasma flowing back and forth across a given radius. These carry with them their specific angular momenta, which, however, vary with radius. The net effect is a torque by one ring on another, since the exchanged angular momentum densities do not cancel exactly. To characterize the unknown velocity of these plasma eddies and the unknown length scale over which they flow, one writes the viscosity as

$$v = \alpha c_s H, \qquad 7.64$$

where $c_s \equiv R_g T/\mu$ is the sound speed. The constant α is expected to be <1. Since $H = c_s/\Omega$ (see equation 7.62), the α-disk prescription of viscosity is sometimes also written in the form

$$v = \frac{\alpha c_s^2}{\Omega}. \qquad 7.65$$

Magnetized disks subject to the MRI derive, instead, their viscous dissipation from the action of the magnetic field on the underlying plasma. Still, we can make a direct connection with the α-disk prescription by making the elegant and simple identification

$$\alpha \equiv \frac{2}{3}\beta_v \beta_P, \qquad 7.66$$

with which we recover equation (7.65) and which now characterizes the *anomalous* viscosity of our magnetized disk.

Disk Equations

We have based all of our discussion thus far on the application of equations (7.19)–(7.22) to the development of the MRI in magnetized disks and the consequent anomalous viscosity to which this gives rise. However, to make further progress and to actually solve for the structure of these disks, we must now incorporate the viscous dissipation as a heating term in the energy equation (7.21) and as a source or sink of momentum in equation (7.20). For simplicity, we will continue to ignore the other possible source of heat—Ohmic dissipation, which converts magnetic field energy directly into thermal energy—since the resistivity in such a system is poorly known (though probably not zero!). In a full three-dimensional magnetohydrodynamic treatment, one does not need to be exclusive like this, but, even there, the methodology calls for experimentation with various resistive strengths in order to gauge the relative importance of diffusive field annihilation compared with viscous heating.

Equation (7.19) is always valid, and under these circumstances, equation (7.22) does not change either. However, viscosity alters the momentum of the orbiting gas, so equation (7.20) must be modified as follows:

$$\frac{d(\rho \mathbf{v})}{dt} + \rho \vec{\nabla} \Phi + \rho \mathbf{v}(\vec{\nabla} \cdot \mathbf{v}) = \frac{1}{4\pi}(\mathbf{B} \cdot \vec{\nabla})\mathbf{B} - \vec{\nabla}\left(P + \frac{B^2}{8\pi}\right)$$

$$+ 2\,\hat{e}^j\,\nabla^i\,(S_{ij}\,\rho\,v), \qquad 7.67$$

where P is now understood to represent the nonmagnetic pressure $\rho R_g T/\mu + P_{rad}$, and the unit vector $\hat{e}^j$ is attached to the j-coordinate. In addition,[38]

$$S_{ij} \equiv \frac{1}{2}\left(\frac{\partial v_i}{\partial x^j} + \frac{\partial v_j}{\partial x^i} - \frac{2}{3}\delta_{ij}\frac{\partial v_k}{\partial x^k}\right). \qquad 7.68$$

When the flow is Keplerian, one can easily show that $S_{R\phi} = -(3/4)\,\Omega$. Also, for Sagittarius A*, the radiation pressure P_{rad} is dominated by radio waves in the Rayleigh-Jeans limit, so

$$P_{rad} = \frac{8\pi}{9}kT\left(\frac{\nu_m}{c}\right)^3. \qquad 7.69$$

In this expression, ν_m is the frequency below which radiation is highly absorbed, meaning that the optical depth from the emitter to infinity is >1.

Before we move on to equation (7.21) for the internal energy of the gas, let us further consider the consequences of equation (7.67) and see how the overall energy budget of the disk is affected by the anomalous viscosity. The total energy (excluding internal) includes kinetic, gravitational, and magnetic terms. A common procedure for getting the energy conservation equation is to project equation (7.67)

---

[38] See, e.g., Landau and Lifshitz (1987).

onto the vector **v** and equation (7.22) onto the vector **B**. The first step yields

$$\frac{\partial}{\partial t}\left[\rho\left(\frac{1}{2}v^2+\Phi\right)\right]=-\vec{\nabla}\cdot\left[\rho\mathbf{v}\left(\frac{1}{2}v^2+\Phi\right)\right]-\mathbf{v}\cdot\vec{\nabla}P$$

$$+2\mathbf{v}\cdot\hat{e}^j\,\nabla^i\,(S_{ij}\,\rho\,v)-\frac{1}{4\pi}\mathbf{v}\cdot[\mathbf{B}\times(\vec{\nabla}\times\mathbf{B})],\qquad 7.70$$

and the second produces

$$\frac{\partial}{\partial t}\left(\frac{B^2}{8\pi}\right)=\frac{1}{4\pi}\mathbf{B}\cdot[\vec{\nabla}\times(\mathbf{v}\times\mathbf{B})]\qquad\qquad 7.71$$

(remembering that $\eta=0$ here). And now adding these two equations gives us an expression for the conservation of kinetic, gravitational, and magnetic energy:

$$\frac{\partial}{\partial t}\left[\frac{B^2}{8\pi}+\rho\left(\frac{1}{2}v^2+\Phi\right)\right]=-\vec{\nabla}\cdot\left[-2\nu\rho\mathbf{v}\cdot\tilde{\mathbf{S}}+\frac{1}{4\pi}\mathbf{B}\times(\mathbf{v}\times\mathbf{B})\right.$$

$$\left.+\rho\mathbf{v}\left(\frac{1}{2}v^2+\Phi\right)\right]-2\nu\rho S^2-\mathbf{v}\cdot\vec{\nabla}P,\qquad 7.72$$

where in tensor notation $\mathbf{v}\cdot\tilde{\mathbf{S}}\equiv\hat{e}^j\,v^i\,S_{ij}$ (and j is not repeated).

In steady state, the total energy density does not change, so the left-hand side is zero. Of the remaining terms, the divergence of the bracketed term gives the net outflux or influx of energy density per unit volume, and the last term represents the so-called PdV work. Thus, we infer the heating term due to viscous dissipation to be

$$\Gamma\equiv 2\nu\rho S^2.\qquad\qquad 7.73$$

From equation (7.72), we therefore see that (in steady state)[39]

$$\Gamma=-\vec{\nabla}\cdot\left[-2\nu\rho\mathbf{v}\cdot\tilde{\mathbf{S}}+\frac{1}{4\pi}\mathbf{B}\times(\mathbf{v}\times\mathbf{B})+\rho\mathbf{v}\left(\frac{1}{2}v^2+\Phi\right)\right]-\mathbf{v}\cdot\vec{\nabla}P.\quad 7.74$$

---

[39] As it turns out, the divergence of the viscous flux in this expression is insignificant compared with the other terms, so it may be neglected in the actual implementation of the heating rate. This argument was made by Balbus, Gammie, and Hawley (1994).

Knowing the heating rate Γ, together with the cooling rate Λ, which arises primarily from synchrotron, bremsstrahlung, and inverse Compton emission, we can now return to equation (7.21) for the internal energy of the gas and update this expression with the additional energy source and sink:

$$\frac{\partial(\rho\epsilon)}{\partial t} = -\vec{\nabla}\cdot(\mathbf{v}\rho\epsilon) + \Gamma - \Lambda - P\vec{\nabla}\cdot\mathbf{v}, \qquad 7.75$$

where $\rho\epsilon = \xi nkT + 3P_{rad}$ is the thermal plus radiation energy density.[40] In the fully ionized but nonrelativistic limit, $\xi = 3$, whereas in the relativistic electron (though not proton) limit, $\xi = 9/2$.

We are now almost ready to solve equations (7.19), (7.22), (7.72), and (7.75) in order to determine the structure of the compact magnetized disk surrounding Sagittarius A*. However, there is still a very important ingredient missing—the set of boundary conditions that fix the temperature, density, and accretion rate at the disk's outer edge. Of course, we have a rough idea of how large these quantities should be from the large-scale simulations described earlier in this chapter. Unfortunately, we do not (yet) have the means to *directly* divine the values of these parameters with greater precision. Nonetheless, we can still make use of other pertinent data, primarily the spectrum shown in figure 3.1, to infer these conditions indirectly.

Calculation of Sagittarius A*'s mm and X-ray Spectrum

The flux density (at Earth) produced by the Keplerian disk is given by

$$F_{\nu_0} = \frac{1}{D^2}\int I_{\nu'}\left(1 - \frac{r_s}{r}\right)^{3/2} dA, \qquad 7.76$$

where $D \approx 8.0$ kpc is the distance to the galactic center, ν_0 is the observed frequency at infinity,[41] and ν' is the frequency measured by a stationary observer in the Schwarzschild frame. (For simplicity, we will assume the

---

[40]Note that, for simplicity, we here ignore the possible contribution to the internal energy from nonthermal particles. We already know, however, that a pure thermal distribution of charges cannot fully account for all of Sagittarius A*'s spectrum (e.g., see figure 3.1). We will discuss the role played by nonthermal particles later in this chapter.

[41]In this subsection, ν will denote the frequency and should not be confused with the kinematic viscosity we introduced earlier.

metric for a nonspinning black hole.) The frequency transformations are given by (cf. equations 5.114 and 5.115)

$$\nu_0 = \nu' \left(1 - \frac{r_S}{r}\right)^{1/2} \qquad\qquad 7.77$$

and

$$\nu' = \nu \frac{(1 - v_\phi^2/c^2)^{1/2}}{1 - (v_\phi/c)\cos\theta}, \qquad\qquad 7.78$$

where ν is the frequency measured in the comoving frame and θ is the angle between the azimuthal velocity v_ϕ and the line of sight. Since the radial velocity will always be much smaller than v_ϕ, we can ignore this component in the transformation equations. Thus, $\cos\theta = \sin i \cos\phi$, where i is the inclination angle of the axis perpendicular to the Keplerian flow and ϕ is the azimuth of the emitting element. When the Doppler shift is included, the blueshifted region is located primarily near $\phi = 0$ while the redshifted region is at $\phi = \pi$. The other quantities that are necessary for an evaluation of the flux density are the area element

$$dA = \frac{r\cos i}{(1 - r_S/r)^{1/2}} \, dr \, d\phi \qquad\qquad 7.79$$

and the specific intensity[42]

$$I_{\nu'} = B'_{\nu'} (1 - e^{-\tau}), \qquad\qquad 7.80$$

where

$$B'_{\nu'} = \left[\frac{(1 - v_\phi^2/c^2)^{1/2}}{1 - (v_\phi/c)\cos\theta}\right]^3 B_\nu \qquad\qquad 7.81$$

and

$$\tau = \int \kappa'_{\nu'} \, ds = \kappa_\nu \frac{2H}{\cos i} \frac{1 - (v_\phi/c)\cos\theta}{(1 - v_\phi^2/c^2)^{1/2}} \qquad\qquad 7.82$$

[42] A good reference text for the definition of all these quantities and their usage is the book *Radiative Processes in Astrophysics*, by Rybicki and Lightman (1979).

is the optical depth. The frequency-dependent quantity κ_ν is the absorption coefficient, and B_ν is the Planck function defined in equation (3.1).

When the optical depth $\tau \ll 1$, Kirchoff's law allows us to write

$$I_{\nu'} \approx B'_{\nu'} \tau = \epsilon_\nu \frac{2H}{\cos i} \left[\frac{(1 - v_\phi^2/c^2)^{1/2}}{1 - (v_\phi/c) \cos \theta} \right]^3, \qquad 7.83$$

where $\epsilon_\nu \equiv B_\nu \kappa_\nu$ is now the emissivity.

The presence of a substantial azimuthal component in the magnetic field (see the discussion following equation 7.51) makes it convenient for us to calculate the observed flux directly from the extraordinary and ordinary components of the intensity. The most straightforward approach is to select the symmetry axis of the Keplerian disk as the reference direction. The observed flux densities in the azimuthal and the reference directions are then given, respectively, by

$$F_{1\nu_0} = \int \left(I_{\nu'}^E \mid \cos^2 \phi' \mid + I_{\nu'}^O \mid \sin^2 \phi' \mid \right) \left(1 - \frac{r_S}{r} \right)^{3/2} \frac{dA}{D^2}, \qquad 7.84$$

$$F_{2\nu_0} = \int \left(I_{\nu'}^E \mid \sin^2 \phi' \mid + I_{\nu'}^O \mid \cos^2 \phi' \mid \right) \left(1 - \frac{r_S}{r} \right)^{3/2} \frac{dA}{D^2}, \qquad 7.85$$

where $\phi' + \pi/2$ is the position angle of the magnetic field vector within the emitting element with azimuth ϕ, so that $\cot \phi' = \cot \phi \, \cos i$, and $I_{\nu'}^E$ and $I_{\nu'}^O$ are the specific intensities for the extraordinary and ordinary waves, respectively. For thermal synchrotron radiation, the emissivities may be written in the form[43]

$$\epsilon^E = \frac{\sqrt{3}e^3}{8\pi m_e c^2} B \sin \theta' \int_0^\infty N(E) \left[F(x) + G(x) \right] dE \qquad 7.86$$

[43]For pedagogical reasons, we here simply assume a thermal distribution for the electrons and write the emissivities accordingly (see Pacholczyk 1970). However, in a more realistic environment, such a distribution is modified to reflect the dominant acceleration mechanism(s). Later in this chapter and especially in chapter 8, we will examine the impact on $N(E)$ by the acceleration of particles scattering stochastically off the turbulent magnetic field. We will find that under such action the particle distribution evolves away from a pure Maxwellian (see equation 8.12) but is still close enough to it that the current simple analysis is reasonable.

and

$$\epsilon^O = \frac{\sqrt{3}e^3}{8\pi m_e c^2} B \sin\theta' \int_0^\infty N(E)\,[F(x) - G(x)]\,dE, \qquad 7.87$$

where $N(E)$ is the electron distribution function at energy E and

$$\cos\theta' = \frac{\cos\theta - v_\phi/c}{1 - (v_\phi/c)\cos\theta}, \qquad 7.88$$

$$x = \frac{4\pi\nu m_e^3 c^5}{3eB\sin\theta' E^2}, \qquad 7.89$$

$$F(x) = x \int_x^\infty K_{5/3}(z)\,dz, \qquad 7.90$$

$$G(x) = x K_{2/3}(x). \qquad 7.91$$

In these expressions, $K_{5/3}$ and $K_{2/3}$ are the corresponding modified Bessel functions. The total flux density produced by synchrotron emission in the compact magnetized disk is the sum of the contributions from these two emissivities.

We should point out here that a separation of the emissivity into its extraordinary and ordinary components is more than mere convenience. This type of analysis is essential for the development of a clear theoretical understanding of Sagittarius A*'s polarization properties, summarized in figure 3.2. However, we will defer the interpretation of the polarization data to chapter 9, where we will also discuss the future polarimetric imaging of this source.

To complete this discussion, we note that this actually is not the full extent of the radiation field produced by the Keplerian region. In addition, plasma expelled from the accretion flow itself contributes to Sagittarius A*'s overall (primarily radio) spectrum. We shall treat the latter in greater detail later in this chapter; for the moment, we must also consider the contribution from inverse Compton scattering, which boosts some of the mm radiation emitted according to equations (7.86) and (7.87) into X-rays.

Fortunately, the conditions prevalent in this region (notably, the temperature $T \sim 10^{10}\text{–}10^{11}$ K of the electrons and the characteristic frequency $\nu \sim 3 \times 10^{11}$ Hz of the seed photons) allow us to treat the electron-photon scattering in the Thomson regime; that is, for this application, we do not need to worry about the Klein-Nishina energy dependence of the cross section. For simplicity, we may also treat the electrons as locally isotropic, since their thermal energy is close to virial, which means that the upscattered radiation may also be handled isotropically.

Without loss of generality, one therefore calculates the photon number density function n_ϵ from the angle-integrated intensity (where $n_\epsilon \, d\epsilon$ represents the number density of photons with energies between ϵ and $\epsilon + d\epsilon$ and depends primarily on the radius r). Under these conditions, a single electron with Lorentz factor γ emits photons with energies between ϵ_s and $\epsilon_s + d\epsilon_s$ at a rate[44]

$$\frac{dN}{dt \, d\epsilon_s}(\epsilon_s, \gamma) = \frac{3c\sigma_T}{16} \frac{\epsilon_s}{\beta \gamma^2} \int_{-1}^{1} d\mu_s \int_{\epsilon_l}^{\epsilon_u} d\epsilon \frac{n_\epsilon}{\epsilon^2} g(\mu', \mu_s'), \qquad 7.92$$

where $\mu = \cos\theta$ and $\mu_s = \cos\theta_s$ represent the lab-frame (unprimed) propagation angles (with respect to the electron's direction of motion) of the incident and scattered photons, respectively. Special relativistic transformations allow us to write

$$\epsilon_u = \epsilon_s \gamma^2 (1 - \beta \mu_s)(1 + \beta) \qquad 7.93$$

and

$$\epsilon_l = \epsilon_s \frac{(1 - \beta \mu_s)}{(1 + \beta)}. \qquad 7.94$$

The function $g(\mu', \mu_s')$ takes into account the angular dependence of the cross section and is given in terms of the rest-frame (primed) angles by the expression

$$g(\mu', \mu_s') = (1 - \mu'^2)(1 - \mu_s'^2) + (1/2)(1 + \mu'^2)(1 + \mu_s'^2). \qquad 7.95$$

---

[44]This subject has been dealt with extensively in the literature. A more detailed description of this process may be found, e.g., in Melia and Fatuzzo (1989) and Coppi and Blandford (1990).

The rest-frame and lab-frame angles are related through the transformation

$$\mu' = \frac{(\mu - \beta)}{(1 - \beta\mu)}.$$ 7.96

Finally, conservation of energy requires that

$$\mu = \left[1 - \left(\frac{\epsilon_s}{\epsilon}\right)(1 - \beta\mu_s)\right]\left(\frac{1}{\beta}\right).$$ 7.97

With the upscattered photons emitted isotropically, the total Comptonized photon production rate may then be found by integrating over all the electrons in the emission region, yielding

$$\frac{dN_{tot}}{dt\, d\epsilon_s} = \int_{gas} dV \int_0^\infty dE\, n_e\, f(E)\left(\frac{dN}{dt\, d\epsilon_s}\right),$$ 7.98

where n_e is the (radially dependent) number density of electrons and $f(E)$, with $E = \gamma m_e c^2$, is the relativistic Maxwell-Boltzmann distribution function

$$f(E) = \frac{m_e n_e}{4\pi kTc\, K_2(m_e c^2/kT)}\, \exp(-E/kT),$$ 7.99

with K_2 the modified Hankel function of order 2. Note that the temperature (and therefore the Maxwell-Boltzmann distribution function) is also radially dependent.

The X-ray spectrum arising from inverse Compton scattering appears to emerge primarily as a variable component, especially during a flare (see chapter 4). We shall therefore defer further discussion of this process to chapter 8.

Structure of the Disk

With the specification of the ordinary and extraordinary emissivities, we now have a complete set of equations describing the structure of the disk and the spectrum it creates. In solving these expressions, one may perform a χ^2-minimization for a range of initial conditions in order to produce consistent spectral and polarimetric fits to the data, examples of which are shown in figure 7.9. Though the solutions shown here are not unique, the parameter values to which these calculated curves correspond are typical for Sagittarius A* in its current state.

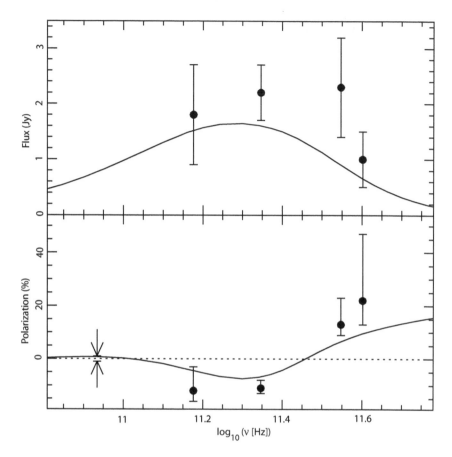

Figure 7.9 Spectra showing total flux density and polarization from Sagittarius A*. The high-frequency data points are from Aitken et al. (2000), and the curves are from the solution of the equations discussed in the text. The limit at 84 GHz is from Bower et al. (1999a). At lower frequencies, the linear polarization is consistent with zero. In this figure, "negative" polarization corresponds to the polarization vector being aligned with the spin axis of the black hole, whereas "positive" is for a perpendicular configuration. When this vector flips by 90°, the polarization crosses zero. (Image from Bromley, Melia, and Liu 2001)

The bottom panel shows the fractional polarization, defined as

$$P_{\nu0} = \frac{F_{1\nu0} - F_{2\nu0}}{F_{1\nu0} + F_{2\nu0}} \qquad\qquad 7.100$$

(see equations 7.84 and 7.85). We should point out here that, whereas one may reasonably ignore general relativistic effects in solving the dynamics equations for the plasma to obtain a first-order solution to the problem, the same cannot be said for the calculation of the intensity and flux. This is due to the fact that light-bending and area-amplification effects are critical to the correct rendering of the disk's appearance at infinity, and this bears directly on the inferred source emissivity. We shall return to this discussion—and provide a more detailed description of how these so-called ray-tracing calculations are carried out—in chapter 9, where we also consider future imaging observations of Sagittarius A*.

The simulations that produce good fits to the spectrum "begin" with a gas inflow rate of $\sim 3.6 \times 10^{16}$ g s^{-1}, at a circularization radius of $\sim 8 r_S$. The radiating flow extends down to $\sim 2 r_S$, at which point the gas settles into a plunging orbit toward the event horizon.

The physical conditions inferred from these fits have profiles like those shown in figure 7.10. The temperature—already close to its virial value following the radiatively inefficient infall toward the circularization radius—ranges from $\sim 10^{10}$ to 10^{11} K in the Keplerian region, while the particle density varies between $\sim 2 \times 10^7$ and 4×10^8 cm^{-3}. The magnetic field intensity increases from about 4 to 20 G. The reader will notice that these conditions were in essence presaged by earlier considerations of the large-scale (presumably convectively unstable) flow and by arguments based on the lack of any significant Faraday depolarization in Sagittarius A*'s spectrum. As far as inclination angles of the disk are concerned, modestly low values, around 30°, give good fits to the data, though higher inclination angles are not strongly excluded.

It is rather straightforward to understand the polarization characteristics seen here. In the optically thick region (below $\sim 1.6 \times 10^{11}$ Hz), the specific intensity of the extraordinary and ordinary waves is almost isotropic in the comoving frame because the optical depth τ (equation 7.82) is very large. Even with the inclusion of the Doppler effect (equation 6.18), the emissivity of the source is relatively independent of position angle. But the optical depths are different for the two waves, as indicated by equation (7.82), and the specific intensity of the extraordinary wave is slightly larger than that of the ordinary wave. From equations (7.86) and (7.87), we see that the second component is larger than the first, and the percentage polarization is therefore negative according to the definition of P_{v_0} in equation (7.100). With an increase in frequency, the extraordinary amplitude becomes even larger (relative to that of the ordinary wave), and so the percentage polarization increases.

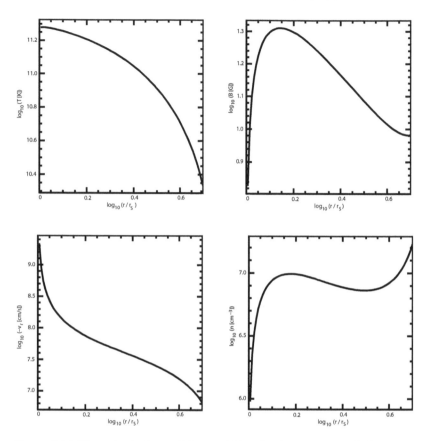

Figure 7.10 The physical conditions within the inner $\sim 5 r_S$ of the compact magnetized disk surrounding Sagittarius A*: (upper left) temperature profile, (upper right) the magnetic field, (lower left) the radial component of velocity, and (lower right) the particle number density. Later in the chapter, we shall study the results of a more detailed three-dimensional magnetohydrodynamic simulation of this model. (Image from Melia, Liu, and Coker 2001)

However, in the optically thin region, the specific intensity is given by equation (7.83). The synchrotron emissivity is very sensitive to the angle between the line of sight and the magnetic field vector **B**; synchrotron radiation is beamed into a plane perpendicular to **B** in the comoving frame. With the inclusion of the Doppler effect, the radiation is beamed into a cone, and the dominant contribution comes from the blueshifted region, which has an azimuth of about zero. Therefore, since the extraordinary wave is more intense than the ordinary wave and since the integrals in equations (7.86) and (7.87) are dominated by radiation from the emitting element with an azimuth of about zero, the first component is larger than the second. In this case, the fractional polarization becomes positive.

In other words, the optically thick emission is dominated by emitting elements on the near and far sides of the black hole, for which the extraordinary wave has a polarization direction parallel to the reference axis. In contrast, the dominant contribution in the thin region comes from the blueshifted emitter to the side of the black hole, where the extraordinary wave has a polarization direction mostly perpendicular to this axis. The sharp decrease in polarization at still higher frequencies is due to the diluting effects of Comptonized emission, which begins to dominate over synchrotron emission at that point.

Magnetohydrodynamic Disk Simulations

The next level in sophistication beyond the analytical treatment we have been describing here has already been broached with recent three-dimensional magnetohydrodynamic simulations of the disk accretion in Sagittarius A*.[45] These calculations discretize the equations we have derived in the previous subsections and evolve them on a spatial grid with sufficient resolution[46] to bring out the essential features of this problem, for example, the wavelengths associated with the fastest-growing modes in the MRI (see the discussion preceding equation [7.41]).

These calculations make use of the pseudo-Newtonian gravitational potential[47] to mimic the dynamically important marginally stable circular orbit of the full Schwarzschild metric at $r = 3r_S$ (see figure 6.2). In this prescription,

$$\Phi = -\frac{GM}{(r - r_S)}, \qquad 7.101$$

[45] See Hawley and Balbus (2002).

[46] These simulations use $128 \times 32 \times 128$ grid zones in cylindrical coordinates (R, ϕ, z). The radial grid extends from $R = 1.5r_S$ to $R = 220r_S$. There are 36 equally spaced zones inside $R = 15r_S$ and 92 zones increasing logarithmically in size outside this radius. In the z-direction, there are 50 zones equally spaced between $-10r_S$ and $10r_S$, with the remainder of the zones logarithmically stretched to the z-boundaries at $\pm 60r_S$. The azimuthal domain is uniformly gridded over $\pi/2$ in angle. The radial and vertical boundary conditions are simple zero-gradient outflow conditions, with which no flow is permitted into the computational domain. The ϕ-boundary is periodic.

[47] This formalism was introduced by Paczyński and Wiita (1980).

with which the specific angular momentum of a circular orbit is

$$l_{\text{circ}} = \frac{(GMr)^{1/2}r}{(r - r_S)}.$$

7.102

In addition, these simulations ignore radiation transport and energy losses, which are, by assumption, dynamically unimportant in the radiatively inefficient system we have here. And the angular momentum is transported by Maxwell and Reynolds stresses arising from magnetic field and velocity correlations in the MRI-induced turbulence.

The general features emerging from these calculations include three principal flow components (see figure 7.11): (1) a hot, rotationally supported disk extending down to the marginally stable orbit, (2) an extended low-density coronal backflow enveloping the disk, and (3) a distinctive jetlike outflow perpendicular to the disk in the vicinity of the black hole. Near the equator, thermal pressure dominates over the pressure due to the magnetic field (i.e., the MRI-generated magnetic field is somewhat below equipartition), whereas the surrounding region is strongly magnetized.

Although this simulation lacks a formal treatment of the energetics required for a detailed calculation of Sagittarius A*'s spectrum, one can nonetheless estimate the synchrotron emissivity as a function of position within the disk. The result of this analysis, summarized in figure 7.12, generally confirms the analytical work of the previous subsections. The highest peak frequencies are $\sim 2.5 \times 10^{11}$ Hz, produced in the very inner portion of the disk, where the particle density is $\sim 10^8$ cm^{-3}, compared with the value $\sim 10^7$ cm^{-3} inferred earlier. In addition, the gas temperature here varies between 4×10^{10} and 10^{11} K, compared with the typical value of $\sim 10^{11}$ K we inferred above. And a similarly favorable comparison may be made of the MRI-generated magnetic fields in the analytical and computational treatments.

All in all, the spectral and polarimetric guidance provided by the observations has greatly influenced how we view the environment within $\sim 10 r_S$ of Sagittarius A*. The black hole at the center of our Galaxy is fed by a hot, tenuous plasma that settles into Keplerian rotation, winding its way inwards under the action of an anomalous viscosity from the MRI-induced turbulence. The mm/sub-mm spectral bump is produced in this region, as is the variable (synchrotron) IR and (inverse Compton) X-ray flux (see chapter 8).

One can easily show, however, that Sagittarius A*'s radio spectrum cannot arise from such a compact volume. In the next section, we shall

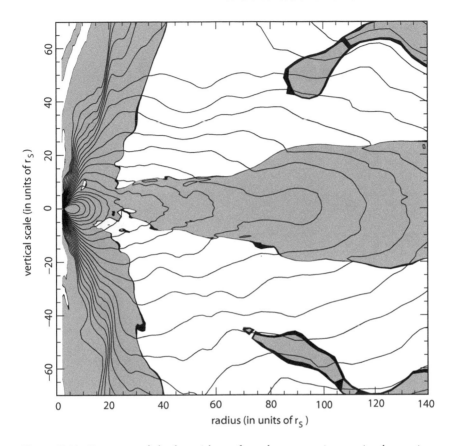

Figure 7.11 Contours of the logarithm of total pressure (magnetic plus gas) at the end of a three-dimensional magnetohydrodynamic simulation of accretion within $\sim 100 r_S$ of the black hole. Cylindrical radius is shown in the horizontal direction (in units of r_S), and the z-coordinate is along the vertical axis. The shaded regions overlaid on top of the contours reveal where gas pressure exceeds magnetic. Note that gas pressure dominates in the disk, in the hot inner torus, and along the funnel wall where matter is being expelled (see §7.6). However, magnetic pressure dominates in the bulk of the coronal envelope atop the disk. (Image from Hawley and Balbus 2002)

see why the cm waves from Sagittarius A* appear to originate from a region well beyond the inner disk. The matter expelled along the funnel in figure 7.11 may be a source for this emission, though it is not yet clear whether energetic particles can be produced in numbers sufficient to power the observed synchrotron emission. On the other hand, relativistic electrons diffusing through the enveloping corona after being accelerated

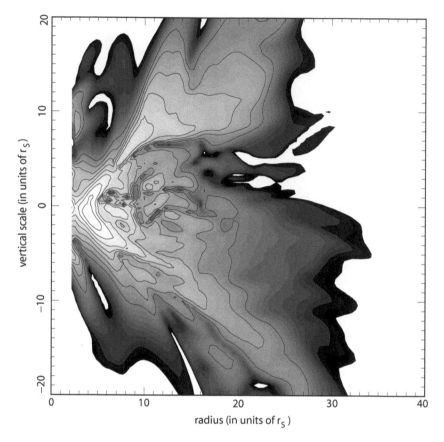

Figure 7.12 Contour map and grayscale of peak frequencies for synchrotron emission at the end of the simulation shown in figure 7.11. The horizontal axis shows the (cylindrical) radius (in units of r_S), and the vertical one gives the z-coordinate. Note the spatial scale here is smaller than in the previous figure. The contours are equally spaced in log frequency, with a peak at 2.5×10^{11} Hz. The highest frequencies emerge from the inner edge of the torus and hence from the smallest spatial scale. (Image from Hawley and Balbus 2002)

stochastically by the turbulent magnetic field may contribute to—and perhaps dominate—this radiation.

7.6 EXPULSION OF MATTER

Within five or six years of Sagittarius A*'s discovery, one could already argue that the radio source at the galactic center cannot be confined within a compact region.[48] For example, the Compton limit, introduced at the beginning of chapter 3, made it difficult to produce the observed radio luminosity from a compact volume consistent with a bound static system.

Gravitational confinement requires $GM\rho/r \gg u$, where ρ is the mass density for a proton number equal to that of the synchrotron-emitting electrons and u is the energy density of the magnetized plasma emitting the radiation. For a black hole mass $M \sim 3.6 \times 10^6 \ M_\odot$, this condition is met when $r \ll 10^{13}$ cm, which is already much smaller than Sagittarius A*'s observed size at cm wavelengths (see figure 2.4 and the related discussion in chapter 2). More important, the observed flux emitted by such a small volume would require a brightness temperature exceeding 10^{12} K, at which runaway inverse-Compton scattering would greatly deplete the pool of radio photons.

Chandra's measurement of Sagittarius A*'s 2–10 keV flux (see figure 3.5) has strengthened this conclusion by providing the means to study its radio and X-ray emissivities simultaneously and self-consistently. For this argument, Sagittarius A*'s steady X-ray luminosity $(2.4^{+3.0}_{-0.6} \times 10^{33}$ ergs s$^{-1})$ must be viewed as an upper limit to the radiation produced on any scale smaller than $\sim 0.''5$, since the X-ray counterpart to the radio source appears to be resolved with a size comparable to the Bondi-Hoyle accretion radius r_{acc} (see equation 7.1).[49]

Using the latest observations, it is not difficult to show that the radio emission from Sagittarius A* cannot be produced via *thermal* synchrotron processes within a bounded accretion flow. This reasoning is based on the fact that with the temperature limited from above by its virial value the required radio emission can be produced by thermal synchrotron only with a sufficiently strong magnetic field, which, however, should not exceed its equipartition value. Attempts to produce the observed radio spectrum in this way result in high electron number

---

[48]This assessment first appeared in a paper by Reynolds and McKee (1980). Additional arguments, based on more recent spectral measurements (Liu and Melia 2001), will be presented later in this subsection.

[49]This almost certainly means that Sagittarius A*'s steady X-ray luminosity is at least partially produced by the large-scale Bondi-Hoyle accretion inflow, as opposed to the X-rays emitted during a flare that, as we shall see in chapter 8, appear to originate from the mm/sub-mm radiation zone. See Melia (1992a, 1994) and Quataert (2002) for a more complete analysis of Sagittarius A*'s quiescent luminosity. A full discussion of Sagittarius A*'s observed quiescent flux appears in Baganoff et al. (2003).

densities, which in turn violate the observed X-ray flux with excessive bremsstrahlung emission.

A fully ionized bounded hydrogen gas has a temperature

$$T(r) < T_{\text{virial}}(r) \approx 1.8 \times 10^{12} \left(\frac{r_S}{r}\right) \text{ K,} \qquad 7.103$$

where

$$3kT_{\text{virial}} \equiv \frac{GM(m_p + m_e)}{r}, \qquad 7.104$$

in terms of the proton (m_p) and electron (m_e) masses. The thermal emission is blackbody limited, so to produce a 0.53 Jy radio flux density at 1.36 GHz with a thermal source at the galactic center, we must have

$$T(r) \left(\frac{r}{r_S}\right)^2 > 3.4 \times 10^{15} \text{ K.} \qquad 7.105$$

Combining this with the virial temperature limit, we infer that $r > 1$, $900 r_S$, for which $T \sim 10^9$ K. The critical point is that the ratio of cyclosynchrotron emissivity ϵ_s at 1.36 GHz to the bremsstrahlung emissivity ϵ_b at 10^{18} Hz implies a lower limit to the particle number density.[50] At 10^{18} Hz, $\epsilon_b = 6.0 \times 10^{-42} n^2$ ergs s^{-1} Hz^{-1} cm^{-3}, whereas at 1.36 GHz, $\epsilon_s = 4.2 \times 10^{-17} nM(x)$, where $M(x) = 0.1746 \ \exp(-1.8899 \, x^{1/3}) \, x^{-1/6}$ when $x = 1.1 \times 10^4/B$ is much bigger than 1. If the magnetic field is in equipartition (a conservative upper limit), then $B^2 = 1.0 \times 10^{-5} n$, and from the condition $\epsilon_b < 2.8 \times 10^{-8} \epsilon_s$, we estimate that $B > 8.1$ G. However, the electron number density n_e must then be $> 6.6 \times 10^6$ cm^{-3}, and for a source size $r \sim 1,900 r_S$ at the galactic center, the bremsstrahlung X-ray flux density at 10^{18} Hz is 3.9×10^{-5} Jy, greatly exceeding the observed limit.

Clearly, the *thermal* radiative efficiency of a bounded accretion flow is not sufficient to produce Sagittarius A*'s cm radiation. Could this long wavelength emission be due to nonthermal particles embedded within the thermal plasma? The answer is presumably yes, and this would enhance the impact of ρ over u, but the diffusion of relativistic particles through the nonrelativistic plasma would still allow the former to escape and create an expanding halo. We shall discuss a concrete example of how

[50] See Melia (1992a) for ϵ_b and Mahadevan, Narayan, and Yi (1996) for ϵ_s.

this might happen below, based on the stochastic acceleration of particles by the turbulent magnetic field.

But let us first consider a phenomenological approach to modeling Sagittarius A*'s radio spectrum with outflowing particles, which has focused primarily on identifying the range of physical parameters required to account for the observations. The limitation of this approach is that many aspects of the spectral model follow from simple scaling laws, which are not very sensitive to the detailed dynamics, and one may generate a wide array of spectral configurations with the adjustment of a few free parameters. In addition, the early attempts at this type of analysis[51] invoked the presence of a large fossil disk, which has since been invalidated by the observations—a result that has reaffirmed the lack of physically motivated guidance for this type of scaling. However, this work still has value in providing context for the more detailed modeling that will follow our preliminary discussion.

In this analysis,[52] one postulates a symmetric bipolar jet of magnetized relativistic protons and electrons, expelled from a nozzle near the black hole. This plasma accelerates under the action of its own pressure gradient and expands sideways at sound speed. For simplicity, the gravitational potential is ignored since its influence is relatively negligible. Assuming an adiabatic outflow and conservation of momentum and particle number, one can then obtain the simple scaling laws that relate the magnetic field, particle density, Lorentz factor, and jet radius as functions of the distance from the nozzle and of the initial conditions at the base. For simplicity, the spectrum produced by this configuration is then calculated using the transfer equation for source-only emission, in which the intensity is given as $I_\nu(\tau_\nu) = (1 - \exp[-\tau_\nu])S_\nu$, where S_ν is the source function.[53]

Initial values must be assigned to six free parameters: the radius of the nozzle, the location of the sonic point (at which the outflow velocity equals the sound speed), the magnetic field, the electron number density, the inclination angle of the jet relative to the plane of the sky, and the characteristic Lorentz factor of the electron distribution. In addition, the stringent IR upper limits (2.8 mJy in H-band, 2.7 mJy in K_s-band, and 6.4 mJy in L'-band) and rather steep spectral power-law index

---

[51] See Falcke, Mannheim, and Biermann (1993).

[52] There are several papers on this topic; a recent treatment may be found in Falcke and Markoff (2000).

[53] See Rybicki and Lightman (1979).

(-2.1 ± 1.4 between 1.6 and 3.8 μm; see chapter 3) require either a steep cutoff in a power-law lepton distribution, with index $p = 2$ and a turnover at roughly five times the initial Lorentz factor, or a highly peaked function, such as a relativistic Maxwell-Boltzmann distribution (equation 7.99).

Again, we emphasize that the primary purpose of this analysis is to infer empirically—by fitting directly to the data—the characteristic size and physical state of the region emitting the radio waves, so this large number of unknowns should not be taken as an indication that our view of Sagittarius A* is overly compliant and poorly constrained. Nonetheless, one should keep in mind the primary assumption in this type of fitting; that is, there exists a well-defined nozzle close to the black hole, where a relativistic plasma is collimated with equipartition between the particles and magnetic field. This is not a trivial point, since the nozzle would be responsible for the dominant mm/sub-mm bump in Sagittarius A*'s spectrum (see figures 3.1 and 7.9). We now know, of course, that the periodic modulations seen during the IR and X-ray flares from this object manifest themselves in the same region where the mm/sub-mm flux is produced (see chapter 4 and the more detailed discussion in chapter 8), suggesting emission in a Keplerian plasma, rather than some well-defined stable nozzle.

These caveats notwithstanding, we obtain a reasonable accounting of Sagittarius A*'s radio spectrum with a nozzle no bigger than $\sim 15 r_S$ and a sonic radius of similar dimensions. The electrons, with characteristic Lorentz factor ~ 50, must have a number density $\sim 10^6$ cm^{-3}, and the (equipartition) magnetic field is inferred to be ~ 20 G. The challenge now is to identify the physical mechanism responsible for creating this expanding halo of relativistic particles surrounding Sagittarius A*.

Stochastic Acceleration and the Outflow of Particles

As we attempt to move beyond a description of accretion onto Sagittarius A* and consider secondary processes that may accelerate some of the infalling particles away from black hole, we run into a theoretical problem that has cut across a broad swath of modern astrophysics. Conceivably, a fraction of the captured plasma may be re-energized— and its trajectory redirected—by one or more of the following influences: hydrodynamic instabilities, electromagnetic acceleration, Fermi shock acceleration, and stochastic particle acceleration by a turbulent magnetic field. Current work on Sagittarius A* seems to be focusing progressively more and more on the last of these mechanisms as the principal agent creating the relativistic halo.[54] Certainly, the ingredients necessary for this physical process to operate—a compact region, a turbulent magnetic field, and energetic seed particles—all seem to be in place. We shall therefore briefly summarize the theory of stochastic acceleration and describe how it may eventually lead to a more comprehensive understanding of Sagittarius A*'s radio emission.

The stochastic acceleration of particles interacting resonantly with plasma waves or turbulence generated via an MHD dissipation process[55] was originally invoked to explain how magnetic reconnection can energize particles during solar flares. In the hot, tenuous environment surrounding Sagittarius A*, the time required for the most energetic particles to escape from the acceleration site is short enough to produce a significant outflux (quantified below) of relativistic particles, forming a nonthermal halo. On an even larger scale, this efflux of energetic electrons and protons may interact with the ambient medium to produce an additional signature at gamma ray energies, which we already touched on at the end of chapter 3 and will revisit later in this section.

In statistical mechanics, Liouville's theorem asserts that the phase-space distribution function is constant along the trajectories of the system; that is, the density of system points, in the vicinity of any given locus traveling through phase space, is constant with time. In a dynamical configuration, with coordinates q_i and conjugate momenta p_i, for $i = 1, \ldots, d$, where d is the number of degrees of freedom, the phase-space distribution function $f(t, q_1, q_2, \ldots q_d, p_1, p_2, \ldots p_d)$ (or $f[t, \mathbf{q}, \mathbf{p}]$ in shorthand notation) gives the number of particles $f(t, \mathbf{q}, \mathbf{p}) \, d^d q \, d^d p$ in the infinitesimal phase-space volume $d^d q \, d^d p$. Liouville's equation is a mathematical formulation of this statement and may be written in the form:

$$\frac{df}{dt} = \frac{\partial f}{\partial t} + \sum_{i=1}^{d} \left(\frac{\partial f}{\partial q_i} \dot{q}_i + \frac{\partial f}{\partial p_i} \dot{p}_i \right) = 0. \qquad 7.106$$

[54] Further details on this work may be found in Liu, Petrosian, and Melia (2004) and Liu, Melia, and Petrosian (2006a). For additional discussion concerning the acceleration of protons to relativistic energies near the black hole, see also Aharonian and Neronov (2005).

[55] See, e.g., Miller and Ramaty (1987), Hamilton and Petrosian (1992), and Petrosian and Liu (2004).

If the system is spatially uniform, as is usually assumed in cases like this, then the partial derivatives with respect to q_i vanish. In addition, the forces acting on the particles are known or may be inferred indirectly, and they may be used to replace the derivatives $\dot{p}_i$.

Of course, in a real situation, particles do escape the system at a rate dictated by the characteristic escape time τ_{esc}, which is generally energy dependent; others may be introduced into the system at a rate $Q(t, \mathbf{p})$ (in units of particles per unit time, per unit momentum). In addition, the total function $f(t, \mathbf{q}, \mathbf{p})$ may be difficult to evaluate exactly. Instead, the collisionless Liouville's equation may be expanded to second order in perturbed quantities—here, the momenta p_i or, equivalently, the magnitude of momentum $p = |\mathbf{p}|$ and the corresponding direction cosine μ, relative to a specified direction (such as the global magnetic field).

The resulting expression is variously called the *kinetic*, or diffusion, or Fokker-Planck, equation. In this formalism, the equation that evolves the distribution function for energetic charged particles

$$\phi(t, p, \mu) \equiv \int d^3x \, f(t, \mathbf{q}, \mathbf{p}), \qquad 7.107$$

due to gyro-resonant interactions with small-amplitude plasma waves superimposed on a uniform magnetic field, is[56]

$$\frac{\partial \phi}{\partial t} = \frac{1}{p^2} \frac{\partial}{\partial p} \left(p^2 D_{pp} \frac{\partial \phi}{\partial p} \right) + \frac{1}{p^2} \frac{\partial}{\partial p} \left(p^2 D_{p\mu} \frac{\partial \phi}{\partial \mu} \right) + \frac{\partial}{\partial \mu} \left(D_{\mu p} \frac{\partial \phi}{\partial p} \right)$$
$$+ \frac{\partial}{\partial \mu} \left(D_{\mu\mu} \frac{\partial \phi}{\partial \mu} \right) + \frac{1}{p^2} \frac{\partial}{\partial p} \left(p^2 \dot{p}_L \phi \right) + Q(t, p). \qquad 7.108$$

This equation has been ensemble averaged over the statistical properties of the plasma waves in accordance with quasi-linear theory. The various quantities appearing here[57] are defined as follows: p is the relativistic unit momentum $\gamma v/c$, where v is the particle speed and γ is the Lorentz factor; μ is the cosine of the pitch angle; $\dot{p}_L$ accounts for the energy loss term due to processes (such as bremsstrahlung and synchrotron) not directly associated with gyro-resonant particle-wave interactions; and

---

[56] Among the early authors to carry out this procedure were Kennel and Engelmann (1966), Hall and Sturrock (1967), and Lerche (1968). See also Schlickeiser (1989).

[57] Note that in this section we define p to be the momentum per unit mass and dimensionless in velocity, not to be confused with the usual definition $p = \gamma mv$, used elsewhere in this book.

$Q(t, p)$ is the source term introduced previously. The Fokker-Planck, or diffusion, coefficients D_{pp}, $D_{p\mu}$, $D_{\mu p}$, and $D_{\mu\mu}$, which arise from the perturbative expansion of the particle distribution function and Boltzmann equation and which represent the forces experienced by the particles through the derivatives p_i, depend on the properties of the wave turbulence (i.e., the wave mode and polarization), on the angle of wave propagation to the ambient magnetic field, and on the power spectrum, including the ratio of the turbulent wave energy to the background magnetic energy.

For most applications of this equation (including the present case), it is safe to assume that the rate of pitch-angle scattering is much larger than the rate of energy diffusion and the rate of particle escape from the system. In other words, one assumes that scatterings changing the direction of **p** occur more frequently than those changing the magnitude p. Then the particle distribution can be assumed to be isotropic, and the pitch angle may be eliminated by integrating equation (7.108) with respect to μ.

Writing

$$F(t, p) = \int_{-1}^{+1} \phi(t, p, \mu)\, d\mu \qquad\qquad 7.109$$

and representing the scattering loss of particles via pitch-angle diffusion as a loss term $F(t, p)/\tau_{\text{esc}}$, we can rewrite equation (7.108) in the form

$$\frac{\partial F(t, p)}{\partial t} = \frac{1}{p^2}\frac{\partial}{\partial p}\left(p^2 \langle D_{pp}\rangle \frac{\partial F(t, p)}{\partial p}\right) + \frac{1}{p^2}\frac{\partial}{\partial p}\left(p^2 \dot{p}_L F[t, p]\right)$$
$$- \frac{F(t, p)}{\tau_{\text{esc}}} - \frac{1}{2}Q(t, p), \qquad\qquad 7.110$$

where $\langle D_{pp}\rangle$ now represents an average of D_{pp} over μ.

The final step in our derivation of the energy kinetic equation for particles interacting with the plasma waves involves the conversion of p into the kinetic energy variable $E = \gamma - 1$. This can be done by putting

$$N(t, E)\, dE = 4\pi p^2 F(t, p)\, dp \qquad\qquad 7.111$$

and

$$\frac{\partial}{\partial p} = \frac{dE}{dp}\frac{\partial}{\partial E}. \qquad\qquad 7.112$$

211

With the additional relations $\beta = v/c$, $p^2 = E(E+2)$, and $p\,dp = (E+1)\,dE$, equation (7.110) transforms into (the so-called diffusion-convection equation)

$$\frac{\partial N}{\partial t} = \frac{\partial}{\partial E}\left[D_{EE}\frac{\partial N}{\partial E} - (A - \dot{E}_L)N\right] - \frac{N}{\tau_{esc}} + S(t, E), \qquad 7.113$$

where $\dot{E}_L = \beta\,\dot{p}_L$,

$$D_{EE} = \beta^2\,\langle D_{pp}\rangle, \qquad 7.114$$

$$A(E) = \frac{1}{p^2}\frac{d}{dp}\left[p^2\beta\langle D_{pp}\rangle\right], \qquad 7.115$$

and

$$S(t, E) = \frac{2\pi p^2}{\beta}Q(t, p). \qquad 7.116$$

In a complete treatment of stochastic acceleration by a turbulent magnetic field, one would also need to specify the temporal evolution of the wave spectrum $W(t, \mathbf{k})$, which represents the spectral energy density of the magnetic field as a function of wavenumber $\mathbf{k}$. This quantity accounts for all of the particle-wave interaction physics, since the energy diffusion coefficient D_{EE} depends on W and on A, which here may also be written $A = (D_{EE}/E)(2 - \gamma^{-2})/(1 + \gamma^{-1})$. Often, the wave spectrum reaches equilibrium and attains a power-law distribution with an index q determined by the (turbulent) cascade and leakage terms in steady state. In addition, the relativistic particles themselves reach steady state (with $\partial N/\partial t = 0$) when the acceleration terms D_{EE} and A dominate. Their spectrum is characterized by an index $\delta = q - 1$ and a cutoff at high γ, where energy losses and particle escape dominate over the rate of acceleration.

In writing down a practical form for $W(t, \mathbf{k})$, one assumes that the turbulence is isotropic, homogeneous, and stationary, with the result that (ignoring the time variable)

$$W(k) = \frac{q-1}{k_{min}}\left(\frac{k_{min}}{k}\right)^q W_{tot}, \qquad 7.117$$

where

$$W_{\text{tot}} = \int_{k_{\min}}^{\infty} W(k)\, dk.$$ 7.118

One can derive from this[58] the Fokker-Planck coefficients D_{EE} and $A(E)$. When $2 < q \leq 4$,

$$D_{EE} = D_0 \frac{[E(E+2)]^{(q-1)/2}}{E+1}$$ 7.119

and

$$A(E) = q D_0 [E(E+2)]^{(q-3)/2},$$ 7.120

where

$$D_0 = \frac{\pi(q-1)^2}{q^2(q^2-4)} \left(\frac{ck_{\min}}{\Omega_e}\right)^{q-1} \frac{f_{\text{turb}}\,\Omega_e}{\psi^4}.$$ 7.121

The other new quantities we have defined are the plasma parameter $\psi = \omega_{pe}/\Omega_e$, the nonrelativistic gyrofrequency $\Omega_e = eB/m_e c$, the ratio $f_{\text{turb}} = 8\pi\, W_{\text{tot}}/B^2$ of turbulent energy to magnetic energy, and the plasma frequency $\omega_{pe} = (4\pi n_e e^2/m_e)^{1/2}$, in terms of the electron number density n_e. Note that $f_{\text{turb}} \sim 1$ in the absence of an external large-scale magnetic field.

However, for $1 < q < 2$,

$$D_{EE} = D_0' \frac{[E(E+2)]^{1/2}}{E+1}$$ 7.122

and

$$A(E) = 2 D_0' [E(E+2)]^{-1/2},$$ 7.123

where

$$D_0' \approx \frac{\pi(q-1)}{8} \left(\frac{ck_{\min}}{\Omega_e}\right)^{q-1} \left(\frac{m_e}{m_p}\right)^{(2-q)/2} \frac{f_{\text{turb}}\,\Omega_e}{\psi^{(2+q)}}$$ 7.124

to within a factor of order 1.

[58] Readers interested in seeking a detailed derivation of these quantities should consult the original work published in Hamilton and Petrosian (1992) and Schlickeiser (1997).

The acceleration of particles at a given energy is dominated by interactions with certain specific wave modes, akin to the resonance condition encountered in a harmonic oscillator. Charges with Lorentz factor γ, velocity v, and pitch angle cosine μ couple strongly with the wave satisfying resonance condition $\omega = k_{\parallel} v \mu + \gamma^{-1}$, where ω and $k_{\parallel}$ are the wave frequency and wavevector parallel to the underlying magnetic field (keeping only the first harmonic of the gyrofrequency). Electrons with higher energies resonate with waves with smaller wave numbers (k) corresponding to larger spatial scales.

We can learn a great deal about the behavior of equation (7.113), the solution $N(E)$, and the wave spectrum $W(t, \mathbf{k})$ by simply examining the pertinent timescales associated with the particle-wave interaction. The first of these is the light transit time (with speed c) for an acceleration region of size R:

$$\tau_{\mathrm{tr}} \equiv \frac{R}{c}. \qquad 7.125$$

The pitch-angle scattering timescale (see equation 7.108) is

$$\tau_{\mathrm{sc}} = \frac{1}{D_{\mu\mu}}, \qquad 7.126$$

which may be written

$$\tau_{\mathrm{sc}} = \gamma^{2-q} \tau_p \qquad 7.127$$

in terms of τ_p, a characteristic timescale in the plasma, *defined* by the expression

$$\tau_p^{-1} \equiv \frac{\pi}{2} \Omega_e \, f_{\mathrm{turb}} (q - 1) k_{\mathrm{min}}^{(q-1)}. \qquad 7.128$$

On their random-walk spatial diffusion, the charges must scatter a number of times $\sim [R/\langle l \rangle]^2$ in order to escape, where $\langle l \rangle$ is the mean free path. Thus,

$$\tau_{\mathrm{esc}} = \frac{R^2}{\beta^2 c^2 \tau_{\mathrm{sc}}}. \qquad 7.129$$

Another important timescale to consider is that associated with acceleration, which, according to equation (7.113) is

$$\tau_{ac} = \frac{E^2}{D_{EE}}.$$ 7.130

This may itself be written as

$$\tau_{ac} = \frac{2\tau_p}{\beta_A{}^2}\gamma^{2-q},$$ 7.131

where

$$\beta_A \equiv \frac{B/c}{(8\pi m_p n_e)^{1/2}}$$ 7.132

is the Alfvén velocity divided by c. Finally, in Sagittarius A*, where the photon energy density is at least one order of magnitude lower than the magnetic field energy density,[59] the radiative loss of relativistic electrons is dominated by synchrotron processes, with a timescale

$$\tau_{loss} = \frac{E}{\dot{E}_L} = \frac{(\gamma - 1)m_e^2 c^3}{4\pi e^4 n_e (\ln \Lambda/\beta + 4\beta^2\gamma^2/9\psi^2)},$$ 7.133

where $\ln \Lambda \simeq 20$ and the first and the second terms in the denominator give the Coulomb collision and the synchrotron loss rate, respectively.

The particle distribution is sensitive to four primary quantities: the size R of the acceleration zone; the turbulent energy density, represented by f_{turb}; the particle number density n_e; and the magnetic field B. To date, the most reliable measurement of Sagittarius A*'s intrinsic size has been made with closure amplitude imaging using the VLBA.[60] At 7 mm wavelength, the source is confined to a region roughly 24 Schwarzschild radii in extent, though the size is variable on a timescale of hours to weeks and sometimes reaches $\sim 60 r_S$. Correspondingly, Sagittarius A*'s average flux density at this wavelength is ≈ 1.5 Jy. The stochastic acceleration corresponding to the 7 mm wavelength emission must therefore be

[59] One may easily confirm this by comparing the photon energy density associated with the mm/sub-mm flux shown in figure 3.1 to the magnetic energy density $B^2/8\pi$ inferred from the simple spectral fitting discussed above.

[60] See figure 2.4. The latest observations have been reported by Bower et al. (2004).

confined to a region of known size (i.e., $R \sim 20 r_S$) and must produce a particle distribution that correctly accounts for the measured radiative output at this wavelength. The other parameters are well constrained from our previous discussion. For example, $f_{turb} \approx 1$ in the absence of an external large-scale magnetic field. The source of electrons is apparently the hot flow accreting toward the black hole, with a temperature of $\sim 10^{10}$–10^{11} K. Thus, the injected electrons have a mean Lorentz factor $\gamma_{in} \sim 10$, corresponding to a temperature of $\sim 6 \times 10^{10}$ K.

The accelerated electrons have a relatively flat spectrum below this energy and a power-law distribution above it. The distribution cuts off when the synchrotron cooling time (equation 7.133) becomes shorter than the acceleration time (equation 7.131). With $\tau_{loss} \approx 7.8 \times 10^8/(B^2 \gamma)$ seconds, one therefore infers a high-energy cutoff

$$\gamma_{max} \approx 1.7 \times 10^5 B^{-2} (20 r_S/R). \qquad 7.134$$

This provides us with sufficient information to delimit the range of B. Since γ_{max} should be larger than γ_{in}, we immediately infer that

$$B < 130 \left(\frac{20 r_S}{R}\right)^{1/2} \left(\frac{10}{\gamma_{in}}\right)^{1/2} \text{G.} \qquad 7.135$$

The corresponding maximum frequency of synchrotron emission is

$$\nu_{max} \approx 5.0 \times 10^{17} B^{-3} \left(\frac{20 r_S}{R}\right)^2 \text{Hz.} \qquad 7.136$$

In order to avoid producing IR emission in excess of the observed limit (see chapter 3), we require $\nu_{max} \leq 10^{14}$ Hz, which provides the lower limit

$$B \geq 17 \left(\frac{20 r_S}{R}\right)^{2/3} \text{G.} \qquad 7.137$$

An additional very important observational constraint emerges from Sagittarius A*'s total radio luminosity $L_{radio} \approx 5 \times 10^{34}$ ergs s^{-1} and another from its luminosity $L_{7mm} \approx 5 \times 10^{33}$ ergs s^{-1} at 7 mm (estimated from the flux density quoted above). Since the source is known to be in transition from self-absorbed to optically thin across the mm/sub-mm bump (see figure 3.1) and optically thin at ν_{max} in order to be consistent with the 7 mm image (see the discussion following figure 2.4), we may estimate the luminosity associated with this emission using the simple scaling[61]

$$\nu_{max} L_\nu(\nu_{max}) \approx 1.1 \times 10^{31} n_e B^{0.4} \left(\frac{3V}{4\pi R^3}\right)$$

$$\times \left(\frac{R}{20r_S}\right)^{2.2} \left(\frac{\gamma_{in}}{10}\right)^{1.2} \text{ergs s}^{-1}, \qquad 7.138$$

where V is the emitting volume. This luminosity must be less than L_{radio}, but greater than L_{7mm}, from which we gather that the plasma must be strongly magnetized (with $\beta_A > 1$) in the halo region where the $\lambda > 7$ mm radiation is produced. The acceleration and escape times (equations 7.129 and 7.130) resulting from these physical conditions are shown in figure 7.13, for a specific q and particle number density n_e (≈ 500 cm^{-3}) or, alternatively, characterized by the Alfvén velocity β_{AC} (equation 7.132). The distinction between electrons and protons is due entirely to their different masses.

Actually, this mass-dependent response to stochastic acceleration produces an additional observational signature—beyond simply accounting for Sagittarius A*'s radio halo, for which this work was originally designed. The predicted proton ejection rate may prove to be relevant to new high-energy observations of the galactic center (see also the discussion at the end of chapter 3). From the timescales shown in figure 7.13, we see that low-energy protons have an escape time much longer than their acceleration. Thus, most of these particles are energized efficiently up to about $E_{min} \sim 300$ Mev, above which the acceleration and escape times become comparable and the protons settle into a power-law distribution. This spectrum may be described as

$$N_p(E) = (\delta - 1) n_e E_{min}^{\delta-1} E^{-\delta}, \qquad 7.139$$

assuming that the energy cutoff is much higher than the "turnover" energy E_{min}. The ensuing proton luminosity for $E_p \geq E_o = 0.1$ TeV is then

$$L_p = \frac{\delta - 1}{\delta - 2} \frac{V n_e E_{min}^{\delta-1} E_o^{\delta-2}}{\tau_{esc}} \approx 3.2 \times 10^{34} n_e \left(\frac{E_{min}}{300 \text{ MeV}}\right)^{1.2}$$

$$\times \left(\frac{0.1 \text{ TeV}}{E_o}\right)^{0.2} \left(\frac{3V}{4\pi R^3}\right) \left(\frac{R}{20 r_S}\right)^2 \text{ergs s}^{-1}. \qquad 7.140$$

[61] See Pacholczyk (1970).

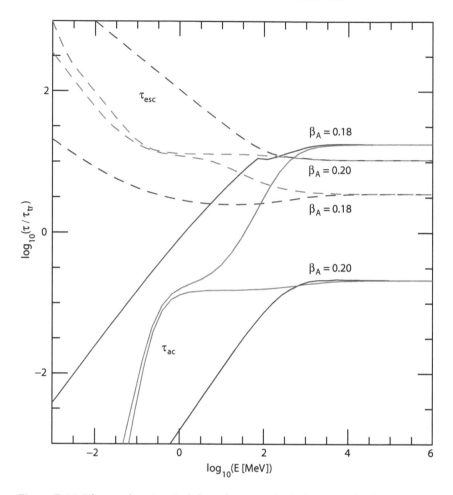

Figure 7.13 The acceleration (solid) and escape (dashed) timescales for electrons (thin) and protons (thick), in plasmas with $\beta_A = 0.18$ and 2.0 (indicated in the figure). The timescales are shown in units of the light transit time $\tau_{tr} = R/c$. The turbulence is assumed to be in energy equipartition with the magnetic field and $q = 2$. In a weakly magnetized plasma ($\beta_A < 1.0$), the waves can accelerate low-energy electrons to a Lorentz factor $\beta_A m_p / m_e$, where m_p / m_e is the proton to electron mass ratio, causing a sharp rise in the electron acceleration time with energy beyond 1 MeV. The acceleration of nonrelativistic electrons is dominated by high-frequency electromagnetic waves. In a strongly magnetized plasma ($\beta_A > 1.0$), the acceleration time of relativistic electrons is almost energy independent. (Image from Liu, Melia, and Petrosian 2006b)

For the electron density $n_e \sim 500$ cm^{-3} required to fit Sagittarius A*'s observed 7 mm flux, we infer an *escaping*, relativistic proton luminosity $L_p \approx 1.6 \times 10^{37}$ ergs s^{-1}, injected into the medium surrounding the black hole.

Once they are out of the acceleration site, these particles diffuse through the ambient medium (described earlier in this chapter and depicted in figure 7.2), random-walking several parsecs away from the galactic center. But on their way out, the protons undergo a series of interactions, including $pN \to pN$, m_{meson}, and $m_{N\overline{N}}$, where N is either a p or a neutron n, m_{meson} denotes the energy-dependent multiplicity of mesons (mostly pions), and $m_{N\overline{N}}$ is the multiplicity of nucleon/antinucleon pairs (both increasing functions of energy). Since $m_{N\overline{N}}/m_{\text{meson}} < 10^{-3}$ at low energy and even smaller at higher energies,[62] one may safely ignore the antinucleon production events. The charge exchange interaction ($p \to n$) occurs around 40% of the time at accelerator energies, and this fraction is predicted to be only very weakly energy dependent. Other possible interactions of accelerated p's—all potentially important for cooling—are $p\gamma \to p\pi^0\gamma$, $p\gamma \to n\pi^+\gamma$, $p\gamma \to e^+e^-p$, and $pe \to eNm_{\text{meson}}$.

The (differential) π^0 emissivity resulting from an isotropic distribution of relativistic protons $N_p(E)$ (in units of cm^{-3} eV^{-1}) interacting with cold (fixed-target) ambient hydrogen of density n_e is given by the expression

$$Q_{\pi^0}^{pp}(E_{\pi^0}) = cn_e \int_{E^{th}(E_{\pi^0})} dE \, N_p(E) \frac{d\sigma(E_{\pi^0}, E)}{dE_{\pi^0}}, \qquad 7.141$$

where $E^{th}(E_{\pi^0})$ is the minimum proton energy required to produce a pion with total energy E_{π^0} (determined through kinematical considerations) and $[Q_{\pi^0}^{pp}] = $ pion ss^{-1} cm^{-3} eV^{-1}. The resulting gamma ray emissivity is then given by the expression

$$Q_\gamma(E_\gamma) = 2 \int_{E_{\pi^0, \min}(E_\gamma)} dE_{\pi^0} \frac{Q_{\pi^0}^{pp}(E_{\pi^0})}{(E_{\pi^0}^2 - m_{\pi^0}^2)^{1/2}}, \qquad 7.142$$

where $E_{\pi^0, \min}(E_\gamma) = E_\gamma + m_{\pi^0}^2/(4E_\gamma)$.

---

[62] See Crocker et al. (2005).

At proton energies $E > 5\,\text{GeV}$—above the Δ resonance-affected region—the differential cross section is approximated by the scaling form[63]

$$\frac{d\sigma(E, E_{\pi^0})}{dE_{\pi^0}} = \frac{\sigma_0}{E_{\pi^0}} h_{\pi^0}(x_0), \qquad 7.143$$

where $x_0 \equiv E_{\pi^0}/E$, $\sigma_0 = 32$ mbarn, and

$$h_{\pi^0}(x_0) = 0.67(1 - x_0)^{3.5} + 0.5e^{-18x_0}. \qquad 7.144$$

This scaling form properly takes into account the large pion multiplicities that occur at high energies.

Thus, for the parent proton distribution governed by equation (7.139), we can write the neutral pion emissivity due to proton-proton scattering as

$$Q_{\pi^0}^{pp}(E_{\pi^0}) = N_p(E_{\pi^0})\,\sigma_0\,n_e c\,H^0(\delta) \qquad 7.145$$

and, consequently, the photon emissivity due to the decay of these π^0's as

$$Q_\gamma(E_\gamma) \simeq c n_e \int_{E_\gamma}^\infty dE_{\pi^0} \int_{E_{\pi^0}}^\infty dE\, N_p(E) \frac{d\sigma(E, E_{\pi^0})}{dE_{\pi^0}} \frac{2}{E_{\pi^0}}$$

$$\simeq \frac{2}{\delta} N_p(E_\gamma)\,\sigma_0\,n_e c\,H^0(\delta), \qquad 7.146$$

where in both equations immediately above we employ

$$H^0(\delta) \equiv \int_0^1 dx_0\, x_0^{\delta-2}\, h_{\pi^0}(x_0) = 2\left\{\Gamma(\delta - 1)\left[18^{1-\delta}\right.\right.$$

$$\left.\left. + 15.5865\,\Gamma^{-1}(3.5 + \delta)\right] - E(2 - \delta; 18)\right\}, \qquad 7.147$$

in which $\Gamma(x)$ is the Euler gamma function and $E(n; z)$ is the exponential integral function that satisfies

$$E(n; z) \equiv \int_1^\infty \exp(-zt)/t^n\, dt. \qquad 7.148$$

---

[63] See Blasi and Colafrancesco (1999).

This is the result we have been seeking for a direct comparison with the TeV flux from the galactic center measured by HESS (discussed at the end of chapter 3).

According to equations (7.140) and (7.146), the physical conditions required to explain the 7 mm flux also produce—via neutral pion decay associated with proton-proton scattering in the extended medium—a factor of about four times the measured TeV flux from this region. The slight disparity between theory and observation may be due to anisotropy in the particle diffusion or to the fact that the escaping protons are only partially filtered by the surrounding plasma, that is, that some diffuse farther than ~ 3 pc before scattering with other protons.

One of the remaining issues regarding Sagittarius A*'s radio emission concerns the circular polarization observed longward of ~ 7 mm (see figure 3.2). Unlike the linearly polarized light shortward of ~ 3 mm, which apparently originates on scales $< 10r_S$ from within the Keplerian region, the circular polarization must be a reflection of physical processes occurring in the nonthermal halo, on scales $\sim 20r_S$.

Circular polarization is a common feature of quasars and appears to be generated near synchrotron self-absorbed cores.[64] In the specific source 3C 273, observations of the proper motion of circular-polarization-producing regions show that this phenomenon is intrinsic to the emitter, as opposed to being due to foreground effects.

However, direct emission of circularly polarized light is inadequate to explain the observed levels. In a realistic source, the fraction of polarization is strongly suppressed by the tangled magnetic field and possibly the emissivity of e^+-e^- pairs, which do not contribute to the circular polarization (because they orbit in opposite directions). Of the remaining mechanisms that have been proposed, coherent radiation processes[65] produce polarization in a narrow frequency range, which has now been ruled out by the multiwavelength observations discussed in chapter 3. The mechanism most likely to be responsible for the circular polarization in Sagittarius A* appears to be Faraday conversion,[66] which is related to the Faraday rotation we discussed in §3.2 and then reintroduced in §7.3.

[64] See, e.g., Homan and Wardle (1999).
[65] See, e.g., Benford and Tzach (2000).
[66] A detailed discussion of this process appears in Ruszkowski and Begelman (2002). See also Beckert and Falcke (2002).

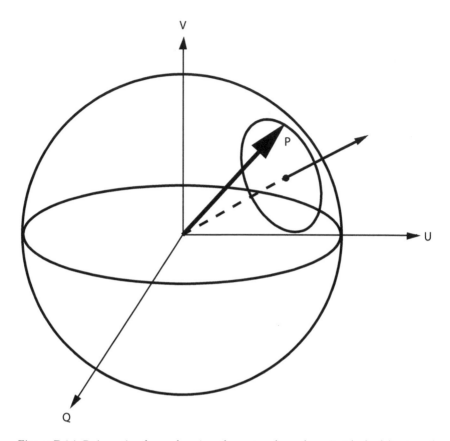

Figure 7.14 Poincaré sphere showing the natural mode axis (dashed line) and the polarization vector (labeled P). Circular polarization corresponds to the poles; the equator represents linear polarization. (Image from Ruszkowski and Begelman 2002)

Faraday rotation is a specific example of a more general phenomenon known as birefringence. When the natural modes are linearly or elliptically polarized (see equations 3.7 and 3.9), birefringence results in the cyclic conversion between linearly and circularly polarized radiation due to the changing phase relationships between the modes along the ray. One often invokes the Poincaré sphere to demonstrate this effect, as seen in figure 7.14. The vector P represents the arbitrary elliptical polarization, with its tip lying on the sphere and characterized by Cartesian coordinates $(Q, U, V)/I$, which are the Stokes parameters defined after equation (3.15). The north and south poles correspond to right- and left-circular polarizations, since for these $\phi_2 - \phi_1 = \pi/2$ and

$|E_{01}| = |E_{02}|$, so $Q = U = 0$. Linear polarization corresponds to points along the equator, for which $\phi_2 - \phi_1 = 0$, so $V = 0$. As the radiation passes through the medium, birefringence causes the tip of the arrow P to rotate about the axis of the natural plasma modes. So, for example, Faraday rotation corresponds to the case in which the natural mode axis is vertical (along V), and the polarization vector P rotates around it.

The radiative transfer of polarized radiation in a turbulent plasma may be solved by adopting transfer equations for a piecewise homogeneous medium with a weakly anisotropic dielectric tensor.[67] Assuming a highly tangled magnetic field B with a very small mean component $\langle B \rangle$, the solutions may be characterized in terms of the ratio $\delta \equiv \langle B \rangle / B$. As the cm wavelength radiation propagates through the birefringent nonthermal halo of Sagittarius A*, the relative phase of the circular polarization modes changes with position, resulting in a conversion from Stokes Q to Stokes U and then finally to Stokes V. However, when δ is too large, Faraday depolarization reduces Stokes Q and U and thus leads to a suppression of circular polarization as well. Circular polarization therefore has an extremum. This behavior is typified by the solution plotted in figure 7.15, where the mean linear and circular polarizations are shown as functions of δ for the case of radiation transfer through a high synchrotron depth.

Circular polarization can exceed linear polarization at higher values of the bias parameter δ, since an excess of the former over the latter requires a significant level of Faraday depolarization. Archival VLA data indicate that the mean circular polarization in Sagittarius A* was stable over more than a decade. As indicated in the figure, this persistence almost certainly points to a nonzero bias in $\langle B \rangle$.

To summarize the last few sections of this chapter, we have shown that Sagittarius A*'s mm/sub-mm spectrum is apparently produced by a hot, magnetized Keplerian region orbiting within $\sim 10 r_S$ of the center. This plasma, however, cannot sustain a brightness temperature high enough to also account for Sagittarius A*'s radio flux longward of ~ 7 mm. Instead, a fraction of the accreting matter is evidently accelerated to relativistic energies and is subsequently ejected, forming a nonthermal halo surrounding the black hole. The 7 mm glow is produced within $\sim 20 r_S$ of Sagittarius A*, where the stochastic acceleration that energizes electrons also functions to produce a steady loss of nonthermal protons diffusing out as far as ~ 2–3 pc from the source.

---

[67] See Sazonov (1969) and Jones and O'Dell (1977a).

223

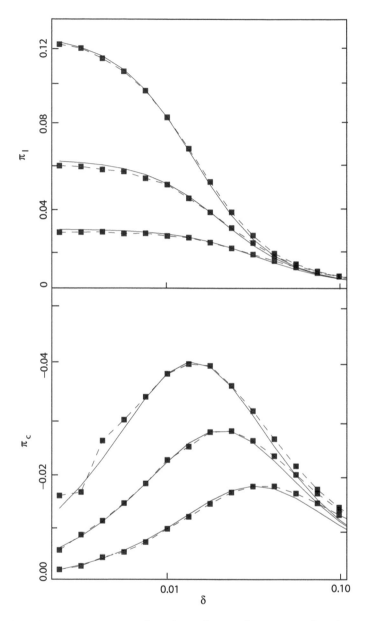

Figure 7.15 Linear (π_l, top panel) and circular (π_c, bottom panel) polarizations for a very large synchrotron depth, that is, a self-absorbed source, and varying number of turbulent zones along the line of sight (10^8, 3×10^5, and 10^5, top to bottom). The solid lines denote analytical results, whereas squares connected by the dashed lines show the results of Monte Carlo simulations. (Image from Ruszkowski and Begelman 2002)

Sagittarius A*'s linear polarization, associated with its mm/sub-mm spectral component, arises within the Keplerian region; the nonthermal halo extending out to $>20r_S$ is optically thin to this radiation. On the other hand, the circularly polarized radiation originates from the extended halo, which is self-absorbed for radio waves longward of $\sim 7\,$mm. It appears that Faraday conversion in this region is responsible for producing the net circular polarization.

This is currently where things stand with regard to our understanding of Sagittarius A*, at least in its nonflaring state. It is a somewhat fragmented picture that, however, should continue to improve as better resources become available to afford theorists the opportunity of modeling the accretion of matter onto the black hole in a more comprehensive manner. But we have not yet exhausted the rich supply of data we may use to unravel the mystery of how this object functions in all its guises. In particular, we have barely broached the subject of variability, not to mention the evidence that Sagittarius A*'s IR and X-ray flux is sometimes modulated with an approximately twenty-minute quasi-period. This will be the focus of chapter 8, where we shall begin to probe the space and time near the marginally stable orbit of this supermassive black hole.

CHAPTER 8

Flares

Compared to Sagittarius A*'s quiescent emission, the IR and X-ray flares reflect a significant (though transient) change in the system's physical state. The biggest events in this category produce fluxes tens of times greater than those from the nonflaring configuration. There is also the suspicion, based primarily on Sagittarius' A*'s IR lightcurve, that perhaps there is no real "steady" state in this source and that all of its emission is associated with transient events spanning a range of durations and amplitude; we just happen to identify the biggest of these as "flares."

Several mechanisms have been proposed to account for this variability in Sagittarius A*, including gravitational lensing,[1] star-disk collisions,[2] and physical processes associated with the black hole's accretion flow.[3] However, the observed short timescale (about forty minutes) for the variation rules out all but the accretion scenario, possibly linked to the ejection of particles, as the likely cause of these fluctuations. Light travel time arguments delimit the emitting region to a size no bigger than about 5 AU, or equivalently about 70 Schwarzschild radii for a $3.6 \times 10^6 \, M_\odot$ black hole. The IR and X-ray cyclic modulations appear to restrict the possibilities even further, since a Keplerian origin for the associated periods seems to be unavoidable.

8.1 FLARE PHYSICS

Flares are likely triggered by some plasma instability or by changes in the dynamics of the accretion flow. For example, the dissipation of angular momentum is different above and below the marginally

---

[1] See, e.g., Alexander and Loeb (2001).

[2] See Nayakshin and Sunyaev (2003).

[3] There exists a much richer literature on this process than for the others. Papers on this topic include Melia (1992a, 1994), Hollywood and Melia (1997), Markoff et al. (2001), Liu and Melia (2002a), and Yuan, Quataert, and Narayan (2003).

stable orbit (MSO) defined in §5.5, and this may lead to a strong gravitational dissipation and acceleration of electrons. A temporary change in the accretion rate through the inner disk may also, through the MRI (see §7.5), produce a magnetic flare that subsequently accelerates electrons and protons to relativistic energies, temporarily enhancing the synchrotron emissivity.[4] In addition, variable IR and X-ray emission may result from the acceleration of electrons into a broken power-law distribution during episodes of magnetic reconnection[5] or during episodic shock acceleration at the base of an outflowing jet.[6]

Before proceeding too far with this, let us first examine the various timescales pertaining to the inner disk to see if these scenarios are physically relevant. The anomalous viscosity $v \equiv (2/3)W_{r\phi}/\Sigma\,\Omega$ was defined in equation (7.55) in terms of the stress $W_{r\phi}$, column density Σ, and angular velocity Ω. The radial velocity

$$v_r \sim \left(\frac{4\beta_P\beta_v}{9}\right)\left(\frac{GM}{R}\right)^{1/2}, \qquad 8.1$$

produced by this v according to equation (7.54), with the gas temperature set equal to its virial value, depends on two parameters: the ratio β_v of the stress to the magnetic field energy density (with a typical value of $\sim 0.1 - 0.2$) and β_P, the ratio of magnetic energy density to thermal pressure P. The MHD simulations discussed in chapter 7 indicate that $\beta_P \sim 0.03$ under most conditions.

The viscous timescale for the gas in the inner few gravitational radii $(r_G \equiv GM/c^2)$ of the disk is therefore

$$\tau_v \equiv \frac{r_G}{v_r} \sim 9.6 \left(\frac{r}{r_G}\right)^{1/2}\left(\frac{0.05}{\beta_P\beta_v}\right) \text{ minutes.} \qquad 8.2$$

Notice that τ_v is consistent with the variability timescale (around ten minutes) during a typical flare when $\beta_P\beta_v \sim 0.05$, which is close to the result seen in the MHD simulations, when one takes into account the various approximations made in these calculations and their limited spatial resolution.

----

[4] Additional discussion on these various possibilities may be found in Liu and Melia (2002a) and Liu, Melia, and Petrosian (2006a).

[5] See Yuan, Quataert, and Narayan (2004).

[6] See Markoff et al. (2001).

By comparison, the dynamical timescale for gas flowing through the inner Keplerian region is

$$\tau_d \equiv \frac{2\pi r}{v_k},$$

8.3

where $v_k = (GM/r)^{1/2}$ is the azimuthal velocity at radius r. Its scaled value $\tau_d \approx 1.3(r/r_G)^{3/2}$ minutes suggests that a nonequilibrium process may be responsible for initiating the injection or depletion of matter through the inner orbits, which then results in an overall fluctuation on a viscous timescale as the disk readjusts. Thus, whereas the duration of a fluctuation probably corresponds to the time required for viscosity to reestablish equilibrium, the overall extent of the flare may be attributable to certain characteristics of the infalling plasma (perhaps its spatial extent or its clumping profile).

Although this simple analysis still leaves open the question of how the particles are actually energized and whether most of their subsequent radiation occurs near the acceleration site itself or farther out in a jet, the association we can make between the viscous timescale and the variability of the observed flares suggests that we may be on the right track with this hypothesis—that a temporary enhancement in $\dot{M}$ through the inner disk gives rise to the outburst. Several important changes are expected to result from the injection of new matter into the system. Chief among these is the intensification of B, which enhances the synchrotron emissivity and, at the same time, alters the disk's anomalous viscosity. Since the MRI apparently saturates at $\beta_P \sim 0.05$, a temporary enhancement in $\dot{M}$ by a factor of ten would be sufficient—when the concomitant factor of around 3 increase in B is taken into account—to produce the observed amplitude of the strongest outbursts seen in Sagittarius A*.

But are the IR and X-ray flares simply spectral variations of the same event, or do they actually occur at different locations (such as disk versus jet) and, possibly, at different times? Though it is reasonable to assume the former, this has yet to be confirmed observationally in any compelling fashion. Future multiwavelength observations will no doubt lay this issue to rest. For now, let us suppose that they do represent simultaneous components of the same spectrum. The fact that the IR photon distribution is so much steeper than that of its X-ray counterpart excludes a direct extrapolation of the same power law from one to the other (see figure 8.1). For example, during the peak of the IR flare[7]

---

[7] See Eisenhauer et al. (2005).

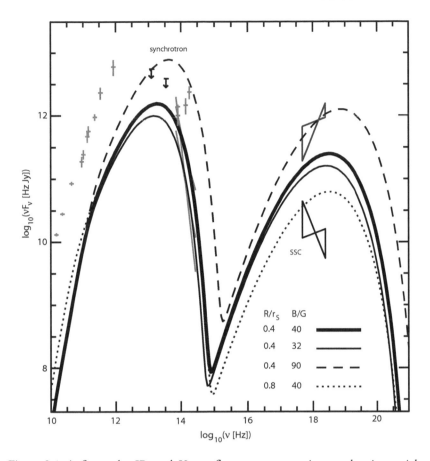

Figure 8.1 A fit to the IR and X-ray flare spectrum using stochastic particle acceleration, as described in the text. The parameter values include a critical Lorentz factor $\gamma_c = 75$ and a total number of radiating electrons $\mathcal{N} = 3.8 \times 10^{42}$. For simplicity, the emitting plasma is assumed to be a uniform sphere, with size R and magnetic field B as indicated. The data include the *Chandra* steady state flux (lower butterfly) and the peak X-ray spectral component of the October 27, 2000, flare (Baganoff et al. 2001). In the IR, the data points represent peak fluxes during individual events, and the crosses (barely visible below the theoretical curves) are the IR spectra measured during the flare of July 15, 2004 (Eisenhauer et al. 2005). There are four theoretical curves: (thick solid) overall fit to the peak IR emission of the July 15, 2004, event; (thin solid) the rising and decay phase spectrum of this IR event; (dashed) a fit to the peak emission of the October 27, 2000, *Chandra* flare; and (dotted) the same as the thick solid curve, except that $R = 0.8r_S$. The data points to the left of the panel correspond to the peak flux of variable radio and mm emission (fluctuating on a timescale of days to weeks) that originates on larger spatial scales (see chapter 7). (Image from Liu, Melia, and Petrosian 2006a)

observed on July 15, 2004, a power-law fit $\nu F_\nu \propto \nu^{-\alpha}$ (with ν the emission frequency) to its power spectrum yields $\alpha = 2.2 \pm 0.3$. During its rising and decaying phases, a similar fit requires an index $\alpha = 3.7 \pm 0.9$.

For this reason, it is thought that the mm/sub-mm–to–IR portion of the spectrum is probably due to synchrotron, whereas the X-rays are produced by synchrotron self-Compton (SSC; see equation 7.98). Incidentally, this constraint is empirically motivated and is independent of the fully fledged theory describing how the emission occurs.

Now, with synchrotron radiation, the power radiated by one relativistic electron per unit frequency is[8]

$$P(\omega) = \frac{\sqrt{3}}{2\pi} \frac{e^3 B \sin \alpha_p}{m_e c^2} F\left(\frac{\omega}{2\omega_c}\right), \qquad 8.4$$

where α_p is the pitch angle (i.e., the angle between the magnetic field and the particle's velocity), and the other symbols have their usual meaning. The function F is defined as

$$F\left(\frac{\omega}{2\omega_c}\right) \equiv \left(\frac{\omega}{\omega_c}\right) \int_{\omega/\omega_c}^{\infty} K_{5/3}(\xi)\, d\xi \qquad 8.5$$

in terms of the modified Bessel function of order $5/3$ and the so-called critical angular frequency

$$\omega_c \equiv \frac{3}{2}\gamma^2 \omega_{\text{gyr}} \sin \alpha_p. \qquad 8.6$$

(The factor $1/2$ inside the argument of F is kept here for consistency with the notation used in its derivation.) Aside from factors of order unity, this expression actually has a very simple physical interpretation. Notice that $\omega_c \sim \gamma^2 \omega_{\text{gyr}}$, where $\omega_{\text{gyr}} \equiv eB/m_e c$ is the classical gyration frequency. This therefore represents a change in frequency (from ω_{gyr} to ω_c) due to inverse Compton scattering (by the gyrating electron) of a virtual photon in the magnetic field, given that two transformations of the frequency are required (see equation 6.18)—one from the laboratory frame into the electron's rest frame and then back into the laboratory frame. Thus, although the single particle spectrum (equation 8.4)

---

[8] A full derivation of this result may be found in the author's book on *Electrodynamics* (2001b).

involves a great number of harmonics, which blend together to form a continuous envelope, the peak of this photon distribution not surprisingly lies very close to $\omega/\omega_c \sim 1$.

It is not difficult to derive from the formula for $P(\omega)$ the photon spectrum produced by power-law electrons with an energy distribution

$$N(\gamma)\,d\gamma = C\gamma^{-p}\,d\gamma, \quad \gamma_1 < \gamma < \gamma_2. \qquad 8.7$$

The proportionality constant C depends on several factors, including the pitch angle and, especially, on the conditions giving rise to the particle acceleration. The total power radiated per unit volume per unit frequency by such a distribution is given by the integral of $N(\gamma)\,d\gamma$, weighted by $P(\omega)$, over all Lorentz factors γ. We therefore have

$$P_{tot}(\omega) = C \int_{\gamma_1}^{\gamma_2} P(\omega)\,\gamma^{-p}\,d\gamma \propto \int_{\gamma_1}^{\gamma_2} F\left(\frac{\omega}{2\omega_c}\right)\gamma^{-p}\,d\gamma. \qquad 8.8$$

A simple change of variables $x \equiv \omega/\omega_c$, with $\omega_c \propto \gamma^2$, transforms this equation into

$$P_{tot}(\omega) \propto \omega^{-(p-1)/2} \int_{x_1}^{x_2} F(x)x^{(p-3)/2}\,dx, \qquad 8.9$$

in which the limits x_1 and x_2 correspond to γ_1 and γ_2, respectively. Unfortunately, both of these limits depend on ω. However, the energy range is often sufficiently wide so that one may safely put $x_1 \to 0$ and $x_2 \to \infty$, in which case the integral is roughly constant,

$$P_{tot}(\omega) \propto \omega^{-(p-1)/2}, \qquad 8.10$$

and we see that the spectral index α of the radiation is related to that of the particle distribution according to

$$\alpha = \frac{p-1}{2}. \qquad 8.11$$

So if the IR flare emission is produced via the synchrotron process, at least a portion of the electron distribution must have the power-law form of equation (8.7) with $p > 7.4$, suggesting an exponential cutoff—presumably dictated by the acceleration process—at $\gamma_c \approx (\nu_{IR}/\nu_{gyr})^{1/2}$, where $\nu_{gyr} = \omega_{gyr}/2\pi$. The X-ray flares, on the other hand, often display a very hard spectrum with $\alpha \approx 0.7 \pm 0.5$, and in the SSC context, this requires a much flatter electron distribution with $p \sim 2$, emitting primarily in the mm/sub-mm range. That is because with SSC the same electrons emitting the synchrotron photons also upscatter them, so the seed photons producing the X-rays in this process must have frequencies $\sim 10^{18}/\gamma_c^2$ Hz, where, as we shall see, $\gamma_c \sim 100$. Phenomenologically, at least, the particle distribution producing the combined IR/X-ray flare spectrum must therefore have the form

$$N(\gamma) = N_0 \gamma^{-p} \exp\left(-\frac{\gamma}{\gamma_c}\right) \qquad \text{(with } p \sim 2\text{)}. \qquad 8.12$$

It turns out that producing the right blend of physical conditions to simultaneously fit both the IR and X-ray flare emissions is not trivial. However, the ideas and conclusions we have just described can be quite probative, given that a synchrotron spectrum, complemented by its self-Comptonized component, critically restricts the viable source configurations. The synchrotron luminosity for an isotropic relativistic distribution of electrons such as the one we have here[9] is

$$\mathcal{L}_{\text{syn}} \approx \frac{16\, e^4}{3 m_e^2 c^3} \mathcal{N} B^2 \gamma_c^2$$

$$\approx 2.0 \times 10^{36} \left(\frac{\mathcal{N}}{10^{43}}\right) \left(\frac{B}{40\,\text{G}}\right)^2 \left(\frac{\gamma_c}{100}\right)^2 \text{ ergs s}^{-1}, \qquad 8.13$$

where $\mathcal{N} = 2 N_0 \gamma_c^3$ is the total number of accelerated electrons. If this radiation field, with energy density U_{syn}, is also isotropic, then the corresponding upscattered X-ray emission is

$$\mathcal{L}_{\text{SSC}} = \frac{U_{\text{syn}}}{U_B} \mathcal{L}_{\text{syn}} \approx \frac{8\pi \mathcal{L}_{\text{syn}}^2}{c A B^2}$$

$$\approx 5.2 \times 10^{35} \left(\frac{\mathcal{L}_{\text{syn}}}{10^{36}\,\text{ergs s}^{-1}}\right)^2 \left(\frac{B}{40\,\text{G}}\right)^{-2} \left(\frac{A}{r_S^2}\right)^{-1} \text{ ergs s}^{-1}, \qquad 8.14$$

where $U_B = B^2/8\pi$ is the magnetic field energy density and A is the surface area of the source. This expression assumes that $\mathcal{L}_{\text{syn}} \approx U_{\text{syn}} c A$.

---

[9] As before, a good resource for this topic is Pacholczyk (1970).

The unstable region in Sagittarius A* is constrained rather well because these two equations contain only three unknowns. For example, if we know B, then we can estimate the relativistic electron number density n_e and source size R. Let us say, for the sake of argument, that the emitter is uniform and spherical. Then $A = 4\pi R^2$ and $\mathcal{N} = 4\pi R^3 n_e/3$. The simultaneous solution to these equations then yields

$$R \approx 0.64 \left(\frac{\mathcal{L}_{\text{syn}}}{10^{36} \text{ ergs s}^{-1}} \right) \left(\frac{\mathcal{L}_{\text{SSC}}}{10^{35} \text{ ergs s}^{-1}} \right)^{-1/2} \left(\frac{B}{40\,\text{G}} \right)^{-1} r_S, \qquad 8.15$$

$$n_e \approx 4.6 \times 10^6 \left(\frac{\mathcal{L}_{\text{syn}}}{10^{36} \text{ ergs s}^{-1}} \right)^{-2} \left(\frac{\mathcal{L}_{\text{SSC}}}{10^{35} \text{ ergs s}^{-1}} \right)^{3/2}$$
$$\times \left(\frac{B}{40\,\text{G}} \right) \left(\frac{\gamma_c}{100} \right)^{-2} \text{ cm}^{-3}. \qquad 8.16$$

It is quite remarkable, really, that without very much theoretical input the flare data may be interrogated to yield such specific information. We also realize, of course, that once the total number of energetic electrons required to produce a bright flare is known, we can set a lower limit[10] on the mass accretion rate $\dot{M}$ and the accretion timescale τ_{acc}, since

$$\tau_{\text{acc}} \dot{M} \sim \mathcal{N} m_p, \qquad 8.17$$

where m_p is the proton mass. If during the instability the accretion time is comparable to the observed rise time of the flare, then we estimate that

$$\dot{M} \sim 0.9 \times 10^{16} \left(\frac{\mathcal{N}}{10^{43}} \right) \left(\frac{\tau_{\text{acc}}}{30\,\text{mins}} \right)^{-1} \text{ g s}^{-1}. \qquad 8.18$$

Given this preliminary analysis of the data, let us now pursue the possibility that the flares in Sagittarius A* are produced by a magnetic event, possibly driven by the aforementioned accretion instability. Much of the groundwork for this discussion was laid out in chapter 7, where we applied the theory of stochastic acceleration to Sagittarius A*'s quiescent radio emission and its apparent coupling to the TeV radiation detected from the inner ~ 1 arcmin of the Galaxy.

---

[10]This is a lower limit because not all of the accreting electrons are necessarily accelerated to relativistic energies.

The solution to equation (7.113) with $E = \gamma m_e c^2$ is expressible in the form of (8.12), the index p set by the acceleration (τ_{acc}) and escape (τ_{esc}) times. For the physical conditions we inferred above, τ_{esc} is longer than the duration of the flare, so most of the radiating electrons cluster near $\gamma_c \sim 100$ (see the discussions following equations 8.6 and 8.11). Sample spectra produced by these particles are shown in figure 8.1 and compared directly with the available data. The thin solid curve corresponds to a calculation with $B = 32$ Gauss and fits the rising and decay phase spectrum of the IR flare of July 15, 2004. The dashed curve corresponds to a flare with $B = 90$ Gauss, which fits the peak flux of the X-ray flare of October 27, 2000. Evidently, the variation in flare characteristics may be attributed primarily to changes in the magnetic field. Note, however, that the optically thin IR emission depends only on the total number of energetic electrons (other than B), whereas the X-ray emission also depends on the source area (inversely, if the total number of radiating particles remains fixed). In this figure, the dotted curve shows the spectrum produced with the area of the emission region increased by a factor of 4 over the previous cases.

One gathers from this that disk instabilities may indeed produce flares in Sagittarius A*, but this theory is still incomplete. One cannot help but notice the small size of the unstable region, typically < 1 Schwarzschild radius in extent. Perhaps the analogy with the Sun is stronger than we think; that is, magnetic loops may be generating most of the particle dynamics. More likely, the physical variables peak toward smaller radii (unlike our simplistic assumption of a uniform emitter), so most of the contribution to the flare flux originates close to the marginally stable orbit. This certainly seems to be suggested by the periodicity detected during some flares, and we shall have much more to say about this in the next section.

8.2 Periodicity

The monotonic decrease of the period evident in both the IR (see, e.g., figure 4.5) and X-ray flares (see, e.g., figure 8.2) suggests that we may be witnessing the evolution of an event moving inwards through the last portion of the accretion disk, very near the marginally stable orbit. As we saw earlier in this chapter, the viscous timescale in the disk $\tau_v = r_G/v_r \approx 9.6\,(r/r_G)^{1/2}$ minutes is approximately 23.5 minutes at $r = 3r_S$, the MSO for a nonrotating (i.e., $a/r_G = 0$) black hole. The MSO

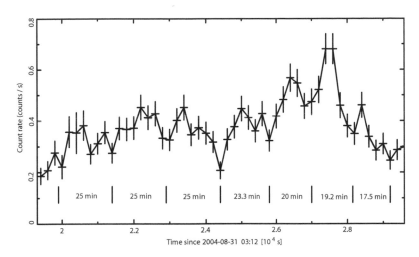

Figure 8.2 Lightcurve of the August 31, 2004, X-ray flare detected with XMM-*Newton*, binned in 200-second intervals. Markers and labels are used to show the gradual decrease in the period from 25 to 17.5 minutes. (Image from Bélanger et al. 2005)

has a progressively smaller radius as a/r_G increases. By comparison, the dynamical timescale $t_d \approx 1.3 \, (r/r_G)^{3/2}$ is roughly nineteen minutes at this radius.

As we noted earlier, the fact that both of these times are comparable to the inferred quasi-period suggests that the transient azimuthal asymmetry giving rise to the modulated flux may be due to either a dynamical event or a viscous process near the MSO.[11] It may be a wave pattern co-rotating with the gas, or perhaps it is a hot spot where matter has fallen in from larger radii, impacting the small disk.[12] In either case, the pattern of modulation over one complete cycle is the result of several relativistic effects, including a Doppler shift, light bending, and area amplification near the black hole's event horizon. We shall return to this topic in the following section.

For the sake of specificity, let us here analyze the (quasi-)periodic modulation seen during the X-ray event of August 31, 2004 (figure 8.2). The radius $\langle r \rangle \approx 2.7 \, r_S$ inferred from the average period in this event

[11] Some earlier considerations on how the timing data may be interrogated for information on the orbital properties at this radius may be found in Melia et al. (2001).
[12] See Falcke and Melia (1997).

(assuming a black hole mass of $3.6 \times 10^6 \, M_\odot$) is interesting for several reasons. First, since this location is very near the MSO of a Schwarzschild black hole, the fact that the viscous timescale is of the same order and slightly larger than the average period means that we could also be observing an event caused by the sudden reconfiguration of magnetic field lines frozen into a blob of matter flowing across the MSO toward the event horizon. Thus, if the shortest observed period (i.e., 17.5 minutes) were to actually correspond to this orbit, then the black hole spin could not be zero since $\langle r \rangle < 3r_S$. Indeed, from equation (6.155) we see that in this case[13] $a/r_G \approx 0.4$ (where $r_G \equiv Gm/c^2$) for a prograde orbit, or ≈ 0.3 in the case of retrograde disk rotation.

Second, the inference that $\langle r \rangle \approx 2.7 r_S$ is in line with the conclusion drawn in chapter 7 that there exists a hot, magnetized disk accreting through the inner $\sim 10 r_S$ of Sagittarius A*. Such a disk, we recall, can produce $\approx 10\%$ linearly polarized light with a position angle that rotates by $\approx 90°$ across the mm/sub-mm bump.

Third, this X-ray flare and others detected earlier by the *Chandra* X-ray Observatory feature strong variability near the middle of the event; in each case the X-ray flux drops by a factor of 40%–500% in 10–15 minutes. Simple light travel time arguments constrain the X-ray emitting region to be no bigger than ~ 17–$34 r_S$. Thus, the tight orbit inferred for the X-ray emitting gas is suggestively similar to that of the mm/sub-mm emitting plasma. The X-rays and mm/sub-mm photons apparently originate from the same medium, the latter due to synchrotron emission, the former via synchrotron self-Compton processes.

Fourth, a size of ≈ 2.4–$2.9 r_S$ for the X-ray emitting region is an affirmation of the measured intrinsic size of Sagittarius A* at $\lambda 7 \, \mathrm{mm}$, which we discussed extensively in the previous chapter. These X-ray orbits complement studies at other wavelengths by confirming the theoretically anticipated size-frequency relationship, in which the most energetic radiation is produced on the smallest spatial scales.[14]

But where exactly is the inner edge of the accretion disk in Sagittarius A*? This question has been asked in a broader context with the help of extensive MHD simulations of the plunging region in a pseudo-Newtonian potential (see equation 7.101) to identify several characteristic inner radii in black hole accretion.[15] Our discussion thus far

---

[13] See also Genzel et al. (2003a).
[14] See Melia, Jokipii, and Narayanan (1992) and Narayan, Yi, and Mahadevan (1995).
[15] See Krolik and Hawley (2002).

has assumed that either the average X-ray period (~ 21.4 minutes) or the last (and smallest) period (~ 17.5 minutes) ought to correspond to the MSO.

For magnetized accretion, however, matter flowing past the marginally stable orbit may still remain "magnetically" coupled with the outer disk even below r_{MSO}, so the constraint $\langle r \rangle < 3r_S$ does not necessarily mean that a/r_G is quite as large as 0.3–0.4. It makes more sense to consider a dynamically more meaningful radius—the so-called stress edge—where plunging matter loses dynamical contact with the material farther out. From a practical standpoint, this location may simply be defined as the surface at which the speed first becomes super-magnetosonic, that is, where it first exceeds the sound speed associated with fluctuations in the magnetic field. The MHD simulations[16] indicate that this transition occurs somewhere between $2.3r_S$ and $3r_S$.

The specific angular momentum $j = r^2\Omega(r)$, written in terms of the orbital angular frequency $\Omega(r)$, continues to decrease below r_{MSO}, though Ω may not necessarily trace its Keplerian value

$$\Omega_K(r) \equiv \left(\frac{GM}{r^3} \right)^{1/2} \qquad 8.19$$

in this region. Without any magnetic coupling across r_{MSO}, matter would retain all the specific angular momentum it had at that last circular orbit, so the accreted value of j, which we shall call j_{in}, would then simply be

$$j_{in} = r_{MSO}^2 \Omega_K(r_{MSO}). \qquad 8.20$$

Instead, the MHD simulations show that $j_{in} \approx 0.95\, j(r_{MSO})$, which suggests that the stress edge r_{stress} is $\sim 2.7r_S$, smaller than r_{MSO} but within the range of values indicated by the location of the trans-magnetosonic surface.

In Sagittarius A*, the IR and X-ray periods do not match exactly, which may be an indication that one (or possibly both) of these measurements is incorrect; it may also mean that the IR and X-ray photons trace slightly different emission regions at the source. As before, we shall study the stress edge using the phenomenology associated with

---

[16]Note that a nonspinning black hole was assumed for all of these calculations, so this discussion pertains solely to the Schwarzschild case, where $a/r_G = 0$.

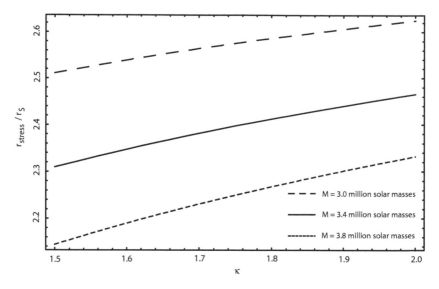

Figure 8.3 The stress edge radius r_{stress}, in units of the Schwarzschild radius r_S, as a function of κ, the exponent in the power-law formulation of the angular frequency $\Omega(r)$. The different curves span the range of possible masses in Sagittarius A*. By comparison, the marginally stable orbit has a radius $3r_S$ when the black hole spin a/r_G is zero.

the X-ray period, though a similar analysis may be made (separately or in parallel) with the IR data as well. Since we do not expect $j(r)$ to necessarily follow its Keplerian value below r_{MSO}, we will adopt the formulation

$$\Omega(r) = \Omega_0 r^{-\kappa}. \qquad 8.21$$

Clearly, $\kappa = 3/2$ for Keplerian rotation, and $\kappa = 2$ in the extreme case of angular momentum conservation. A reasonable result in our analysis would therefore correspond to $3/2 \leq \kappa \leq 2$. At the boundary r_{MSO}, we expect $\Omega = \Omega_K$, so

$$\Omega_0 = cr_G^{1/2} r_{\text{MSO}}^{\kappa-3/2}. \qquad 8.22$$

It is straightforward to calculate r_{stress} from these expressions using the last period (i.e., 17.5 minutes) emerging from the lightcurve in figure 8.2, and the result is plotted as a function of κ in figure 8.3. Note that whereas we previously identified this period with r_{MSO} it here represents instead the last orbit where the matter decouples dynamically from the disk and

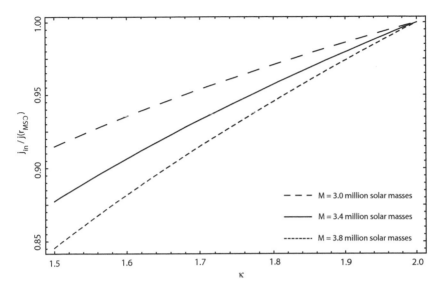

Figure 8.4 The ratio of accreted specific angular momentum j_{in} to the specific angular momentum at the marginally stable orbit, for the same range of black hole mass as that used in figure 8.3.

plunges toward the event horizon. For a mass of $3.6 \times 10^6 \, M_\odot$, r_{stress} falls within the rather narrow range $2.15 r_S$–$2.35 r_S$ for permitted values of κ.

The accreted specific angular momentum corresponding to this is shown in figure 8.4, from which we also note that $0.87 j(r_{MSO}) \leq j_{in} \leq j(r_{MSO})$ for a broad range of κ and mass. The ratio $j_{in}/j(r_{MSO}) = 0.95$ would require $\kappa \approx 1.8 \pm 0.1$, for which $r_{stress} \approx 2.4 r_S$. In principle, both of these results are consistent with the MHD simulations, indicating that the infalling plasma below the MSO remains magnetically coupled to the outer disk, though the dissipation of angular momentum is not quite strong enough in this region to force the gas into Keplerian rotation.

We are only now beginning to appreciate the physics of flaring in Sagittarius A*, so what we think we understand about this source may evolve as the array of observations and simulations expands. If we adopt the (probably overly) simplistic view that the last period corresponds to the MSO, then Sagittarius A*—with a mass of $3.6 \times 10^6 \, M_\odot$—must be spinning at a rate $a/r_G \sim 0.3 - 0.4$. With a more realistic accounting of the magnetic coupling between matter in the plunging region and that beyond the MSO, however, we infer that the onset of the instability probably occurs somewhere near r_{MSO} and that the flaring activity continues as the gas spirals inwards, ending several orbits later when

239

the matter crosses the stress edge at $\sim 2.4 r_S$ and falls toward the event horizon.

Of course, the ultimate significance of a well-understood periodicity in Sagittarius A* is that unless this compact radio source is unrelated to the evident dark matter concentration at the galactic center, we could not avoid the conclusion that the mass $M = 3.6 \times 10^6 \, M_\odot$ must be confined to a region no bigger than $\approx 1/5$ AU. Thus, the reality of a 21.4-minute modulation in Sagittarius A*'s X-ray flux bolsters the case for imaging the black hole's shadow, since it validates the theoretical framework used to predict the shape and size of the dark depression in its emissivity on scales of $30 \, \mu$arcsecs or less.[17] We shall begin developing this theory in the next section and complete our discussion by showing several simulated images in chapter 9.

8.3 GENERAL RELATIVISTIC FLUX MODULATIONS

The fact that we even see a cyclic modulation suggests that relativistic effects cannot be ignored, at least during the flare. In the absence of any shadowing of the disk (e.g., by a binary companion), Newtonian gravity does not modulate the flux even if the emitter lacks azimuthal symmetry. Regardless of whether the disk is tilted, we see it in its entirety from our distant perspective, so its phase is observationally irrelevant.

In general relativity, however, modulations from disk inhomogeneities arise due to several effects, including strong light bending and Doppler shift corrections within ~ 4–$5 r_S$ of the black hole and area amplification in the image projected onto the plane of the sky.[18] We don't actually know yet what the precise structure of the emitter is during a flare, so we cannot calculate its lightcurve with certainty, even taking these effects into account. Nonetheless, we can understand—at least qualitatively— why a periodic modulation arises with the use of a simple toy model.

---

[17]See Hollywood and Melia (1997), Falcke, Melia, and Agol (2000), and Bromley, Melia, and Liu (2001).

[18]The groundwork for interpreting and understanding a system such as this was laid out by Bardeen, Press, and Teukolsky (1972), Cunningham and Bardeen (1973), and Cunningham (1975). Subsequent adaptations of these early developments to a study of variability data for active galactic nuclei (AGN) have been carried out by several authors, most notably Abramowicz et al. (1989), Bao and Stuchlik (1992), and Chakrabarti and Wiita (1994). These workers considered the role played by relativistic corrections in determining specific AGN characteristics, such as the shape of their lightcurves.

For calculational purposes, one may regard the disk as consisting of a relatively large number of concentric rings of emitting particles at source radii r_s and source polar angle $\theta_s = \pi/2$. The enhanced emission associated with the flare is thought of as arising in a "hot" patch propagating as a fixed pattern around the center with an angular velocity Ω_+. For simplicity, let us take this patch to be a wedge; any shape can be incorporated into the calculation when more precise fitting is required.

The observed flux from this arrangement may be calculated with the prescription outlined earlier in equations (7.76)–(7.83), though with the appropriate formulation of the intensity based on the dominant emission process relevant to the waveband of interest. The validity of these procedures rests ultimately upon the fundamental equations governing photon propagation in the Kerr metric.[19] A photon trajectory originating at source coordinates $(t_s, r_s, \theta_s, \phi_s)$ and terminating at observation coordinates $(t_o, r_o, \theta_o, \phi_o)$ may be characterized by the azimuthal component of angular momentum p_ϕ, the energy (at infinity) E, and the polar component of angular momentum

$$p_\theta = (Q + a^2 E^2 \cos^2\theta - p_\phi^2 \cot^2\theta)^{1/2}. \qquad 8.23$$

Here Q is a constant of the motion that arises when the Hamilton-Jacobi equation governing these geodesics (see equation 5.106) is solved by separation of variables, and $a \equiv J/Mc$ is the usual parameter specifying the magnitude of the black hole's spin angular momentum (equation 6.136).

For calculational purposes, the equations to be solved may be simplified considerably by taking the unit of length to be the Schwarzschild radius $r_S = 2GM/c^2 = 1$ and setting $G = c = 1$ in accordance with the convention of "geometrized units" (so that the black hole mass is $M = 1/2$). Introducing the dimensionless variables $\lambda \equiv p_\phi/E$ and $q \equiv Q^{1/2}/E$, one may then write the photon propagation equations as

$$\int_{r_s}^{r_o} \frac{dr}{R(r)^{1/2}} = \int_{\theta_s}^{\theta_o} \frac{d\theta}{\Theta(\theta)^{1/2}}, \qquad 8.24$$

$$\Delta\phi = \int_{r_s}^{r_o} dr \, \frac{\tilde{\Delta}^{-1}\{r[a + \lambda(r-1)]\}}{R(r)^{1/2}} + \lambda \int_{\theta_s}^{\theta_o} d\theta \, \frac{\cos^2\theta}{\Theta(\theta)^{1/2}}, \qquad 8.25$$

[19]These were first set forth by Carter (1968).

241

$$\Delta t = \int_{r_s}^{r_o} dr \, \frac{\tilde{\Delta}^{-1}[(r^2+a^2)^2 - a^2\tilde{\Delta} - a\lambda r]}{R(r)^{1/2}} + a^2 \int_{\theta_s}^{\theta_o} d\theta \, \frac{\cos^2\theta}{\Theta(\theta)^{1/2}}, \qquad 8.26$$

where

$$R(r) = r^4 + (a^2 - \lambda^2 - q^2)r^2 + [(a-\lambda)^2 + q^2]r - a^2q^2, \qquad 8.27$$

$$\Theta(\theta) = q^2 + a^2\cos^2\theta - \lambda^2\cot^2\theta, \qquad 8.28$$

and

$$\tilde{\Delta} = r^2 + a^2 - r. \qquad 8.29$$

For each ring of emitting particles in the disk and each polar observation angle θ_o, there is a superset $\tilde{S}_o$ consisting of an infinite number of sets S_o^n of (λ, q), each of which satisfies equation (8.24), with the index $n\,(=0, 1, 2, \dots)$ indicating the number of times the photon trajectories in the set transit the equatorial plane as they propagate from source to observer. Each set S_o^n corresponds in turn to a set of image coordinates ("impact parameters") (α, β), where

$$\alpha = -\frac{\lambda}{\sin\theta_o}, \qquad 8.30$$

$$\beta = \pm(q^2 - a^2\cos\theta_o^2 - \lambda^2\cot\theta_o^2)^{1/2}, \qquad 8.31$$

and the sign of β is given by $(\partial\theta/\partial r)_o/|(\partial\theta/\partial r)_o|$.

When plotted on the plane of the celestial sphere at infinity (with the origin of coordinates at the center of the black hole and the positive z-axis corresponding to the hole's spin angular momentum vector), the set (α, β) determines the nth-orbit image of a ring of emitting particles as observed at (r_o, θ_o). Having said this, only the direct-orbit $(n=0)$ image contributes significantly to the specific power flux observed at infinity, partly because the higher-orbit rays are occluded by the disk but also because gravitational focusing "squeezes" the higher-orbit images so that their apparent areas at infinity are small compared to that of the *direct-orbit* image (we shall see a very good example of this in color plate 9). For simplicity, therefore, one usually considers only the direct-orbit image for the calculation of the flux.

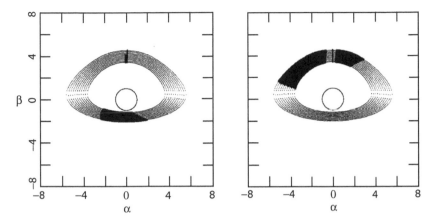

Figure 8.5 Projected image in the plane of the sky of a disk surrounding a Schwarzschild black hole with an inclination angle cosine $\mu = 0.4$. The panel on the left corresponds to the zero of the observational time, showing the "normal" image of a $50°$ hot wedge—the large dark patch at the bottom, or near side, of the image—which is somewhat distorted by light-travel time effects. The two auxiliary wedge images—the small dark patches at the top—are formed by rays that circulate completely about the z-axis of the black hole, oriented in the $+\beta$-direction. The panel on the right gives the appearance of the disk and the hot wedge at observational phase 0.7, corresponding approximately to a maximum in the observed frequency-integrated flux. In this case, the distortion and splitting of the wedge image arise primarily from general relativistic time delays and light bending. (Image from Hollywood et al. 1995)

In the case of a homogeneous, axisymmetric steady-state disk, one needs no additional information to calculate the time-independent specific power flux observed at infinity. On the other hand, if the emitted power varies with time, then one must go on to insert each member of S_o^o into equations (8.25) and (8.26) to obtain the corresponding *azimuthal angular displacements* $\Delta\phi \equiv \phi_o - \phi_s$ and *photon travel times* $\Delta t \equiv t_o - t_s$. Given these quantities and knowledge of the time evolution of the disk in its local reference frame, one may then easily determine the corresponding time evolution of the apparent image of the disk as observed at infinity. By integrating the source intensity function (along with its frequency-shift weighting factors) over a series of "snapshots" of the apparent disk image, one thereby obtains the time-dependent flux.

The appearance of the disk as observed in asymptotically flat space is shown for two different phases in figure 8.5, for an assumed inclination angle cosine $\mu = 0.4$. The zero of source time is chosen so that at that

243

instant, according to an observer fixed in the reference frame of the black hole, the hot wedge is centered on $\phi_s = 0$. The ray emitted with zero angular momentum at $(t_s = 0, r_s = r_{max}, \theta_s = \pi/2, \phi_s = 0)$, where r_{max} is the radius of the outermost disk ring, is then taken as a reference ray. The arrival of this ray at $(t_o, r_o, \theta_o, \phi_o)$ defines the zero of the observer's time. Then the "observational phase" is defined to be the time elapsed, according to the observer, since $t_o = 0$, divided by the period $P_+ \equiv 2\pi/\Omega_+$, modulo unity. The asymmetry in the appearance of the wedge at zero observational phase (the left panel in this figure) arises from light travel time delays of rays from other regions of the wedge relative to the reference ray emitted at $\phi_s = 0$. For comparison, the right panel exhibits the appearance of the same disk at an observational phase of 0.7, corresponding approximately to the observed maximum in the frequency-integrated flux.

It is not too difficult to see why the integrated flux from the configuration shown here should vary with phase. Doppler shifts enhance the intensity when the wedge is moving toward the observer and decrease it on the diametrically opposite side of the disk. In addition, light bending broadens the image of the wedge when it orbits on the far side of the black hole and increases the number of rays reaching the observer relative to the front end. Together, these effects modulate the flux with a period equal to P_+.

The images shown here, though crude by comparison with those routinely produced today, have brought us to the realm of strong-field physics. In the next chapter, we shall follow and develop this lead, and seriously consider the possibility, within this decade, of producing an actual image of Sagittarius A* using global mm arrays.

Chapter 9

Strong Field Physics

We have grown accustomed to the idea of black holes as objects of great influence on their surroundings but not themselves easily detectable. The SO (or S) stars, for example, appear to be sampling Sagittarius A*'s overwhelming gravitational pull within ~ 0.1 pc of the galactic center (see figure 5.9), yet one can hardly forget that this distance from the black hole is still $\sim 3 \times 10^5$ Schwarzschild radii. And when it comes to a black hole's radiative characteristics, its spectrum and luminosity are produced by proxy, via the compression and emissivity of matter accreting towards it or expelled into a surrounding nonthermal halo or jet.[1]

However, this does not mean that we should abandon all hope of ever seeing the signature of strong gravity in the realm ($\sim 3r_S$) where general relativity imprints its indelible mark on space and time. Indeed, the last chapter has taught us that even Sagittarius A*'s disruptive flares apparently contain coded timing information pointing to the existence of a marginally stable orbit in the accreting plasma, an effect not seen in Newtonian gravity.

But this is only the beginning. The mm/sub-mm radiation originates from only a handful of Schwarzschild radii above the event horizon, so it must be subject to significant light bending and area amplification that, given suitable conditions, can lead to a shadow observable at Earth's distance from the galactic center. A *spinning* black hole would produce even stronger effects, including distortions to the shadow, that one might use to measure the spin itself. And with spin, gravity would acquire a dependence on polar angle that can sometimes induce a precession in the disk. We shall consider several of these strong-field consequences, beginning with the impact of frame dragging on the infalling plasma.

---

[1]Hawking radiation from such a massive black hole is so much fainter than the intensity of hot magnetized matter surrounding it that we may safely ignore its contribution to the overall photon flux for the purpose of this discussion.

9.1 SPIN-INDUCED DISK PRECESSION

The oscillation frequency of a perturbation in the perpendicular direction to a circular orbit in the equatorial plane of a Kerr black hole is easily derived from the geodesic equations resulting from the metric in equation (6.145) (see also equation 6.107). Following standard procedure, one writes

$$\frac{dX^\alpha}{dT} = (c, v^r, v^\theta, r\Omega + v^\phi), \qquad\qquad 9.1$$

where v^r, v^θ, and v^ϕ are the velocity components associated with the infinitesimal perturbations and Ω is the coordinate angular velocity of a circular (prograde or retrograde) orbit defined in equation (6.155). From the geodesic equations, one must then derive linearized relations for v^r, v^θ, and v^ϕ, and with the assumption of a cyclic variation

$$v^\theta = v_0 e^{i\omega_\theta t}, \qquad\qquad 9.2$$

there results a dispersionlike equation for the oscillation frequency ω_θ associated with motion in the vertical direction:[2]

$$\omega_\theta^2 = \Omega^2 \left(1 + \frac{3a^2}{r^2} \mp \frac{4ar_G^{1/2}}{r^{3/2}} \right). \qquad\qquad 9.3$$

Thus, for small values of a/r_G, the *nodal precession* frequency $\omega_{p\theta} \equiv \Omega - \omega_\theta$ may be written as

$$\omega_{p\theta} \approx \frac{2acr_G^2}{r^3}. \qquad\qquad 9.4$$

Notice that $\omega_\theta = \Omega$ when $a = 0$, meaning that there is no precession in that case. Notice also that $\omega_{p\theta}$ decreases with radius, which produces differential precession when $a \neq 0$.

Because this frequency depends on r, the final configuration of a disk orbiting a spinning black hole with its angular momentum vector misaligned relative to the spin axis depends on whether the viscous timescale t_v (see equation 8.2) is shorter or longer than the precession period $2\pi/\omega_{p\theta}$. Assuming the disk has negligible inertia, its internal

---

[2] See Okazaki, Kato, and Fukue (1987).

viscosity acts to dissipate the relative precessional motion between adjoining rings, thereby aligning the midplane of the inner region with the equatorial plane of the black hole. On the other hand, the outer portions of the disk remain in their original plane because the Lense-Thirring precession (as this effect is also known) rate drops off sharply as r increases, so the internal pressure and viscous stresses acting inside the disk can limit the effects of differential precession.

This process, in which the Lense-Thirring precession in thin, cold disks flattens the inner region toward the equatorial plane, producing a warped accretion pattern, is known as the Bardeen-Petterson effect.[3]

But Sagittarius A*'s disk is hardly cold. Indeed, plasma falling toward the event horizon in this object heats to near virial temperatures, exceeding 10^{10} K (see chapter 7). Under such circumstances, it is not obvious a priori that the disk should thus succumb to this process. As we shall see, this is more than an academic exercise, for depending on the outcome, a spinning Sagittarius A* may produce an observational signature spawned by the frame-dragging effect predicted in the Kerr metric.[4]

At least three groups of researchers have carried out detailed numerical simulations to examine the conditions under which the Bardeen-Petterson effect may be absent.[5] Since the gravitational influence of interest simulated in these studies occurs several Schwarzschild radii from the black hole, the relativistic effects are included in a post-Newtonian approximation, wherein the momentum equation may be written in the form

$$\frac{d\mathbf{v}}{dt} = -\frac{1}{\rho}\vec{\nabla}P + \mathbf{v} \times \mathbf{h} - \vec{\nabla}\Phi + \mathbf{F}_{\text{visc}}. \qquad 9.5$$

This expression follows directly from equation (7.67), with the inclusion of a new term $\mathbf{v} \times \mathbf{h}$ representing the gravitomagnetic force per unit mass near a rotating black hole.[6] All the other variables have the same physical meaning as previously defined, and $\mathbf{F}_{\text{visc}}$ now represents the viscous force

---

[3] See Bardeen and Petterson (1975).

[4] These ideas were investigated in the context of Sagittarius A* by Liu and Melia (2002b).

[5] All three sets of simulations are based on the smooth-particle hydrodynamics scheme. The first group, Nelson and Papaloizou (2000), explored the general properties of the Bardeen-Petterson effect and were the first to demonstrate its absence under certain conditions. Fragile and Anninos (2005) extended this work to tilted thick disk accretion onto black holes, and Rockefeller, Fryer, and Melia (2005) specifically modeled the disk in Sagittarius A*.

[6] See, e.g., Blandford (1996).

per unit mass (see equation 7.74). The gravitomagnetic force, so named because of its similarity to the $\mathbf{v} \times \mathbf{B}$ terms in the Lorentz force equation of electrodynamics, is a good representation of the influence felt by a particle embedded in the swirling spacetime of a Kerr metric.

In this equation, $\mathbf{h}$ is defined by the expression

$$\mathbf{h} = \frac{2\mathbf{S}}{r^3} - \frac{6(\mathbf{S} \cdot \mathbf{r})\,\mathbf{r}}{r^5}, \qquad 9.6$$

where

$$\mathbf{S} = \frac{G\mathbf{J}}{c^2} \qquad 9.7$$

and

$$\mathbf{J} = ac M \hat{\mathbf{k}} \qquad 9.8$$

is the angular momentum of the black hole (see equation 6.137), with $\hat{\mathbf{k}}$ a unit vector pointing in the z-direction. In addition, the gravitational acceleration due to the black hole is calculated using

$$-\vec{\nabla}\Phi = -\frac{GM}{r^3}\left(1 + \frac{3r_S}{r}\right)\mathbf{r}, \qquad 9.9$$

which produces the correct apsidal precession frequency at large distances from the center.

Counterbalancing the Lense-Thirring effect induced by the gravitomagnetic force are the pressure gradients in the disk and the viscous coupling of neighboring rings within it. To correctly incorporate the latter effect, these simulations make use of the anomalous viscosity arising from the magnetorotational instability described midway through chapter 7 (resulting in equation 7.66).

Whether the forces internal to the disk can effectively counter the Lense-Thirring effect is usually gauged via the Mach number $\mathcal{M} \equiv v/c_s$, which characterizes the heat content of the plasma in terms of the sound speed $c_s \equiv (\partial P/\partial \rho)^{1/2}$. Given that v is dominated by its azimuthal component v_ϕ, we may use equation (7.62) to recast this very useful quantity into the more geometrical phrasing $\mathcal{M} \sim (H/r)^{-1}$. Hotter disks have greater vertical support against collapse to the equatorial plane, so they tend to be thicker and their Mach number is small. The opposite is true of cold disks, which are usually geometrically thin.

Figure 9.1 The three-dimensional arrangement of 326,034 SPH particles for a calculation in which the internal disk coupling is not strong enough to maintain an undistorted configuration. The inner portion of the inclined disk is warped into the equatorial plane of the black hole (aligned horizontally in the image) through the action of the Bardeen-Petterson effect. (Image from Rockefeller, Fryer, and Melia 2005)

Figure 9.1 shows a typical disk configuration produced by the numerical simulations described above in cases where the Mach number is much larger than 1 and the Bardeen-Petterson effect is fully manifested.[7] Specifically, this is a disk with initial outer radius $25\,r_G$, inclination angle $i = 10°$, and Mach number $\mathcal{M} = 12$, orbiting a maximally spinning black hole with $a/r_G = 1$.

In this image, the black hole spin axis is vertical, and we see that the outer portion of the disk has maintained its original 10° inclination relative to this direction. However, due to the Bardeen-Petterson effect, its inner region is warped and now lies in the equatorial plane of the black hole. By this time in the simulation, which corresponds to four Keplerian orbits at $r = 25r_G$, the transition region between the two disk segments occurs at roughly $14\,r_G$.

In contrast, a disk constructed according to the best estimates we have for the conditions at the galactic center does not warp; instead, the entire structure remains tilted out of the equatorial plane and precesses around the spin axis of the black hole. Figure 9.2 shows the three-dimensional arrangement of particles in one particular simulation with outer disk radius $110\,r_G$, inclination angle $i = 30°$, and midplane Mach number $\mathcal{M} \approx 3$, after 85 Keplerian orbits at $60\,r_G$. Evidently, the transition from a warped to nonwarped disk occurs somewhere near $\mathcal{M} \sim 5$. Thus, the

---

[7]This calculation is called E1 in Nelson and Papaloizou (2000). The figure shown here, however, is based on the more elaborate simulation, also labeled E1, of Rockefeller, Fryer, and Melia (2005).

Figure 9.2 The three-dimensional arrangement of 423,626 SPH particles in a simulated disk constructed using the gas temperature and density, constrained by the observations, in the environment near Sagittarius A* (see chapter 7). The disk is relatively thick, with pressure gradients strong enough to maintain coupling between neighboring rings. Here, the Bardeen-Petterson effect is suppressed, and even the innermost portion of the disk remains aligned with the outer disk. There is no warping into the equatorial plane of the black hole. (Image from Rockefeller, Fryer, and Melia 2005)

disk in Sagittarius A* appears to be sufficiently hot and viscous so that the entire structure precesses coherently about the spin axis.

An important observational signature of this effect is the dependence of the precession period on the disk's outer radius for a given value of the black hole spin parameter. This connection arises because the size of the disk provides a measure of the moment of inertia of the structure, whose response to the applied gravitomagnetic torque determines the rate of precession.

Based on these simulations, one might expect to detect a modulation in Sagittarius A*'s apparent emissivity with a period of ~ 50–500 days, arising in a disk of size ~ 20–$30 r_S$. Such a periodicity might emerge in portions of Sagittarius A*'s spectrum produced by an occulted emitter; for example, since its radio emission is produced on scales of ~ 20–$100 r_S$ (see chapter 7), this disk precession may lead to a variable aspect that periodically attenuates the overall radio flux from this region.

It is tantalizing to think of this theoretical interpretation as the basis for the 106-day period discussed in §4.2, if future radio observations confirm its viability. A disk precessing in a nonisotropic gravitational field can probe the spacetime within the inner $\sim 10 r_S$ of a Kerr metric and, more important, may provide the means of directly measuring the black hole's spin.

On the other hand, a nonconfirmation of such a modulation on a ~ 50–500-day timescale would tell us that either no disk is present in Sagittarius A* (which would conflict with the other seemingly compelling evidence we considered in chapters 7 and 8) or that its properties—specifically, the size and orientation—are different from what we now imagine. A failure to detect a long-period modulation in Sagittarius A* could also signal that the geometry of the radio emission region has not yet been properly identified. This is clearly still an open topic, and much work remains to be done in what may turn out to be a very fruitful investigation.

9.2 MICROLENSING

Moving closer to the black hole itself, the novelty that confronts us next in the context of general relativity is the bending of light. Strictly speaking, general relativity does not *predict* that light paths should be bent under the influence of a gravitating body, but the principle of equivalence, upon which the theory is built, holds that special relativity is perfectly valid inside any freely falling frame of reference. Thus, all trajectories, those of massive particles as well as massless, must appear curved to an accelerated observer. The fact that light rays are bent when passing through a gravitational field is therefore an integral component of general relativity.

The geodesic paths produced by this effect—trajectories forged solely under the influence of gravity—are described by equation (6.107), as long as the proper time τ can adequately track the particle's progress along its motion. By definition, however, the interval ds is zero for light (giving rise to the term *null geodesic*) because every observer measures time intervals and distances that preserve the constancy of its speed c. So the proper time for light never advances, $d\tau$ is always zero, and we cannot employ equation (6.107) to evaluate the trajectory of light under any circumstances.

This is really not such a big deal when we realize that τ is but one variable that describes progress along a trajectory. For light, it happens to be an inappropriate one because it never changes. Instead, let us define a more generalized variable λ that does change along the path, thinking of it simply as a quantity that gives a one-to-one correspondence between the spatial coordinates that evolve along with it. Thus, if r changes by a certain amount Δr after λ advances by a known increment $\Delta\lambda$, the corresponding change $\Delta\theta$ will also be correctly specified by the same increment $\Delta\lambda$. What this amounts to is finding differential equations for the coordinates, equivalent to equation (6.107), in terms of λ instead of τ.

In the Cartesian inertial frame (CIF), light rays are straight and

$$\frac{d^2 x^\alpha}{d\lambda^2} = 0. \tag{9.10}$$

Following the derivation of equation (6.98), we then find that the corresponding equations of motion in the accelerated frame must be

$$\frac{d^2 X^\alpha}{d\lambda^2} + \Gamma^\alpha{}_{\nu\mu} \frac{dX^\nu}{d\lambda} \frac{dX^\mu}{d\lambda} = 0. \tag{9.11}$$

For massive particles, $d\tau$ is proportional to $d\lambda$, and λ is then often normalized such that $\lambda = \tau$, in which case we recover the geodesic equations in (6.107). But for light, the normalization of λ must be handled independently of τ.

In a Schwarzschild metric, the ϕ-equation (6.119) becomes

$$r^2 \frac{d\phi}{d\lambda} = \text{constant} \equiv l \tag{9.12}$$

(representing the conserved, though here not specific, angular momentum), and the T-equation (6.120) similarly transforms to

$$\left(1 - \frac{2GM}{c^2 r}\right) \frac{dT}{d\lambda} = E/c^2. \tag{9.13}$$

The variable λ has been normalized in such a way as to make the constant appearing on the right-hand side of (9.13) equal to the conserved energy at infinity, divided by c^2, though here it clearly excludes any rest mass energy. The third equation, for the coordinate r, becomes

$$c^2 \left(\frac{dr}{d\lambda}\right)^2 = E^2 - V_{\text{eff}}^0(r), \qquad 9.14$$

where now

$$V_{\text{eff}}^0(r) = \left(1 - \frac{2GM}{c^2 r}\right) \frac{l^2}{r^2} c^2 \qquad 9.15$$

is the effective (photon) potential.

Thus, eliminating λ from equations (9.12) and (9.14) gives the equation for a photon orbit

$$\frac{d\phi}{dr} = \pm \frac{1}{r^2 \left[(1/b^2) - (1/r^2)\left(1 - 2GM/c^2 r\right)\right]^{1/2}}, \qquad 9.16$$

where we have introduced the *impact* parameter

$$b \equiv \frac{cl}{E}. \qquad 9.17$$

We may simplify this equation somewhat with an appropriate change of variable $u = 1/r$, keeping the root corresponding to $l > 0$:

$$\frac{d\phi}{du} = \left(\frac{1}{b^2} - u^2 + \frac{2GM}{c^2} u^3\right)^{-1/2}. \qquad 9.18$$

Notice that if we neglect the u^3 term—the general relativistic correction—this equation has the simple solution

$$r \sin(\phi - \phi_0) = b, \qquad 9.19$$

where ϕ_0 is a constant of integration. This is, of course, the Newtonian result, in which the photon trajectory is a straight line. Evidently, all of the curvature introduced by the equivalence principle is contained within the factor $(2GM/c^2)u^3$.

For a complete solution, one can always integrate equation (9.18) numerically and find the exact light trajectory for any impact parameter b. But for pedagogical purposes, we will seek an analytic solution in an intermediate regime, where the relativistic correction is small though not entirely negligible, that is, where r is greater than several Schwarzschild radii $(2GM/c^2)$.

Let us introduce yet another change of variable

$$w \equiv u\left(1 - \frac{GM}{c^2}u\right),$$ 9.20

which may also be inverted

$$u = w\left(1 + \frac{Gm}{c^2}w\right) + O\left(\left[\frac{GM}{c^2}\right]^2 u^2\right)$$ 9.21

to demonstrate that terms of order $(r_G/r)^2$ may be ignored. Equation (9.18) then becomes

$$\frac{d\phi}{dw} = \left(1 + \frac{2GM}{c^2}w\right)\left(\frac{1}{b^2} - w^2\right)^{-1/2} + O\left(\left[\frac{GM}{c^2}\right]^2 u^2\right),$$ 9.22

which may be integrated to give

$$\phi = \phi_0 + \frac{2GM}{c^2b} + \arcsin(bw) - \frac{2GM}{c^2}\left(\frac{1}{b^2} - w^2\right)^{1/2}.$$ 9.23

The initial trajectory has $w \to 0$, so the incoming direction is specified by the condition $\phi \to \phi_0$. The photon reaches its smallest radius of approach when $dr/d\lambda = 0$, which, according to equation (9.14), occurs when

$$\left(\frac{E}{l}\right)^2 = \left(1 - \frac{2GM}{c^2r}\right)\frac{1}{r^2}$$ 9.24

or

$$\frac{1}{b} = u\left(1 - \frac{2GM}{c^2}u\right)^{1/2}.$$ 9.25

With the approximation $r_G/r \ll 1$, the photon's radius of closest approach therefore corresponds to $w = 1/b$. At that point,

$$\phi = \phi_0 + \frac{2GM}{c^2b} + \frac{\pi}{2},$$ 9.26

the photon having passed through an angle of $\pi/2 + 2GM/c^2 b$. By symmetry, it passes through an additional angle of the same size as it moves outwards from this radius, so by the time the influence of gravity has subsided, the photon will have passed through a total angle of $\pi + 4GM/c^2 b$.

Of course, if the photon were traveling on a straight line, this angle would be exactly π. The net deflection it incurs as a result of the curvature imposed by gravity is evidently

$$\Delta\phi = \frac{4GM}{c^2 b}.$$

9.27

This angle is small for objects in the solar system, though quite measurable in the case of the Sun. Given $M = 1 \, M_\odot$ and $b = 6.96 \times 10^5$ km, the maximum deflection one would expect for light grazing past the solar limb is $\approx 1''.75$. In 1919, eclipse expeditions were sent to islands off the northeast coast of Brazil and the west coast of Africa to measure this angle for about a dozen stars. The experiment yielded values between about $1''.3$ and $2''.1$, in substantial agreement with Einstein's prediction,[8] a historically important result that hardly went unnoticed by the general public in the 1920s. As we shall see, Sagittarius A*'s much larger mass deflects light rays by a considerably bigger angle, though its much greater distance from us than that of the Sun largely cancels the benefit thereby gained.

The phenomena associated with the deflection of light by massive bodies are now referred to as *gravitational lensing*. Eddington himself noted in the 1920s that under certain conditions there may be multiple light paths connecting an observer to the source, so gravitational lensing can give rise to several images of the same object.

However, not all such images are measurable using the same technique. One often considers gravitational lensing grouped into three separate categories: (1) *Strong* lensing refers to situations in which the lens mass is so large that the multiple images are separated by arcseconds and are easily resolvable. Examples in this category include galaxy clusters. (2) *Weak* lensing only distorts the light from background sources, so the lensed image appears stretched relative to the initial

---

[8] It was Sir Arthur Eddington, a student and strong supporter of relativity theory through the early decades of the twentieth century, who led the expedition to Príncipe, in the Gulf of Guinea (Dyson, Eddington, and Davidson 1920). The second expedition, to Sobral in Brazil, was directed by Andrew Crommelin of the Royal Observatory at Greenwich.

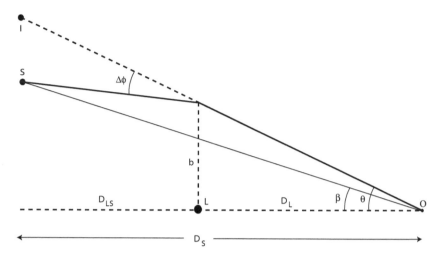

Figure 9.3 Gravitational lensing of a ray originating at the source position S by the black hole creating the lens at L. The observer at O sees the image at position I making an angle θ with respect to the vector pointing from O to L. The general relativistic deflection angle is $\Delta\phi$, and b is the impact parameter.

shape. (3) *Microlensing* produces multiple images separated in the plane of the sky by only microarcseconds. Since current technology cannot yet separate these out, only the amplification due to the lens can be measured.

Let us now see how the images are constructed using the geometric components shown in figure 9.3. In the small-angle approximation,

$$\theta D_S = \beta D_S + \Delta\phi D_{LS}, \qquad\qquad 9.28$$

where D_S and D_{LS} are, respectively, the projected distances between the source and observer and between the source and black hole (which is producing the lens at L). A simple rearrangement gives

$$\beta = \theta - \Delta\phi \frac{D_{LS}}{D_S}, \qquad\qquad 9.29$$

often called the *lens equation*. Thus, noting that $b \approx \theta D_L$ and introducing the definition of $\Delta\phi$ in equation (9.27), we have finally

$$\theta = \frac{\beta}{2} \pm \frac{1}{2}\sqrt{\beta^2 + 4\theta_E^2}, \qquad\qquad 9.30$$

where

$$\theta_E \equiv \sqrt{\frac{4GM}{c^2} \frac{D_{LS}}{D_S D_L}}$$

9.31

is the angular opening of the so-called Einstein ring—the image produced when the source, lens, and observer are all colinear. Under these circumstances (i.e., $\beta = 0$), $|\theta| = \theta_E$ in all directions. But in the more general case (with $\beta \neq 0$), gravitational lensing produces at least two images, one inside the Einstein ring and the other outside it.

Over the past decade, both categories (1) and (3) have been considered for possible lensing experiments using Sagittarius A*. For this object, the Einstein ring radius is

$$\theta_E = \left\{ \frac{4GMr}{c^2(r + R_0)R_0} \right\}^{1/2}$$

$$\approx 21 \left(\frac{M}{3.6 \times 10^6 \, M_\odot} \right)^{1/2} \left(\frac{r}{1 \, \text{pc}} \right)^{1/2} \text{mas},$$

9.32

where $R_0 = 8$ kpc is the distance to the galactic center and r is the distance beyond Sagittarius A* to the object being lensed.

In category (1), the goal is to actually discern both images of a lensed star among the many objects clustered around the black hole.[9] Of course, both the probability of capturing a lensing event and the background confusion depend on the stellar density. At the location of Sagittarius A*, the central cluster distribution is given by equation (5.11), with a core radius $r_c \sim 0.4$ pc and a central density $\rho_0 = 4 \times 10^6 \, M_\odot \, \text{pc}^{-3}$. So the Einstein ring radius is ~ 10 mas for a star lying near r_c behind the central mass, increasing to ~ 200 mas for a star near the far edge of the cluster at ~ 100 pc.

Notice that at the distance to the galactic center the latter angle corresponds to a projected radius $r_E \approx 0.008$ pc, so lensed images will always appear within the cluster's core. Simply glancing at the field would not be sufficient to distinguish lensed images from unlensed stars, but due to the stars' proper motion, the two images (one on either side of Sagittarius A*) would be transient, lasting a few years or less, and would rise and fall in luminosity together.

[9]This was first contemplated by Wardle and Yusef-Zadeh (1992).

This experiment, however, will not be feasible until observational capabilities will have improved considerably beyond current levels. One can show very easily that in order to have a probability close to one of seeing at least one event at any given time, the central surface density of stars should be greater than $1,000$ arcsec^{-2}. This in turn would require an angular resolution of $< 0''.01$ to individually separate the lensed images from the crowded background field of faint stars. Presently, the highest attained spatial resolutions are the diffraction-limited values of $\sim 0''.05$–$0''.15$ used for the proper motion surveys discussed in chapter 5. Thus, although this "strong" lensing experiment will eventually be carried out, it is more likely that a category (3) observation will produce the first results.

In microlensing amplification of faint stars, the amplified—though unresolved—images of these sources rise above the detection threshold and then fade again as they transit behind the black hole. In other words, such events would appear as time-varying sources close to Sagittarius A*. It is therefore intriguing that one or two such variable IR sources have already been detected.[10] These objects brightened to $K \sim 15^m$ before fading away again and appear to have varied on a timescale of about one year.

However, although this behavior is consistent with that of microlensing events, a proper interpretation of these occurrences must include a quantitative assessment of the lensing probability. Given the density of stars in the central cluster and the sampling rate employed thus far, one can easily show that there is only a $\sim 1\%$ probability that a single microlensed event could have been detected thus far.[11] But the advantage of category (3) observations over those in category (1) is that these detection probabilities will increase significantly with improved observational sensitivity and a more frequent monitoring of Sagittarius A*.

This brings us to the end of our discussion concerning microlensing in Sagittarius A*, but we are not yet done with light-bending issues. In the next section, we shall investigate what is arguably the most exciting future development in black hole research—the actual imaging of a black hole's shadow.

---

[10] See Genzel et al. (1997) and Ghez et al. (1998).
[11] See Alexander and Sternberg (1999).

9.3 IMAGING THE SHADOW OF THE BLACK HOLE

The compact radio emission at ~ 1 mm in Sagittarius A* lies in a portion of the spectrum where the intrinsic source size becomes apparent over the scatter-broadening effects of the intervening screen (see figure 2.4 and the discussion following equation 7.138). This radiation, we have concluded, is emitted within a region roughly the size of Earth's orbit about the Sun, corresponding to $\sim 15 r_S$ for a black hole with a mass of $3.6 \times 10^6 \, M_\odot$.

The synchrotron emission at mm/sub-mm wavelengths, moreover, appears to be optically thin, as gauged by the turnover in the spectrum across the "bump" (see figure 7.9). It appears that nature has provided us with a fortuitous set of circumstances, in which the emitting gas falling onto Sagittarius A* becomes transparent at about the same wavelength where blurring effects due to interstellar scattering begin to ebb. So can we reasonably expect to see the black hole directly through the murky medium?

The idea of a black hole creating a dark depression in an otherwise brightly lit background of emission was raised long ago by Bardeen,[12] who described the idealized appearance of such an object in front of a planar-emitting source, showing that it literally would appear as a "black hole." Back then, such a calculation was mere theoretical musing, limited to what-if conjecture. But Sagittarius A* has made the prospect of actually seeing this dark depression—or shadow—an imminent reality.

Much of the mathematical formalism one needs to simulate the image of Sagittarius A* at mm/sub-mm wavelengths has already been introduced in the calculation of its linear polarization (figure 7.9) and the demonstration of flux modulation during a typical IR or X-ray flare (see §8.3). The overall specific intensity $I_\nu^{E,O}$ of an ordinary or extraordinary wave observed at infinity is an integration of the corresponding emissivity $\epsilon^{E,O}$ (see equations 7.86 and 7.87) over the path length along photon geodesics (see equation 9.16 for a Schwarzschild metric). To obtain $I_\nu^{E,O}$ from the accretion disk, one makes the simplifying assumption that the emitter is geometrically thin, with a uniform slab structure in the vertical direction. Other than this approximation, it is straightforward to incorporate all of the relativistic effects into the calculation, including frame dragging, gravitational redshift, light bending, and Doppler boosting.[13]

---

[12] See Bardeen (1974).

[13] A useful second-order geodesic solver is described in Bromley, Chen, and Miller (1997).

The output of this ray-tracing calculation is a pixelized image, with specific intensities at the detector calculated from the relativistic invariant I_ν/ν^3 (see also figure 8.5). Emitter-frame frequencies come from the projection of the emitted photon four-momentum onto the four-velocity of the emitter.[14] Similarly, local emission angles, such as θ' in equations (7.86) and (7.87), follow from projection of the photon four-momentum onto the spacelike components of a tetrad tied to the emitter frame. One assumes Keplerian flow except at small radii, where circular orbits are unstable. In this case the emitters are on firewall trajectories as if perturbed from the minimum stable orbit (at $3\,r_S$ for a black hole with zero angular momentum). Fluxes are then calculated by summing over the pixels in the image, taking into account the physical size of the detector array and the distance to the disk.

The ray-tracing method used to produce the images shown in color plate 9 was tuned for the synchrotron problem in two important ways.[15] First, this code was designed to integrate radiative transfer equations along photon trajectories. This is useful, for example, when absorption of light from high-order images occurs after multiple disk crossings or when one needs to include the redshift of photons as they travel from one region of the emitting gas to another.

Second, this algorithm calculates the polarization fraction in a strongly curved spacetime; this was the essence of the calculation that produced figure 7.9. The specific intensity of each polarization can be evaluated separately in the emitter and detector frames from the relativistic invariant I_ν/ν^3, and the degree of polarization is itself an invariant. The position angle of polarized light may also be calculated from a relativistic invariant related to the parallel transport of a polarization vector along a null ray.[16] The parallel transport operation may be performed quite directly by defining a reference vector at the detector and numerically propagating it along with the null ray itself. In this fashion, a position angle may be mapped consistently from one frame to another, an essential feature for calculating radiative transfer for the ordinary and extraordinary waves that make multiple passes through the disk.

Insofar as the black hole's image is concerned, a closed curve on the sky plane divides a region where geodesics intersect the horizon

---

[14]See Cunningham (1975).

[15]For a complete description of these simulations, refer to Falcke, Melia, and Agol (2000) and Bromley, Melia, and Liu (2001).

[16]See Connors and Stark (1977).

from a complementary region whose geodesics miss it. This "apparent boundary" of the black hole—the perimeter of its shadow—is a circle of radius $\sqrt{27}r_G$ in the Schwarzschild case $(a = 0)$ but has a more flattened shape of similar size for a Kerr black hole, slightly dependent on inclination. Note that the apparent boundary is much larger than the event horizon itself due to the strong bending of light by the black hole.

Some photons may originate in front of the accretor, though still within the apparent boundary, but these experience strong gravitational redshift and a shorter total path length, leading to a smaller integrated emissivity. Photons originating just outside the apparent boundary can orbit the black hole near the circular photon radius several times, adding to the observed intensity. All together, these effects conspire to produce a marked deficit of observed intensity inside the apparent boundary relative to that outside.

However, due to the lingering effects of scatter broadening—even at mm wavelengths—this is not quite what we see on Earth. According to figure 2.4, these effects drop sharply $(\sim \lambda^2)$ toward higher frequencies, and the intrinsic source begins to emerge from the scattered confusion once we cross into the mm portion of the spectrum. But scatter broadening never really dies away completely.

To properly simulate an *observed* image, we must take two additional effects into account: interstellar scattering and the finite telescope resolution achievable from the ground. The first of these may be incorporated by smoothing the image with an elliptical Gaussian with a full width at half maximum (FWHM) of 24.2 μarcsec $\times (\lambda/1.3\,\text{mm})^2$ along the major axis and 12.8 μarcsec $\times (\lambda/1.3\,\text{mm})^2$ along the minor axis.[17] Of course, we do not yet know the spin axis of the black hole, so the position angle of this ellipse is arbitrary. For specificity, one may simply take the position angle to be 90° for the major axis.

The second of these effects, the telescope resolution, may then be added in an idealized form by convolving the smoothed image with a spherical Gaussian point-spread function of FWHM 33.5 μarcsec $\times (\lambda/1.3\,\text{mm})^{-1}(l/8{,}000\,\text{km})^{-1}$, which is the possible resolution of a global interferometer with 8,000 km baselines.[18] When the actual array is available, this calculation can be improved by using the exact point-spread function of the system, which will depend on the number and placement of the participating telescopes.

---

[17] See Lo et al. (1998).
[18] See Krichbaum (1996).

The illustrative panels shown in color plate 9 demonstrate the feasibility of imaging Sagittarius A* polarimetrically at mm/sub-mm wavelengths. The size of the shadow, roughly $5r_S$ in diameter, represents a projected size of $\sim 30\,\mu$arcsec, which is already within a factor of 2–3 times the current VLBI capability. Moreover, forming images that show individual components of polarized light has the added value of providing more specific information on the emitting region.

For example, the bright crescent shape itself and its highly distinctive distribution of polarized light may be understood as a manifestation of the relativistic beaming from material on the incoming side of the disk. For optically thin synchrotron emission, the specific intensity in the emitter frame is highly anisotropic, and the beaming effect causes an alignment between the observed photon trajectory and the direction of maximum emission only on the incoming side. This effect is most apparent in the images that show individual components of polarized light. In the absence of relativity, the extraordinary emission will dominate the flux along the disk spin axis where the photon emission and polarization position angles are at $90°$ to the azimuthal magnetic field. However, the relativistic beaming causes the region of maximum extraordinary emission to shift toward the incoming side of the disk.

The technical methods to achieve the necessary resolution to form these images directly are currently being developed for wavelengths shortward of ~ 1.3 mm. The challenge will be to push this technology even further toward 0.8 or even 0.6 mm VLBI. Depending on how short a wavelength is required, the projected timescale for completing the VLBI array may be less than ten years. Unfortunately, extending this work to much shorter wavelengths than this (where we would be even more certain that Sagittarius A* is optically thin and where scatter broadening is less significant) from the ground is not feasible; at hundreds of microns, Earth's atmosphere becomes optically thick. Such observations must be made from space, rendering the goal of imaging Sagittarius A* much more expensive. Finally, the accretion flow appears to be optically thin to electron scattering at X-ray wavelengths as well, so, in principle, the shadow may also be detectable with space-based X-ray interferometry.

Developments such as these will continue to refine our view of Sagittarius A*, rendering this book, at best, a work in progress. That's the nature of an exciting scientific discipline, in which the sense of imminent discovery is palpable. Many recognize now that Sagittarius A* represents the most compelling case for the existence of black holes and that it may, within the next few years, offer us the means of testing general relativity's predictions in the strong-field limit. That's quite a challenge, but if past history is any guide, even in this we may be underestimating what is yet to come.

References

Abramowicz, M., G. Bao, A. Lanza, and W. Zhang, "X-ray AGN Variability with Hour to Month Timescales," *Proceedings of the 23d ESLAB Symposium* (1989), pp. 871–876.

Aharonian, F. A., A. G. Akhperjanian, K.-M. Aye, A. R. Bazer-Bachi, M. Beilicke, W. Benbow, D. Berge et al., "Very High Energy Gamma Rays from the Direction of Sagittarius A," *Astronomy and Astrophysics* 425 (2004), pp. L13–L17.

Aharonian, F. A., A. G. Akhperjanian, M. Beilicke, K. Bernloehr, H. Bojahr, O. Bolz, H. Boerst et al., "A Search for TeV Gamma-Ray Emission from SNRs, Pulsars and Unidentified GeV Sources in the Galactic Plane in the Longitude Range Between $-2°$ and $85°$," *Astronomy and Astrophysics* 395 (2002), pp. 803–811.

Aharonian, F. A., and A. Neronov, "High-Energy Gamma Rays from the Massive Black Hole in the Galactic Center," *The Astrophysical Journal* 619 (2005), pp. 306–313.

Aitken, D. K., J. Greaves, A. Chrysostomou, T. Jenness, W. Holland, J. H. Hough, D. Pierce-Price, and J. Richer, "Detection of Polarized Millimeter and Submillimeter Emission from Sagittarius A*," *The Astrophysical Journal Letters* 534 (2000), pp. L173–L176.

Alberdi, A., L. Lara, J. M. Marcaide, P. Elosegui, I. I. Shapiro, W. D. Cotton, P. J. Diamond, J. D. Romney, and R. A. Preston, "VLBA Image of Sagittarius A* at $\lambda = 1.35$ cm," *Astronomy and Astrophysics* 277 (1993), p. L1.

Alexander, T., "Stars and Singularities: Stellar Phenomena near a Massive Black Hole" in *The Galactic Black Hole: Lectures on General Relativity and Astrophysics*, The Institute of Physics Publishing, Bristol, UK, 2003, pp. 246–275.

Alexander, T., and A. Loeb, "Enhanced Microlensing by Stars around the Black Hole in the Galactic Center," *The Astrophysical Journal* 551 (2001), pp. 223–230.

Alexander, T., and A. Sternberg, "Near-Infrared Microlensing of Stars by the Supermassive Black Hole in the Galactic Center," *The Astrophysical Journal* 520 (1999), pp. 137–148.

Allen, D. A., A. R. Hyland, and D. J. Hillier, "The Source of Luminosity at the Galactic Centre," *Monthly Notices of the Royal Astronomical Society* 244 (1990), pp. 706–713.

Allen, D. A., and R. H. Sanders, "Is the Galactic Center Black Hole a Dwarf?" *Nature* 319 (1986), pp. 191–194.

Anantharamaiah, K. R., A. Pedlar, R. D. Ekers, and W. M. Goss, "Radio Studies of the Galactic Center, II: The Arc, Threads, and Related Features at 90 Cm (330 MHz)," *Monthly Notices of the Royal Astronomical Society* 249 (1991), pp. 262–281.

Antonucci, R., and R. Barvainis, "Excess 2 Centimeter Emission—A New Continuum Component in the Spectra of Radio-Quiet Quasars," *The Astrophysical Journal Letters* 332 (1988), pp. L13–L17.

Backer, D. C., "Interstellar Scattering of Sgr A*," *AIP Conference Proceedings* 174 (1988), pp. 111–116.

———, "Radio Observations of Sgr A*," *NATO ASI C* 445 (1994), p. 403.

Backer, D. C., and R. A. Sramek, "Apparent Proper Motions of the Galactic Center Compact Radio Source and PSR 1929+10," *The Astrophysical Journal* 260 (1982), pp. 512–519.

———, "Proper Motion of the Compact, Nonthermal Radio Source in the Galactic Center, Sagittarius A*," *The Astrophysical Journal* 524 (1999), pp. 805–815.

Baganoff, F. K., M. W. Bautz, W. N. Brandt, G. Chartas, E. D. Feigelson, G. P. Garmire, Y. Maeda et al., "Rapid X-ray Flaring from the Direction of the Supermassive Black Hole at the Galactic Centre," *Nature* 413 (2001), pp. 45–48.

Baganoff, F. K., Y. Maeda, M. Morris, M. W. Bautz, W. N. Brandt, W. Cui, J. P. Doty et al., "Chandra X-Ray Spectroscopic Imaging of Sagittarius A* and the Central Parsec of the Galaxy," *The Astrophysical Journal* 591 (2003), pp. 891–915.

Bahcall, J. N. and R. A. Wolf, "Star Distribution around a Massive Black Hole in a Globular Cluster," *The Astrophysical Journal* 209 (1976), pp. 214–232.

———, "The Star Distribution around a Massive Black Hole in a Globular Cluster, II: Unequal Star Masses," *The Astrophysical Journal* 216 (1977), pp. 883–907.

Balbus, S. A., C. F. Gammie, and J. F. Hawley, "Fluctuations, Dissipation and Turbulence in Accretion Discs," *Monthly Notices of the Royal Astronomical Society* 271 (1994), pp. 197–201.

Balbus, S. A., and J. F. Hawley, "A Powerful Local Shear Instability in Weakly Magnetized Disks, I: Linear Analysis; II: Nonlinear Evolution," *The Astrophysical Journal* 376 (1991), pp. 214–233.

Balick, B., and R. L. Brown, "Intense Sub-Arcsecond Structure in the Galactic Center," *The Astrophysical Journal* 194 (1974), pp. 265–270.

Bao, G., and Z. Stuchlik, "Accretion Disk Self-Eclipse—X-ray Light Curve and Emission Line," *The Astrophysical Journal* 400 (1992), pp. 163–169.

Bardeen, J. M., "Properties of Black Holes Relevant to their Observation" in *Gravitational Radiation and Gravitational Collapse*, D. Reidel Publishing Company, Dordrecht, Netherlands, 1974, pp. 132–144.

Bardeen, J. M., and J. A. Petterson, "The Lense-Thirring Effect and Accretion Disks around Kerr Black Holes," *The Astrophysical Journal Letters* 195 (1975), pp. L65–L67.

Bardeen, J. M., W. H. Press, and S. A. Teukolsky, "Rotating Black Holes: Locally Nonrotating Frames, Energy Extraction, and Scalar Synchrotron Radiation," *The Astrophysical Journal* 178 (1972), pp. 347–370.

Barret, D., J. F. Olive, L. Boirin, C. Done, G. K. Skinner, and J. E. Grindlay, "Hard X-ray Emission from Low-Mass X-ray Binaries," *The Astrophysical Journal* 533 (2000), pp. 329–351.

Beckert, T., and H. Falcke, "Circular Polarization of Radio Emission from Relativistic Jets," *Astronomy and Astrophysics* 388 (2002), pp. 1106–1119.

Becklin, E. E., I. Gatley, and M. W. Werner, "Far-Infrared Observations of Sagittarius A: The Luminosity and Dust Density in the Central Parsec of the Galaxy," *The Astrophysical Journal* 258 (1982), pp. 135–142.

Bélanger, G., A. Goldwurm, P. Goldoni, J. Paul, R. Terrier, M. Falanga, P. Ubertini et al., "Detection of Hard X-ray Emission from the Galactic Nuclear Region with INTEGRAL," *The Astrophysical Journal Letters* 601 (2004), pp. L163–L166.

Bélanger, G., A. Goldwurm, F. Melia, P. Ferrando, N. Grosso, D. Porquet, R. Warwick, and F. Yusef-Zadeh, "Repeated X-ray Flaring Activity in Sagittarius A*," *The Astrophysical Journal* 635 (2005), pp. 1095–1102.

Benford, G., and D. Tzach, "Coherent Synchrotron Emission Observed: Implications for Radio Astronomy," *Monthly Notices of the Royal Astronomical Society* 317 (2000), pp. 497–500.

Binney, J., and S. Tremaine, *Galactic Dynamics*, Princeton University Press, Princeton, NJ, 1987.

Birkhoff, G., *Relativity and Modern Physics*, Harvard University Press, Cambridge, MA, 1923.

Blaauw, A., C. S. Gum, J. L. Pawsey, and G. Westerhout, "The New I. A. U. System of Galactic Coordinates (1958 Revision)," *Monthly Notices of the Royal Astronomical Society* 121 (1960), pp. 123–131.

Blackman, E. G., "Distinguishing Solar Flare Types by Differences in Reconnection Regions," *The Astrophysical Journal Letters* 484 (1997), pp. L79–L82.

Blandford, R. D., "Observational Tests of General Relativity" in *Gravitational Dynamics*, Cambridge University Press, Cambridge, 1996, pp. 129–140.

Blasi, P., and S. Colafrancesco, "Cosmic Rays, Radio Halos and Nonthermal X-ray Emission in Clusters of Galaxies," *Astroparticle Physics* 12 (1999), pp. 169–183.

Bondi, H., and F. Hoyle, "On the Mechanism of Accretion by Stars," *Monthly Notices of the Royal Astronomical Society* 104 (1944), pp. 273–283.

Born, M., and E. Wolf, *Principles of Optics*, 4th ed., Pergamon Press, New York, 1970.

Bower, G. C., and D. C. Backer, "7 Millimeter VLBA Observations of Sagittarius A*," *The Astrophysical Journal Letters* 496 (1998), pp. L97–L100.

Bower, G. C., D. C. Backer, J.-H. Zhao, M. Goss, and H. Falcke, "The Linear Polarization of Sagittarius A*, I: VLA Spectropolarimetry at 4.8 and 8.4 GHz," *The Astrophysical Journal* 521 (1999a), pp. 582–586.

Bower, G. C., H. Falcke, and D. C. Backer, "Detection of Circular Polarization in the Galactic Center Black Hole Candidate Sagittarius A*," *The Astrophysical Journal Letters* 523 (1999b), pp. L29–L32.

Bower, G. C., H. Falcke, R. M. Hernstein, J.-H. Zhao, W. M. Goss, and D. C. Backer, "Detection of the Intrinsic Size of Sagittarius A* through Closure Amplitude Imaging," *Science* 304 (2004), pp. 704–708.

Bower, G. C., H. Falcke, R. J. Sault, and D. C. Backer, "The Spectrum and Variability of Circular Polarization in Sagittarius A* from 1.4 to 15 GHz," *The Astrophysical Journal* 571 (2002), pp. 843–855.

Bower, G. C., M. C. H. Wright, D. C. Backer, and H. Falcke, "The Linear Polarization of Sagittarius A*, II: VLA and BIMA Polarimetry at 22, 43, and 86 GHz," *The Astrophysical Journal* 527 (1999), pp. 851–855.

Bower, G. C., M. C. H. Wright, H. Falcke, and D. C. Backer, "BIMA Observations of Linear Polarization in Sagittarius A* at 112 GHz," *The Astrophysical Journal Letters* 555 (2001), pp. L103–L106.

———, "Interferometric Detection of Linear Polarization from Sagittarius A* at 230 GHz," *The Astrophysical Journal* 588 (2003), pp. 331–337.

Brandenburg, A., A. Nordlund, R. Stein, and U. Torkelsson, "DynamoGenerated Turbulence and Large-Scale Magnetic Fields in a Keplerian Shear Flow," *The Astrophysical Journal* 446 (1995), pp. 741–754.

Bromley, B., K. Chen, and W. A. Miller, "Line Emission from an Accretion Disk around a Rotating Black Hole: Toward a Measurement of Frame Dragging," *The Astrophysical Journal* 475 (1997), pp. 57–64.

Bromley, B., F. Melia, and S. Liu, "Polarimetric Imaging of the Massive Black Hole at the Galactic Center," *The Astrophysical Journal Letters* 555 (2001), pp. L83–L87.

Brown, R. L., "Precessing Jets in Sagittarius A: Gas Dynamics in the Central Parsec of the Galaxy," *The Astrophysical Journal* 262 (1982), pp. 110–119.

Brown, R. L., K. J. Johnston, and K. Y. Lo, "High Resolution VLA Observations of the Galactic Center," *The Astrophysical Journal* 250 (1981), pp. 155–159.

Brown, R. L., and K. Y. Lo, "Variability of the Compact Radio Source at the Galactic Center," *The Astrophysical Journal* 253 (1982), pp. 108–114.

Brown, W. R., M. J. Geller, S. J. Kenyon, and M. J. Kurtz, "Discovery of an Unbound Hypervelocity Star in the Milky Way Halo," *The Astrophysical Journal Letters* 622 (2005), pp. L33–L36.

Carter, B., "Global Structure of the Kerr Family of Gravitational Fields," *Physical Review* 174 (1968), pp. 1559–1571.

Catchpole, R. M., P. A. Whitelock, and I. S. Glass, "The Distribution of Stars within Two Degrees of the Galactic Center," *Monthly Notices of the Royal Astronomical Society* 247 (1990), pp. 479–490.

Chakrabarti, S. K., and P. J. Wiita, "Variable Emission Lines as Evidence of Spiral Shocks in Accretion Disks around Active Galactic Nuclei," *The Astrophysical Journal* 434 (1994), pp. 518–522.

Chan, K., S. H. Moseley, S. Casey, J. P. Harrington, E. Dwek, R. Loewenstein, F. Varosi, and W. Glaccum, "Dust Composition, Energetics, and Morphology of the Galactic Center," *The Astrophysical Journal* 483 (1997), pp. 798–810.

Chatterjee, P., L. Hernquist, and A. Loeb, "Dynamics of a Massive Black Hole at the Center of a Dense Stellar System," *The Astrophysical Journal* 572 (2002), pp. 371–381.

Chevalier, R., "The Galactic Center Wind," *The Astrophysical Journal Letters* 397 (1992), pp. L39–L42.

Claussen, M. J., D. A. Frail, W. M. Goss, and R. A. Gaume, "Polarization Observations of 1720 MHz OH Masers Toward the Three Supernova Remnants W28, W44, and IC 443," *The Astrophysical Journal* 489 (1997), pp. 143–159.

Coker, R. F., and F. Melia, "Hydrodynamical Accretion onto Sagittarius A* from Distributed Point Sources," *The Astrophysical Journal Letters* 488 (1997), pp. L149–L152.

Combi, J. A., G. E. Romero, and P. Benaglia, "The Gamma Ray Source 2EGS J1703-6302: A New Supernova Remnant in Interaction with an HI Cloud?" *Astronomy and Astrophysics* 333 (1998), pp. L91–L94.

Combi, J. A., G. E. Romero, P. Benaglia, and J. L. Jonas, "Detection of a New, Low-Brightness Supernova Remnant Possibly Associated with EGRET Sources," *Astronomy and Astrophysics* 366 (2001), pp. 1047–1052.

Connors, P. A., and R. F. Stark, "Observable Gravitational Effects on Polarised Radiation Coming From Near a Black Hole," *Nature* 269 (1977), pp. 128–129.

Coppi, P. S., and R. D. Blandford, "Reaction Rates and Energy Distributions for Elementary Processes in Relativistic Pair Plasmas," *Monthly Notices of the Royal Astronomical Society* 245 (1990), pp. 453–507.

Crocker, R. M., M. Fatuzzo, R. Jokipii, F. Melia, and R. R. Volkas, "The AGASA/SUGAR Anisotropies and TeV Gamma Rays from the Galactic Center: A Possible Signature of Extremely High-Energy Neutrons," *The Astrophysical Journal* 622 (2005), pp. 892–909.

Cunningham, C. T., "The Effects of Redshifts and Focusing on the Spectrum of an Accretion Disk around a Kerr Black Hole," *The Astrophysical Journal* 202 (1975), pp. 788–802.

Cunningham, C. T., and J. M. Bardeen, "The Optical Appearance of a Star Orbiting an Extreme Kerr Black Hole," *The Astrophysical Journal* 183 (1973), pp. 237–264.

Czerny, B., and M. Elvis, "Constraints on Quasar Accretion Disks from the Optical/Ultraviolet/Soft X-ray Big Bump," *The Astrophysical Journal* 321 (1987), pp. 305–320.

Davidson, J. A., M. W. Werner, X. Wu, D. F. Lester, P. M. Harvey, M. Joy, and M. Morris, "The Luminosity of the Galactic Center," *The Astrophysical Journal* 387 (1992), pp. 189–211.

Davies, R. D., D. Walsh, and R. S. Booth, "The Radio Source at the Galactic Nucleus," *Monthly Notices of the Royal Astronomical Society* 177 (1976), pp. 319–333.

Dehnen, W., and J. J. Binney, "Local Stellar Kinematics from HIPPARCOS Data," *Monthly Notices of the Royal Astronomical Society* 298 (1998), pp. 387–394.

Doeleman, S., Z.-Q. Shen, A. E. E. Rogers, G. C. Bower, M. C. H. Wright, J.-H. Zhao, D. C. Backer et al., "Structure of Sagittarius A* at 86 GHz Using VLBI Closure Quantities," *The Astronomical Journal* 121 (2001), pp. 2610–2617.

Downes, D., and A. Maxwell, "Radio Observations of the Galactic Center Region," *The Astrophysical Journal* 146 (1966), pp. 653–665.

Duschl, W. J., and H. Lesch, "The Spectrum of Sgr A* and Its Variability," *Astronomy and Astrophysics* 286 (1994), pp. 431–436.

Dyson, F. W., A. S. Eddington, and C. Davidson, "A Determination of the Deflection of Light by the Sun's Gravitational Field, from Observations Made at the Total Eclipse of May 29, 1919," *Philosophical Transactions of the Royal Society* 220 (1920), pp. 291–333.

Ebisawa, K., Y. Maeda, H. Kaneda, and S. Yamauchi, "Origin of the Hard X-ray Emission from the Galactic Plane," *Science* 293 (2001), pp. 1633–1635.

Eckart, A., and R. Genzel, "Observations of Stellar Proper Motions near the Galactic Centre," *Nature* 383 (1996), pp. 415–417.

———, "Stellar Proper Motions in the Central 0.1 Parsec of the Galaxy," *Monthly Notices of the Royal Astronomical Society* 284 (1997), pp. 576–598.

Eckart, A., T. Ott, and R. Genzel, "The Sgr A* Stellar Cluster: New NIR Imaging and Spectroscopy," *Astronomy and Astrophysics* 352 (1999), pp. L22–L25.

Eisenhauer, F., R. Genzel, T. Alexander, R. Abuter, T. Paumard, T. Ott, A. Gilbert et al., "SINFONI in the Galactic Center: Young Stars and Infrared Flares in the Central Light-Month," *The Astrophysical Journal* 628 (2005), pp. 246–259.

Ekers, R. D., W. M. Goss, U. J. Schwarz, D. Downes, and D. H. Rogstad, "A Full Synthesis Map of Sgr A at 5 GHz," *Astronomy and Astrophysics* 43 (1975), pp. 159–166.

Ekers, R. D., J. H. van Gorkom, U. J. Schwarz, and W. M. Goss, "The Radio Structure of Sgr A," *Astronomy and Astrophysics* 122 (1983), pp. 143–150.

Epstein, R. I., "Synchrotron Sources: Extension of Theory for Small Pitch Angles," *The Astrophysical Journal* 183 (1973), pp. 593–610.

Esposito, J. A., S. D. Hunter, G. Kanbach, and P. Sreekumar, "EGRET Observations of Radio-Bright Supernova Remnants," *The Astrophysical Journal* 461 (1996), pp. 820–827.

Falcke, H., "Radio Variability of SGR A* at Centimeter Wavelengths" in *The Central Parsecs of the Galaxy*, ASP Conference Series 186, Astronomical Society of the Pacific, San Francisco, 1999, pp. 113–117.

Falcke, H., W. M. Goss, H. Matsuo, P. Teuben, J.-H. Zhao, and R. Zylka, "The Simultaneous Spectrum of Sgr A* from λ20 cm to λ1 mm and the Nature of the mm-Excess," *The Astrophysical Journal* 499 (1998), pp. 731–734.

Falcke, H., K. Mannheim, and P. L. Biermann, "The Galactic Center Radio Jet," *Astronomy and Astrophysics* 278 (1993), pp. L1–L4.

Falcke, H., and S. Markoff, "The Jet Model for Sagittarius A*: Radio and X-ray Spectrum," *Astronomy and Astrophysics* 362 (2000), pp. 113–118.

Falcke, H., and F. Melia, "Accretion Disk Evolution with Wind Infall, I: General Solution and Application to Sagittarius A*," *The Astrophysical Journal* 479 (1997), pp. 740–751.

Falcke, H., F. Melia, and E. Agol, "Viewing the Shadow of the Black Hole at the Galactic Center," *The Astrophysical Journal Letters* 528 (2000), pp. L13–L16.

Fatuzzo, M., and F. Melia, "A Kinship between the EGRET Supernova Remnants and Sagittarius A East," *The Astrophysical Journal* 596 (2003), pp. 1035–1043.

Feast, M., F. Pont, and P. Whitelock, "The Cepheid Period-Luminosity ZeroPoint from Radial Velocities and HIPPARCOS Proper Motions," *Monthly Notices of the Royal Astronomical Society* 298 (1998), pp. L43–L44.

Feast, M., and P. Whitelock, "Galactic Kinematics of Cepheids from HIPPARCOS Proper Motions," *Monthly Notices of the Royal Astronomical Society* 291 (1997), pp. 683–693.

Feng, Y. X., S. N. Zhang, X. Sun, P. Durouchoux, W. Chen, and W. Cui, "Evolution of Iron Kα: Line Emission in the Black Hole Candidate GX339-4," *The Astrophysical Journal* 553 (2001), pp. 394–398.

Figer, D. F., S. S. Kim, M. Morris, E. Serabyn, R. M. Rich, and I. S. McLean, "Hubble Space Telescope/NICMOS Observations of Massive Stellar Clusters near the Galactic Center," *The Astrophysical Journal* 525 (1999), pp. 750–758.

Fleming, T. P., J. M. Stone, and J. F. Hawley, "The Effect of Resistivity on the Nonlinear Stage of the Magnetorotational Instability in Accretion Disks," *The Astrophysical Journal* 530 (2000), pp. 464–477.

Fleysher, R., and the Milagro Collaboration, "Search for Diffuse TeV GammaRay Emission from the Galactic Plane, Using the Milagro Gamma-Ray Telescope," *Bulletin of the American Astronomical Society* 34 (2002), p. 676.

Fox, J. G., "Evidence against Emission Theories," *American Journal of Physics* 33 (1965), pp. 1–17.

———, "Constancy of the Velocity of Light," *Journal of the Optical Society* 57 (1967), pp. 967–968.

Fragile, P. C., and P. Anninos, "Hydrodynamic Simulations of Tilted ThickDisk Accretion onto a Kerr Black Hole," *The Astrophysical Journal* 623 (2005), pp. 347–361.

Frail, D. A., P. J. Diamond, J. M. Cordes, and H. J. van Langevelde, "Anisotropic Scattering of OH/IR Stars toward the Galactic Center," *The Astrophysical Journal Letters* 427 (1994), pp. L43–L46.

Frank, J., A. King, and D. Raine, *Accretion Power in Astrophysics*, Cambridge University Press, Cambridge, 2002.

Fromerth, M. J., F. Melia, and D. A. Leahy, "A Monte Carlo Study of the 6.4 keV Emission at the Galactic Center," *The Astrophysical Journal* 547 (2001), pp. 129–132.

Gaisser, T. K., R. J. Protheroe, and T. Stanev, "Gamma Ray Production in Supernova Remnants," *The Astrophysical Journal* 492 (1998), pp. 219–227.

Gatley, I., T. J. Jones, A. R. Hyland, R. Wade, and T. R. Geballe, "The Spatial Distribution and Velocity Field of the Molecular Hydrogen Line Emission from the Center of the Galaxy," *Monthly Notices of the Royal Astronomical Society* 222 (1986), pp. 299–306.

Geballe, T. R., K. Krisciunas, J. A. Bailey, and R. Wade, "Mapping of Infrared Helium and Hydrogen Line Profiles in the Central Few Arcseconds of the Galaxy," *The Astrophysical Journal Letters* 370 (1991), pp. L73–L76.

Geballe, T. R., R. Wade, K. Krisciunas, I. Gatley, and M. C. Bird, "The Broad-Line Region at the Center of the Galaxy," *The Astrophysical Journal* 320 (1987), pp. 562–569.

Genzel, R., "The Nuclear Star Cluster of the Milky Way" in *Dynamics of Star Clusters and the Milky Way*, ASP Conference Series 228, Astronomical Society of the Pacific, San Francisco, 2001, pp. 291–311.

Genzel, R., A. Eckart, T. Ott, and F. Eisenhauer, "On the Nature of the Dark Mass in the Centre of the Milky Way," *Monthly Notices of the Royal Astronomical Society* 291 (1997), pp. 219–234.

Genzel, R., C. Pichon, A. Eckart, O. E. Gerhard, and T. Ott, "Stellar Dynamics in the Galactic Center: Proper Motions and Anisotropy," *Monthly Notices of the Royal Astronomical Society* 317 (2000), pp. 348–374.

Genzel, R., R. Schödel, T. Ott, A. Eckart, T. Alexander, F. Lacombe, D. Rouan, and B. Aschenbach, "Near-IR Flares from Accreting Gas around the Supermassive Black Hole in the Galactic Center," *Nature* 425 (2003a), pp. 934–937.

Genzel, R., R. Schödel, T. Ott, F. Eisenhauer, R. Hofmann, M. Lehnert, A. Eckart et al. "The Stellar Cusp around the Supermassive Black Hole in the Galactic Center," *The Astrophysical Journal* 594 (2003b), pp. 812–832.

Genzel, R., N. Thatte, A. Krabbe, H. Kroker, and L. E. Tacconi-Garman, "The Dark Mass Concentration in the Central Parsec of the Milky Way," *The Astrophysical Journal* 472 (1996), pp. 153–172.

Genzel, R., D. M. Watson, C. H. Townes, H. L. Dinerstein, D. Hollenbach, D. F. Lester, M. Werner, and J. W. V. Storey, "Far-Infrared Spectroscopy of the Galactic Center: Neutral and Ionized Gas in the Central 10 Parsecs of the Galaxy," *The Astrophysical Journal* 276 (1984), pp. 551–559.

Gerhard, O., "The Galactic Center HeI Stars: Remains of a Dissolved Young Cluster?" *The Astrophysical Journal Letters* 546 (2001), pp. L39–L42.

Gezari, D., "Mid-Infrared Emission amid Luminous Sources in the Central Parsec" in *The Galactic Center*, ASP Conference Series 102, Astronomical Society of the Pacific, San Francisco, 1996, p. 491.

Gezari, S., A. M. Ghez, E. E. Becklin, J. Larkin, I. S. McLean, and M. Morris, "Adaptive Optics Near-Infrared Spectroscopy of the Sagittarius A* Cluster," *The Astrophysical Journal* 576 (2002), pp. 790–797.

Ghez, A., E. E. Becklin, G. Duchene, S. Hornstein, M. Morris, S. Salim, and A. Tanner, "Full Three-Dimensional Orbits for Multiple Stars on Close Approaches to the Central Supermassive Black Hole," *Astronomische Nachrichten* 324 (2003a), pp. 527–533.

Ghez, A., G. Duchene, K. Matthews, S. D. Hornstein, A. Tanner, J. Larkin, M. Morris et al., "The First Measurement of Spectral Lines in a ShortPeriod Star Bound to a Galaxy's Central Black Hole: A Paradox of Youth," *The Astrophysical Journal Letters* 586 (2003b), pp. L127–L131.

Ghez, A., B. L. Klein, M. Morris, and E. E. Becklin, "High Proper Motion Stars in the Vicinity of Sagittarius A*: Evidence for a Supermassive Black Hole at the Center of Our Galaxy," *The Astrophysical Journal* 509 (1998), pp. 678–686.

Ghez, A., M. Morris, E. E. Becklin, A. Tanner, and T. Kremenek, "The Accelerations of Stars Orbiting the Milky Way's Central Black Hole," *Nature* 407 (2000), pp. 349–351.

Ghez, A. M., S. A. Wright, K. Matthews, D. Thompson, D. Le Mignant, A. Tanner, S. D. Hornstein, M. Morris, E. E. Becklin, and B. T. Soifer, "Variable Infrared Emission from the Supermassive Black Hole at the Center of the Milky Way," *The Astrophysical Journal Letters* 602 (2004), pp. L159–L162.

Goldwurm, A., E. Brion, P. Goldoni, P. Ferrando, F. Daigne, A. Decourchelle, R. S. Warwick, and P. Predehl, "A New X-Ray Flare from the Galactic Nucleus Detected with the XMM-*Newton* Photon Imaging Cameras," *The Astrophysical Journal* 584 (2003), pp. 751–757.

Goss, W. M., R. Brown, and K. Y. Lo, "The Discovery of Sgr A*," *Astronomische Nachrichten* S1 (2003), pp. 497–504.

Gould, A., and S. V. Ramirez, "Non-Acceleration of Sagittarius A*: Implications for Galactic Structure," *The Astrophysical Journal* 497 (1998), pp. 713–716.

Güsten, R., R. Genzel, M. C. H. Wright, D. T. Jaffe, J. Stutzki, and A. I. Harris, "Aperture Synthesis Observations of the Circumnuclear Ring in the Galactic Center," *The Astrophysical Journal* 318 (1987), pp. 124–138.

Gwinn, C. R., R. M. Danen, T. K. Tran, J. Middleditch, and L. M. Ozernoy, "The Galactic Center Radio Source Shines below the Compton Limit," *The Astrophysical Journal Letters* 381 (1991), pp. L43–L46.

Hall, D. E., and P. A. Sturrock, "Diffusion, Scattering, and Acceleration of Particles by Stochastic Electromagnetic Fields," *Physics of Fluids* 10 (1967), pp. 2620–2628.

Hall, D. N. C., S. G. Kleinmann, and N. Z. Scoville, "Broad Helium Emission in the Galactic Center," *The Astrophysical Journal Letters* 260 (1982), pp. L53–L57.

Haller, J. W., M. J. Rieke, G. H. Rieke, P. Tamblyn, L. Close, and F. Melia, "Stellar Kinematics and the Black Hole in the Galactic Center," *The Astrophysical Journal* 456 (1996), pp. 194–205.

Hamilton, R. J., and V. Petrosian, "Stochastic Acceleration of Electrons, I: Effects of Collisions in Solar Flares," *The Astrophysical Journal* 398 (1992), pp. 350–358.

Hanson, R. B., "Lick Northern Proper Motion Program, II: Solar Motion and Galactic Rotation," *Astronomical Journal* 94 (1987), pp. 409–415.

Hartman, R. C., D. L. Bertsch, S. D. Bloom, A. W. Chen, P. Deines-Jones, J. A. Esposito, C. E. Fichtel et al., "The Third EGRET Catalog of High-Energy Gamma-Ray Sources," *The Astrophysical Journal Supplement Series* 123 (1999), pp. 79–202.

Hawley, J. F., "Global Magnetohydrodynamical Simulations of Accretion Tori," *The Astrophysical Journal* 528 (2000), pp. 462–479.

Hawley, J. F., and S. A. Balbus, "A Powerful Local Shear Instability in Weakly Magnetized Disks, II: Nonlinear Evolution," *The Astrophysical Journal* 376 (1991), pp. 223–233.

———, "A Powerful Local Shear Instability in Weakly Magnetized Disks, III: Long-Term Evolution in a Shearing Sheet; IV: Nonaxisymmetric Perturbations," *The Astrophysical Journal* 400 (1992), pp. 595–621.

———, "The Dynamical Structure of Nonradiative Black Hole Accretion Flows," *The Astrophysical Journal* 573 (2002), pp. 738–748.

Hawley, J. F., C. F. Gammie, and S. A. Balbus, "Local Three-Dimensional Magnetohydrodynamic Simulations of Accretion Disks," *The Astrophysical Journal* 440 (1995), pp. 742–763.

———, "Local Three-Dimensional Simulations of an Accretion Disk Hydromagnetic Dynamo," *The Astrophysical Journal* 464 (1996), pp. 690–703.

Hay, H. J., J. P. Schiffer, T. E. Cranshaw, and P. A. Egelstaff, "Measurement in the Red Shift in an Accelerated System Using the Mössbauer Effect in Fe^{57}," *Physical Review Letters* 4 (1960), p. 165.

Hills, J. G., "Hypervelocity and Tidal Stars from Binaries Disrupted by a Masive Galactic Black Hole," *Nature* 331 (1988), pp. 687–689.

Hollywood, J. M., and F. Melia, "The Effects of Redshifts and Focusing on the Spectrum of an Accretion Disk in the Galactic Center Black Hole

Candidate Sgr A*," *The Astrophysical Journal Letters* 443 (1995), pp. L17–L20.

———, "General Relativistic Effects on the Infrared Spectrum of Thin Accretion Disks in Active Galactic Nuclei: Application to Sagittarius A*," *The Astrophysical Journal Supplement* 112 (1997),pp. 423–455.

Hollywood, J. M., F. Melia, L. M. Close, D. W. McCarthy, and T. A. DeKeyser, "General Relativistic Flux Modulations in the Galactic Center Black Hole Candidate Sgr A*," *The Astrophysical Journal Letters* 448 (1995), pp. L21–L24.

Homan, D. C., and J. F. C. Wardle, "Detection and Measurement of ParsecScale Circular Polarization in Four AGNs," *The Astronomical Journal* 118 (1999), pp. 1942–1962.

Hornstein, S. D., A. M. Ghez, A. Tanner, M. Morris, E. E. Becklin, and P. Wizinowich, "Limits on the Short-Term Variability of Sagittarius A* in the Near-Infrared," *The Astrophysical Journal Letters* 577 (2002), pp. L9–L13.

Hoskin, M. A., "The 'Great Debate': What Really Happened," *Journal of Historical Astronomy* 7 (1976), pp. 169–182.

Hoyle, F., and R. A. Lyttleton, "On the Accretion Theory of Stellar Evolution," *Monthly Notices of the Royal Astronomical Society* 101 (1941), pp. 227–237.

Hughes, P. A., H. D. Aller, and M. F. Aller, "Polarized Radio Outbursts in BlLacertae—Part Two—the Flux and Polarization of a Piston-Driven Shock," *The Astrophysical Journal* 298 (1985), pp. 301–315.

Igumenshchev, I. V., and R. Narayan, "Three-Dimensional Magnetohydrodynamic Simulations of Spherical Accretion," *The Astrophysical Journal* 566 (2002), pp. 137–147.

Ipser, J. R., and R. H. Price, "Accretion onto Pregalactic Black Holes," *The Astrophysical Journal* 216 (1977), pp. 578–590.

Jackson, J. M., N. Geis, R. Genzel, A. I. Harris, S. Madden, A. Poglitsch, G. J. Stacey, and C. H. Townes, "Neutral Gas in the Central 2 Parsecs of the Galaxy," *The Astrophysical Journal* 402 (1993), pp. 173–184.

Jackson, J. M., M. H. Heyer, T. Paglione, and A. Bolatto, "HCN and CO in the Central 630 Parsecs of the Galaxy," *The Astrophysical Journal Letters* 456 (1996), pp. L91–L95.

Jauncey, D. L., A. K. Tzioumis, R. A. Preston, D. L. Meier, R. Batchelor, J. Gates, P. A. Hamilton et al., "Radio Structure at 8.4 GHz in Sagittarius A, the Compact Radio Source at the Galactic Center," *The Astronomical Journal* 98 (1989), pp. 44–48.

Jeans, J. H., "On the Theory of Star-Streaming and the Structure of the Universe," *Monthly Notices of the Royal Astronomical Society* 76 (1915), pp. 70–84.

Jin, L., "Damping of the Shear Instability in Magnetized Disks by Ohmic Diffusion," *The Astrophysical Journal* 457 (1996), pp. 798–804.

Jokipii, J. R., "Diffusive Shock Acceleration—Acceleration Rate, MagneticField Direction, and the Diffusion Limit" in *Particle Acceleration in Cosmic*

Plasmas, AIP Conference Proceedings 264, Bartol Research Institute, Newark, NJ, 1992, pp. 137–147.

Jones, T. W., and S. L. O'Dell, "Transfer of Polarized Radiation in SelfAbsorbed Synchrotron Sources, I: Results for a Homogeneous Source," *The Astrophysical Journal* 214 (1977a), pp. 522–539.

———, "Transfer of Polarized Radiation in Self-Absorbed Synchrotron Sources, II: Treatment of Inhomogeneous Media and Calculation of Emergent Polarization," *The Astrophysical Journal* 215 (1977b), pp. 236–246.

Kato, S., "Trapped One-Armed Corrugation Waves and QPOs," *Publications of the Astronomical Society of Japan* 42 (1990), pp. 99–113.

Kellerman, K. I., and I. I. K. Pauliny-Toth, "The Spectra of Opaque Radio Sources," *The Astrophysical Journal Letters* 155 (1969), pp. L71–L78.

Kennel, C. F., and F. Engelmann, "Velocity Space Diffusion from Weak Plasma Turbulence in a Magnetic Field," *Physics of Fluids* 9 (1966), pp. 2377–2388.

Kerr, F. J., "Review of Galactic Constants," *Monthly Notices of the Royal Astronomical Society* 221 (1986), pp. 1023–1038.

Kerr, F. J., and D. Lynden-Bell, "Review of Galactic Constants," *Monthly Notices of the Royal Astronomical Society* 221 (1986), pp. 1023–1038.

Kerr, R. P., "Gravitational Field of a Spinning Mass as an Example of Algebraically Special Metrics," *Physical Review Letters* 11 (1963), pp. 237–238.

Kosack, K., H. M. Badran, I. H. Bond, P. J. Boyle, S. M. Bradbury, J. H. Buckley, D. A. Carter-Lewis et al., "TeV Gamma-Ray Observations of the Galactic Center," *The Astrophysical Journal Letters* 608 (2004), pp. L97–L100.

Kowalenko, V., and F. Melia, "Towards Incorporating a Turbulent Magnetic Field in an Accreting Black Hole Model," *Monthly Notices of the Royal Astronomical Society* 310 (1999), pp. 1053–1061.

Koyama, K., Y. Maeda, T. Sonobe, T. Takeshima, Y. Tanaka, and S. Yamauchi, "ASCA View of Our Galactic Center: Remains of Past Activities in X-Rays?" *Publications of the Astronomical Society of Japan* 48 (1996), pp. 249–255.

Krabbe, A., R. Genzel, S. Drapatz, and V. Rotaciuc, "A Cluster of He I Emission-Line Stars in the Galactic Center," *The Astrophysical Journal Letters* 382 (1991), pp. L19–L22.

Krabbe, A., R. Genzel, A. Eckart, F. Najarro, D. Lutz, M. Cameron, H. Kroker et al., "The Nuclear Cluster of the Milky Way: Star Formation and Velocity Dispersion in the Central 0.5 Parsec," *The Astrophysical Journal Letters* 447 (1995), pp. L95–L99.

Krichbaum, T. P., "Millimeter-VLBI with a Large Millimeter Array: Future Possibilities" in *Science with Large Millimeter Arrays*, Springer-Verlag, Berlin, 1996, pp. 95–102.

Krichbaum, T. P., A. Witzel, and J. A. Zensus, "VLBI Observations of the Galactic Center Source Sgr A* at 86 GHz and 215 GHz" in *The Central Parsecs of the Galaxy*, ASP Conference Series 186, Astronomical Society of the Pacific, San Francisco, 1999, pp. 89–97.

Krolik, J., and J. Hawley, "Where Is the Inner Edge of an Accretion Disk around a Black Hole?" *The Astrophysical Journal* 573 (2002), pp. 754–763.

Lacy, J. H., F. Baas, C. H. Townes, and T. R. Geballe, "Observations of the Motion and Distribution of the Ionized Gas in the Central Parsec of the Galaxy," *The Astrophysical Journal Letters* 227 (1979), pp. L17–L20.

Lacy, J. H., C. H. Townes, T. R. Geballe, and D. J. Hollenbach, "Observations of the Motion and Distribution of the Ionized Gas in the Central Parsec of the Galaxy, II," *The Astrophysical Journal* 241 (1980), pp. 132–146.

Lacy, J. H., C. H. Townes, and D. J. Hollenbach, "The Nature of the Central Parsec of the Galaxy," *The Astrophysical Journal* 262 (1982), pp. 120–134.

Landau, L. D., and E. M. Lifshitz, *Fluid Mechanics*, Butterworth-Heinemann, Oxford, UK, 1987.

Laor, A., and H. Netzer, "Massive Thin Accretion Disks, I: Calculated Spectra," *Monthly Notices of the Royal Astronomical Society* 238 (1989), pp. 897–916.

LaRosa, T. N., N. E. Kassim, and T. J. W. Lazio, "A Wide-Field 90 Centimeter VLA Image of the Galactic Center Region," *The Astronomical Journal* 119 (2000), pp. 207–240.

Latvakoski, H. M., G. J. Stacey, G. E. Gull, and T. L. Hayward, "Kuiper Widefield Infrared Camera Far-Infrared Imaging of the Galactic Center: The Circumnuclear Disk Revealed," *The Astrophysical Journal* 511 (1999), pp. 761–773.

Laun, F., and D. Merritt, "Brownian Motion of Black Holes in Dense Nuclei," *The Astrophysical Journal* (2006), in press.

Lee, M.-H., and J. Goodman, "Adiabatic Growth of a Black Hole in a Rotating Stellar System," *The Astrophysical Journal* 343 (1989), pp. 594–601.

Legg, M. P. C., and K. C. Westfold, "Elliptic Polarization of Synchrotron Radiation," *The Astrophysical Journal* 154 (1968), pp. 499–514.

Lerche, I., "Quasilinear Theory of Resonant Diffusion in a Magneto-Active Relativistic Plasma," *Physics of Fluids* 11 (1968), pp. 1720–1727.

Liu, S., and F. Melia, "New Constraints on the Nature of Radio Emission in Sagittarius A*," *The Astrophysical Journal Letters* 561 (2001), pp. L77–L80.

———, "An Accretion-Induced X-ray Flare in Sagittarius A*," *The Astrophysical Journal Letters* 566 (2002a), pp. L77–L80.

———, "Spin-Induced Disk Precession in the Supermassive Black Hole at the Galactic Center," *The Astrophysical Journal Letters* 573 (2002b), pp. L23–L26.

Liu, S., F. Melia, and V. Petrosian, "Stochastic Electron Acceleration during the NIR and X-ray Flares in Sagittarius A*," *The Astrophysical Journal* 636 (2006a), pp. 798–803.

———, "Stochastic Acceleration in the Galactic Center HESS Source." *The Astrophysical Journal* (2006b), in press.

Liu, S., V. Petrosian, and F. Melia, "Electron Acceleration around the Supermassive Black Hole at the Galactic Center," *The Astrophysical Journal Letters* 611 (2004), pp. L101–L104.

REFERENCES

Lo, K. Y., D. C. Backer, R. D. Ekers, K. I. Kellermann, M. Reid, and J. M. Moran, "On the Size of the Galactic Centre Compact Radio Source Diameter Less Than 20 A.U.," *Nature* 315 (1985), pp. 124–126.

Lo, K. Y., D. C. Backer, K. I. Kellermann, M. Reid, J.-H. Zhao, W. M. Goss, and J. M. Moran, "High-Resolution VLBA Imaging of the Radio Source Sgr A* at the Galactic Centre," *Nature* 362 (1993), pp. 38–40.

Lo, K. Y., and M. J. Claussen, "High-Resolution Observations of Ionized Gas in Central 3 Parsecs of the Galaxy: Possible Evidence for Infall," *Nature* 306 (1983), pp. 647–651.

Lo, K. Y., M. H. Cohen, A. S. C. Readhead, and D. C. Backer, "Multiwavelength VLBI Observations of the Galactic Center," *The Astrophysical Journal* 249 (1981), pp. 504–512.

Lo, K. Y., R. T. Schilizzi, M. H. Cohen, and H. N. Ross, "VLBI Observations of the Compact Radio Source in the Center of the Galaxy," *The Astrophysical Journal* 202 (1975), pp. L63–L65.

Lo, K. Y., Z.-Q. Shen, J.-H. Zhao, and P. T. P. Ho, "Intrinsic Size of Sgr A*: 72 Schwarzschild Radii," *The Astrophysical Journal Letters* 508 (1998), pp. L61–L64.

Lynden-Bell, D., and M. J. Rees, "On Quasars, Dust and the Galactic Centre," *Monthly Notices of the Royal Astronomical Society* 152 (1971), pp. 461–475.

Macquart, J.-P., and D. Melrose, "Scintillation-Induced Circular Polarization in Pulsars and Quasars," *The Astrophysical Journal* 545 (2000), pp. 798–806.

Maeda, Y., F. K. Baganoff, E. D. Feigelson, M. Morris, M. W. Bautz, W. N. Brandt, D. N. Burrows et al., "A Chandra Study of Sagittarius A East: A Supernova Remnant Regulating the Activity of Our Galactic Center?" *The Astrophysical Journal* 570 (2002), pp. 671–687.

Maeda, Y., K. Koyama, M. Sakano, T. Takeshima, and S. Yamauchi, "A New Eclipsing X-Ray Burster near the Galactic Center: A Quiescent State of the Old Transient A1742-289," *Publications of the Astronomical Society of Japan* 48 (1996), pp. 417–423.

Mahadevan, R., R. Narayan, and I. Yi, "Harmony in Electrons: Cyclotron and Synchrotron Emission by Thermal Electrons in a Magnetic Field," *The Astrophysical Journal* 465 (1996), pp. 327–337.

Malkan, M. A., and W. L. W. Sargent, "The Ultraviolet Excess of Seyfert 1 Galaxies and Quasars," *The Astrophysical Journal* 254 (1982), pp. 22–37.

Markoff, S., H. Falcke, F. Yuan, and P. L. Biermann, "The Nature of the 10 ksec X-ray Flare in Sagittarius A*," *Astronomy and Astrophysics* 379 (2001), pp. L13–L16.

Markoff, S., F. Melia, and I. Sarcevic, "On the Nature of the EGRET Source at the Galactic Center," *The Astrophysical Journal Letters* 489 (1997), pp. L47–L50.

Marscher, A. P., and W. K. Gear, "Models for High-Frequency Radio Outbursts in Extragalactic Sources, with Application to the Early 1983 Millimeter-to-Infrared Flare of 3C 273," *The Astrophysical Journal* 298 (1985), pp. 114–127.

Mayer-Hasselwander, H. A., D. L. Bertsch, B. L. Dingus, A. Eckart, J. A. Esposito, R. Genzel, R. C. Hartman et al., "High-Energy Gamma Ray Emission from the Galactic Center," *Astronomy and Astrophysics* 335 (1998), pp. 161–172.

McGinn, M. T., K. Sellgren, E. E. Becklin, and D. N. B. Hall, "Stellar Kinematics in the Galactic Center," *The Astrophysical Journal* 338 (1989), pp. 824–840.

McNamara, D. H., J. B. Madsen, J. Barnes, and B. F. Ericksen, "The Distance to the Galactic Center," *Publications of the Astronomical Society of the Pacific* 112 (2000), pp. 202–216.

Melia, F., "An Accreting Black Hole Model for Sagittarius A*," *The Astrophysical Journal Letters* 387 (1992a), pp. L25–L28.

———, "The Nucleus of M31," *The Astrophysical Journal Letters* 398 (1992b), pp. L95–L98.

———, "An Accreting Black Hole Model for Sagittarius A*, 2: A Detailed Study," *The Astrophysical Journal* 426 (1994), pp. 577–585.

———, "X-rays from the Edge of Infinity," *Nature* 413 (2001a), pp. 25–26.

———, *Electrodynamics*, The University of Chicago Press, Chicago, 2001b.

Melia, F., B. Bromley, S. Liu, and C. Walker, "Measuring the Black Hole Spin in Sgr A*," *The Astrophysical Journal Letters* 554 (2001), pp. L37–L40.

Melia, F., and R. Coker, "Stellar Gas Flows into a Dark Cluster Potential at the Galactic Center," *The Astrophysical Journal* 511 (1999), pp. 750–760.

Melia, F., and H. Falcke, "The Supermassive Black Hole at the Galactic Center," *Annual Reviews of Astronomy and Astrophysics* 39 (2001), pp. 309–352.

Melia, F., and M. Fatuzzo, "Particle Dynamics and Gamma Ray Emission in the Magnetospheres of Neutron Stars with Accretion Disks," *The Astrophysical Journal* 346 (1989), pp. 378–390.

Melia, F., M. Fatuzzo, F. Yusef-Zadeh, and S. Markoff, "A Self-Consistent Model for the Broadband Spectrum of Sagittarius A East at the Galactic Center," *The Astrophysical Journal Letters* 508 (1998), pp. L65–L69.

Melia, F., J. R. Jokipii, and A. Narayanan, "A Determination of the Mass of Sagittarius A* from Its Radio Spectral and Source Size Measurements," *The Astrophysical Journal Letters* 395 (1992), pp. L87–L90.

Melia, F., and V. Kowalenko, "Magnetic Field Dissipation in Converging Flows," *Monthly Notices of the Royal Astronomical Society* 327 (2001), pp. 1279–1287.

Melia, F., S. Liu, and R. Coker, "A Magnetic Dynamo Origin for the Submillimeter Excess in Sagittarius A*," *The Astrophysical Journal* 553 (2001), pp. 146–157.

Melrose, D. B., *Electromagnetic Processes in Dispersive Media*, Cambridge University Press, Cambridge, 1991.

Menten, K. M., M. J. Reid, A. Eckart, and R. Genzel, "The Position of Sgr A*: Accurate Alignment of the Radio and Infrared Reference Frames at the Galactic Center," *The Astrophysical Journal Letters* 475 (1997), pp. L111–L114.

Miller, J., and R. Ramaty, "Ion and Relativistic Electron Acceleration by Alfvén and Whistler Turbulence in Solar Flares," *Solar Physics* 113 (1987), pp. 195–200.

Milosavljević, M., and D. Merritt, "Formation of Galactic Nuclei," *The Astrophysical Journal* 563 (2001), pp. 34–62.

Mirabel, I. F., and L. F. Rodriguez, "Microquasars in the Milky Way," *Sky and Telescope* 103 (2002), pp. 32–40.

Miyoshi, M., J. Moran, J. Herrnstein, L. Greenhill, N. Nakai, P. Diamond, and M. Inoue, "Evidence for a Black Hole from High Rotation Velocities in a Sub-parsec Region of NGC 4258," *Nature* 373 (1995), p. 127.

Morris, M., "Massive Star Formation near the Galactic Center and the Fate of the Stellar Remnants," *The Astrophysical Journal* 408 (1993), pp. 496–506.

Morris, M., and E. Serabyn, "The Galactic Center Environment," *Annual Reviews of Astronomy and Astrophysics* 34 (1996), pp. 645–702.

Murakami, H., K. Koyama, and Y. Maeda, "Chandra Observations of Diffuse X-rays from the Sagittarius B2 Cloud," *The Astrophysical Journal* 558 (2001), pp. 687–692.

Najarro, F., D. J. Hillier, R. P. Kudritzki, A. Krabbe, R. Genzel, D. Lutz, S. Drapatz, and T. R. Geballe, "The Nature of the Brightest Galactic Center He I Emission Line Star," *Astronomy and Astrophysics* 285 (1994), pp. 573–584.

Nandra, K., I. M. George, R. F. Mushotzky, T. J. Turner, and T. Yaqoob, "ASCA Observations of Seyfert 1 Galaxies, Data Analysis, Imaging, and Timing," *The Astrophysical Journal* 476 (1997), pp. 70–82.

Narayan, R., and W. B. Hubbard, "Theory of Anisotropic Refractive Scintillation—Application to Stellar Occultations by Neptune," *The Astrophysical Journal* 325 (1988), pp. 503–518.

Narayan, R., I. Yi, and R. Mahadevan, "Explaining the Spectrum of Sagittarius A* with a Model of an Accreting Black Hole," *Nature* 374 (1995), pp. 623–625.

Nayakshin, S., and F. Melia, "Big Blue Bump and Transient Active Regions in Seyfert Galaxies," *The Astrophysical Journal Letters* 484 (1997), pp. L103–L106.

Nayakshin, S., and R. Sunyaev, "Close Stars and an Inactive Accretion Disc in Sgr A*: Eclipses and Flares," *Monthly Notices of the Royal Astronomical Society* 343 (2003), pp. L15–L19.

Nelson, R. P., and C. B. Papaloizou, "Hydrodynamic Simulations of the Bardeen-Petterson Effect," *Monthly Notices of the Royal Astronomical Society* 315 (2000), pp. 570–586.

Okazaki, A., S. Kato, and J. Fukue, "Global Trapped Oscillations of Relativistic Accretion Disks," *Publications of the Royal Astronomical Society of Japan* 39 (1987), pp. 457–473.

Olling, R. P., and M. R. Merrifield, "Refining the Oort and Galactic Constants," *Monthly Notices of the Royal Astronomical Society* 297 (1998), pp. 943–952.

Ozernoy, L. M., R. Genzel, and V. V. Usov, "Colliding Winds in the Stellar Core at the Galactic Centre: Some Implications," *Monthly Notices of the Royal Astronomical Society* 288 (1997), pp. 237–244.

Pacholczyk, A. G., *Radio Astrophysics*, W. H. Freeman and Company, Boston, 1970.

———, "Circular Repolarization in Compact Radio Sources," *Monthly Notices of the Royal Astronomical Society* 163 (1973), pp. 29p–34p.

Paczyński, B., and P. J. Wiita, "Thick Accretion Disks and Supercritical Luminosities," *Astronomy and Astrophysics* 88 (1980), pp. 23–31.

Parker, E. N., *Cosmical Magnetic Fields: Their Origin and Their Activity*, Clarendon Press, Oxford, UK, 1979.

Pavlinsky, M. N., S. A., Grebenev, and R. A. Sunyaev, "X-ray Images of the Galactic Center Obtained with ART-P/GRANAT: Discovery of New Sources, Variability of Persistent Sources, and Localization of X-ray Bursters," *The Astrophysical Journal* 425 (1994), pp. 110–121.

Pedlar, A., K. R. Anantharamaiah, R. D. Ekers, W. M. Goss, J. H. van Gorkom, U. J. Schwarz, and J.-H. Zhao, "Radio Studies of the Galactic Center, I: The Sagittarius A Complex," *The Astrophysical Journal* 342 (1989), pp. 769–784.

Peebles, P. J. E., "Star Distribution near a Collapsed Object," *The Astrophysical Journal* 178 (1972), pp. 371–376.

Petrosian, V., and S. Liu, "Stochastic Acceleration of Electrons and Protons, I: Acceleration by Parallel-Propagating Waves," *The Astrophysical Journal* 610 (2004), pp. 550–571.

Petschek, H. E., "Magnetic Field Annihilation" *The Physics of Solar Flares*, Proceedings of the AAS-NASA Symposium, NASA, Washington, DC, 1964, p. 425.

Phinney, E. S., "Manifestations of a Massive Black Hole in the Galactic Center," *AIP Conference Proceedings* 136 (1989), pp. 543–553.

Piddington, J. H., and H. C. Minnett, "Observations of Galactic Radiation at Frequencies of 1210 and 3000 Mc/s," *Australian Journal of Scientific Research* A4 (1951), pp. 459–475.

Pohl, M. "The Galactic Center Arc as Source of High-Energy Gamma Rays," *Astronomy and Astrophysics* 317 (1997), pp. 441–447.

Portegies Zwart, S. F., S. L. W. McMillan, and O. Gerhard, "The Origin of IRS 16: Dynamically Driven In-Spiral of a Dense Star Cluster to the Galactic Center?" *The Astrophysical Journal* 593 (2003), pp. 352–357.

Predehl, P., and J. Trümper, "ROSAT Observation of the Sgr A Region," *Astronomy and Astrophysics* 290 (1994), pp. L29–L32.

Price, S. D., M. P. Egan, S. J. Carey, D. R. Mizuno, and T. A. Kuchar, "Midcourse Space Experiment of the Galactic Plane," *The Astronomical Journal* 121 (2001), pp. 2819–2842.

Quataert, E., "A Thermal Bremsstrahlung Model for the Quiescent X-ray Emission from Sagittarius A*," *The Astrophysical Journal* 575 (2002), pp. 855–859.

Quataert, E., and A. Gruzinov, "Constraining the Accretion Rate onto Sagittarius A* Using Linear Polarization," *The Astrophysical Journal* 545 (2000), pp. 842–846.

Quinlan, G. D., L. Hernquist, and S. Sigurdsson, "Models of Galaxies with Central Black Holes: Adiabatic Growth in Spherical Galaxies," *The Astrophysical Journal* 440 (1995), pp. 554–564.

Rayner, D. P., R. P. Norris, and R. J. Sault, "Radio Circular Polarization of Active Galaxies," *Monthly Notices of the Royal Astronomical Society* 319 (2000), pp. 484–496.

Readhead, A. C. S., "Equipartition Brightness Temperature and the Inverse Compton Catastrophe," *The Astrophysical Journal* 426 (1994), pp. 51–59.

Reid, M. J., "The Distance to the Center of the Galaxy," *Annual Reviews of Astronomy and Astrophysics* 31 (1993), pp. 345–372.

Reid, M. J., K. M. Menten, R. Genzel, T. Ott, R. Schödel, and A. Eckart, "The Position of Sagittarius A*, II: Accurate Positions and Proper Motions of Stellar SiO Masers near the Galactic Center," *The Astrophysical Journal* 587 (2003), pp. 208–220.

Reid, M. J., A.C.S. Readhead, R. C. Vermeulen, and R. N. Treuhaft, "The Proper Motion of Sgr A*, I: First VLBA Results," *The Astrophysical Journal* 524 (1999), pp. 816–823.

Reynolds, S. P., and C. F. McKee, "The Compact Radio Source at the Galactic Center," *The Astrophysical Journal* 239 (1980), pp. 893–897.

Rieke, G. H., and M. J. Lebofsky, "Comparison of Galactic Center with Other Galaxies," *AIP Conference Proceedings* 83 (1982), pp. 194–203.

Rieke, G. H., and M. J. Rieke, "Ionization of the Mass-Loss Wind of the M Supergiant IRS 7 by the Ultraviolet Flux in the Galactic Center," *The Astrophysical Journal Letters* 344 (1989), pp. L5–L8.

Rockefeller, G., C. L. Fryer, and F. Melia, "Spin-Induced Disk Precession in Sagittarius A*," *The Astrophysical Journal* 635 (2005), pp. 336–340.

Rockefeller, G., C. L. Fryer, F. Melia, and M. S. Warren, "Diffuse X-rays from the Inner 3 Parsecs of the Galaxy," *The Astrophysical Journal* 604 (2004), pp. 662–670.

Rogers, A. E. E., S. Doeleman, M. C. H. Wright, G. C. Bower, D. C. Backer, S. Padin, J. A. Philips, et al., "Small-Scale Structure and Position of Sagittarius A* from VLBI at 3 Millimeter Wavelength," *The Astrophysical Journal Letters* 434 (1994), pp. L59–L62.

Romani, R., R. Narayan, and R. Blandford, "Refractive Effects in Pulsar Scintillation," *Monthly Notices of the Royal Astronomical Society* 220 (1986), pp. 19–49.

Ruffert, M., and F. Melia, "Hydrodynamical 3D Bondi-Hoyle Accretion onto the Galactic Center Black Hole Candidate Sgr A*," *Astronomy and Astrophysics* 288 (1994), pp. L29–L32.

Ruszkowski, M., and M. C. Begelman, "Circular Polarization from Stochastic Synchrotron Sources," *The Astrophysical Journal* 573 (2002), pp. 485–495.

Rybicki, G. B., and A. P. Lightman, *Radiative Processes in Astrophysics*, Wiley-Interscience, New York, 1979.

Salpeter, E. E., "Accretion of Interstellar Matter by Massive Objects," *The Astrophysical Journal* 140 (1964), pp. 796–800.

Sanders, D. B., E. S. Phinney, G. Neugebauer, B. T. Soifer, and K. Matthews, "Continuum Energy Distribution of Quasars—Shapes and Origins," *The Astrophysical Journal* 347 (1989), pp. 29–51.

Sanders, R. H., and K. G. Prendergast, "The Possible Relation of the 3-kiloparsec Arm to Explosions in the Galactic Nucleus," *The Astrophysical Journal* 188 (1974), pp. 489–500.

Sard, R. D., *Relativistic Mechanics*, Benjamin, New York, 1970.

Sault, R. J., and J.-P. Macquart, "Confirmation and Analysis of Circular Polarization from Sagittarius A*," *The Astrophysical Journal Letters* 526 (1999), pp. L85–L88.

Sazonov, V. N., "Generation and Transfer of Polarized Synchrotron Radiation," *Soviet Astronomy* 13 (1969), pp. 396–402.

Schlickeiser, R., "Cosmic Ray Transport and Acceleration, I: Derivation of the Kinetic Equation and Application to Cosmic Rays in Static Cold Media," *The Astrophysical Journal* 336 (1989), pp. 243–293.

———, "Gamma Ray Evidence for Galactic In-Situ Electron Acceleration," *Astronomy and Astrophysics* 319 (1997), pp. L5–L8.

Schödel, R., T. Ott, R. Genzel, A. Eckart, N. Mouawad, and T. Alexander, "Stellar Dynamics in the Central Arcsecond of Our Galaxy," *The Astrophysical Journal* 596 (2003), pp. 1015–1034.

Schödel, R., T. Ott, R. Genzel, R. Hofmann, M. Lehnert, A. Eckart, N. Mouawad et al., "A Star in a 15.2-Year Orbit around the Supermassive Black Hole at the Centre of the Milky Way," *Nature* 419 (2002), pp. 694–696.

Sellgren, K., M. T. McGinn, E. E. Becklin, and D. N. Hall, "Velocity Dispersion and the Stellar Population in the Central 1.2 Parsecs of the Galaxy," *The Astrophysical Journal* 359 (1990), pp. 112–120.

Serabyn, E., and M. Morris, "Sustained Star Formation in the Central Stellar Cluster of the Milky Way," *Nature* 382 (1996), pp. 602–604.

Shapiro, S. L., "Accretion onto Black Holes: The Emergent Radiation Spectrum, II: Magnetic Effects," *The Astrophysical Journal* 185 (1973), pp. 69–82.

REFERENCES

Shields, G. A., "Thermal Continuum from Accretion Disks in Quasars," *Nature* 272 (1978), pp. 706–708.

Shvartsman, V. F., "Halos around "Black Holes," *Soviet Astronomy* 15 (1971), p. 37.

Sidoli, L., T. Belloni, and S. Mereghetti, "A Catalogue of Soft X-ray Sources in the Galactic Center Region," *Astronomy and Astrophysics* 368 (2001), pp. 835–844.

Skinner, G. K., "X-ray Observations of the Galactic Center," *IAU Symposium* 136, Kluwer Academic Publishers, Dordrecht, Netherlands, 1989, pp. 567–580.

Snowden, S. L., R. Egger, M. J. Freyberg, D. McCammon, P. P. Plucinsky, W. T. Sanders, J. H. M. Schmitt, J. Truemper, and W. Voges, "ROSAT Survey Diffuse X-ray Background Maps, II," *The Astrophysical Journal* 485 (1997), pp. 125–135.

Stone, J. M., J. F. Hawley, C. F. Gammie, and S. A. Balbus, "Three-Dimensional Magnetohydrodynamical Simulations of Vertically Stratified Accretion Disks," *The Astrophysical Journal* 463 (1996), pp. 656–673.

Stone, J. M., and J. E. Pringle, "Magnetohydrodynamical Non-radiative Accretion Flows in Two Dimensions," *Monthly Notices of the Royal Astronomical Society* 322 (2001), pp. 461–472.

Tamblyn, P., and G. H. Rieke, "IRS 16—The Galaxy's Central Wimp?" *The Astrophysical Journal* 414 (1993), pp. 573–579.

Tanaka, Y., K. Koyama, Y. Maeda, and T. Sonobe, "Unusual Properties of X-ray Emission near the Galactic Center," *Publications of the Astronomical Society of Japan* 52 (2000), pp. L25–L30.

Townes, C. H., J. H. Lacy, T. R. Geballe, and D. J. Hollenbach, "The Center of the Galaxy," *Nature* 301 (1983), pp. 661–666.

Tsuchiya, K., R. Enomoto, L. T. Ksenofontov, M. Mori, T. Naito, A. Asahara, G. V. Bicknell et al., "Detection of Sub-TeV Gamma Rays from the Galactic Center Direction by CANGAROO-II," *The Astrophysical Journal Letters* 606 (2004), pp. L115–L118.

Van Hoven, G., "Solar Flares and Plasma Instabilities—Observations, Mechanisms, and Experiments," *Solar Physics* 49 (1976), pp. 95–116.

van Langevelde, H. J., D. A. Frail, J. M. Cordes, and P. J. Diamond, "Interstellar Scattering Toward the Galactic Center as Probed OH/IR stars," *The Astrophysical Journal* 396 (1992), pp. 686–695.

Wang, Q. D., E. V. Gotthelf, and C. C. Lang, "A Faint Discrete Source Origin for the Highly Ionized Iron Emission from the Galactic Centre," *Nature* 415 (2002), pp. 148–150.

Wardle, M., and F. Yusef-Zadeh, "Gravitational Lensing by a Massive Black Hole at the Galactic Center," *The Astrophysical Journal Letters* 387 (1992), pp. L65–L68.

Warren, M. S., and J. K. Salmon, "A Parallel Hashed Oct-Tree N-Body Algorithm," *IEEE Computer Society Press* (1993), pp. 12–21.

———, "A Portable Parallel Particle Program," *Computer Physics Communications* 87 (1995), pp. 266–290.

Watson, M. G., R. Willingale, P. Hertz, and J. E. Grindlay, "An X-ray Study of the Galactic Center," *The Astrophysical Journal* 250 (1981), pp. 142–154.

Weiler, K. W., and I. de Pater, "A Catalog of High Accuracy Circular Polarization Measurements," *The Astrophysical Journal Supplement* 52 (1983), pp. 293–327.

Weinberg, S., *Gravitation and Cosmology: Principles and Applications of the General Theory of Relativity*, Wiley-Interscience, New York, NY, 1972.

Wright, M. C. H., and D. C. Backer, "Flux Density of Sagittarius A* at $\lambda = 3$ mm," *The Astrophysical Journal* 417 (1993), pp. 560–564.

Young, P., "Numerical Models of Star Clusters with a Central Black Hole, I: Adiabatic Models," *The Astrophysical Journal* 242 (1980), pp. 1232–1237.

Yu, Q., and S. Tremaine, "Ejection of Hypervelocity Stars by the (Binary) Black Hole in the Galactic Center," *The Astrophysical Journal* 599 (2003), pp. 1129–1138.

Yuan, F., E. Quataert, and R. Narayan, "Nonthermal Electrons in Radiatively Inefficient Flow Models of Sagittarius A*," *The Astrophysical Journal* 598 (2003), pp. 301–312.

———, "On the Nature of the Variable Infrared Emission from Sgr A*," *The Astrophysical Journal* 606 (2004), pp. 894–899.

Yusef-Zadeh, F., W. Cotton, M. Wardle, F. Melia, and D. A. Roberts, "Anisotropy in the Angular Broadening of Sagittarius A* at the Galactic Center," *The Astrophysical Journal Letters* 434 (1994), pp. L63–L66.

Yusef-Zadeh, F., C. Law, M. Wardle, Q. D. Wang, A. Fruscione, C. C. Lang, and A. Cotera, "Detection of X-ray Emission from the Arches Cluster near the Galactic Center," *The Astrophysical Journal* 570 (2002), pp. 665–670.

Yusef-Zadeh, F., and F. Melia, "The Bow Shock Structure of IRS 7—Wind-Wind Collision near the Galactic Center," *The Astrophysical Journal Letters* 385 (1992), pp. L41–L44.

Yusef-Zadeh, F., and M. Morris, "Structural Details of the Sagittarius A Complex—Evidence for a Large-Scale Poloidal Magnetic Field in the Galactic Center Region," *The Astrophysical Journal* 320 (1987), pp. 545–561.

Yusef-Zadeh, F., M. Morris, and D. Chance, "Large, Highly Organized Radio Structures near the Galactic Centre," *Nature* 310 (1984), pp. 557–561.

Yusef-Zadeh, F., D. A. Roberts, and J. Biretta, "Proper Motions of Ionized Gas at the Galactic Center: Evidence for Unbound Orbiting Gas," *The Astrophysical Journal Letters* 499 (1998), pp. L159–L162.

Yusef-Zadeh, F., D. A. Roberts, W. M. Goss, D. A. Frail, and A. J. Green, "Detection of 1720 MHz Hydroxil Masers at the Galactic Center: Evidence for Shock-Excited Gas and Milligauss Fields," *The Astrophysical Journal Letters* 466 (1996), pp. L25–L29.

Yusef-Zadeh, F., M. Wardle, and P. Parastaran, "The Nature of the Faraday Screen toward the Galactic Center Nonthermal Filament G359.54+0.18," *The Astrophysical Journal Letters* 475 (1997), pp. L119–L122.

Zhao, J.-H., G. C. Bower, and W. M. Goss, "Radio Variability of Sagittarius A*—A 106 Day Cycle," *The Astrophysical Journal Letters* 547 (2001), pp. L29–L32.

Zhao, J.-H., and W. M. Goss, "Radio Continuum Structure of IRS 13 and Proper Motions of Compact HII Components at the Galactic Center," *The Astrophysical Journal Letters* 499 (1998), pp. L163–L167.

Zhao, J.-H., R. M. Herrnstein, G. C. Bower, W. M. Goss, and S. M. Liu, "A Radio Outburst Nearly Coincident with the Large X-ray Flare from Sgr A* on 2002-10-03," *The Astrophysical Journal Letters* 603 (2004), pp. L85–L88.

Zylka, R., P. G. Mezger, D. Ward-Thompson, W. J. Duschl, and H. Lesch, "Anatomy of the Sagittarius A Complex, 4: SGR A* and the Central Cavity Revisited," *Astronomy and Astrophysics* 297 (1995), pp. 83–97.

Index

γ-ray luminosity of Sgr A East, 23
1A 1742-294, 66
1E 1740.7–2942, 15, 65
1E 1742.5–2859, 58
1E 1743.1–2843, 66
2002-10-03 X-ray flare, 75
3C 273, 221
3EG J1746–2852, 22, 23
7 mm wavelength radiation, 215, 216,
   221, 236

accelerated frame, 132, 143
acceleration time, 215
accreted angular momentum, 164, 179
accretion disk, 45, 53, 72, 84, 164, 179,
   186, 226, 234, 236
accretion disk flares, 226
accretion filaments, 174
accretion onto Sagittarius A*, 156, 163,
   165
accretion rate, 162, 179
accretion shock, 163
active galactic nuclei (AGN), 2, 33, 43, 49,
   54, 72, 74, 155, 240
Advanced CCD Imaging Spectrometer
   (ACIS), 11
advected currents, 172
AG Car, 103
Aharonian, Felix, 69, 209
air Cerenkov, 69
Alexander, Tal, 106, 107, 110, 226, 228
Alfvén velocity, 173, 181, 215, 217
Allegheny Observatory, 26
ambient magnetic field, 172
Ampère's law, 171, 185
Andromeda, 73
angular broadening, 37
angular momentum barrier, 146
angular momentum conservation, 149
angular momentum dissipation, 226
angular resolution, 5, 58, 66, 68
anomalous resistivity, 173

anomalous viscosity, 186, 189, 191, 202,
   227, 228
antinucleon, 219
apoapse, 109
Ar XVII, 15
area amplification, 235
ART-P (telescope), 59
artificial viscosity, 171
ASCA, 10, 13, 59
asymptotic giant branch (AGB), 86, 101
atmospheric turbulence, 55
AU, 112
Australia Telescope Compact Array, 54
AX J1745.6-2901, 60, 68
AXAF, 60
azimuthal velocity, 193, 228

Baganoff, Fred, 60, 62, 63, 71, 73, plate 8,
   157, 205, 228
Bahcall, John, 94, 97
Balbus, Steve, 180, 183, 187, 191, 201,
   204
Balick, Bruce, 2, 3
balloon instruments, 58
bandwidth, 50, 51
Bardeen-Petterson effect, 247, 249, 250
Bardeen, James, 154, 240, 247, 249, 250,
   259
BeppoSAX, 11
Berkeley, plate 6
Big Blue Bump (BBB), 45
Big Dipper, 1
BIMA, 53
Bio-Bio river, 8
bipolar outflow, 156, 207
birefringence, 221
Birkhoff's theorem, 141, 142
Birkhoff, George, 141
BL Lac, 3
blackbody, 42, 45, 76
blackbody limit, 46, 206
black hole growth rate, 97

black hole image, 28, 41, 185, 195, 199, 215, 240, 258, 259, 261, 262
black hole spin, 54, 84, 149–151, 154, 155, 193, 197, 236, 238, 239, 246
Boltzmann equation, 166, 211
Bondi-Hoyle accretion, 62, 162, 164, 170, 172, 205
bremsstrahlung emission, 16, 24, 59, 63, 69, 160, 163, 179, 192, 206, 211
brightness temperature, 42, 205
Bromley, Benjamin, 78, plate 9, 155, 197, 235, 240
Brown, Robert, 2–4, 44
burst spectrum, 76

Ca XIX, 15
Canes Venatici, 1
CANGAROO, 69
Cartesian coordinates, 124, 125, 135
Cartesian Inertial Frame (CIF), 137–140, 143
central stellar cluster, 86, 100, 101
central stellar cluster evolution, 1, 74, 107, 109, 112
central stellar distribution, 91
Cepheid variable, 25, 26, 34
Cerenkov radiation, 69
chain rule of differentiation, 123
Chandra X-ray Observatory, 8, 13, 16, 18, 19, 23, 24, 60, 62, 63, 68, 71, 72, 75, 79, plate 8, 163, 205, 236
Chandrasekhar, Subrahmanyan, 60
Chevalier, Roger, 158
chirping, 80
Christoffel symbols, 139, 140, 145
circular polarization, 46, 47, 49, 52, 167, 221, 222
circularization radius, 164, 179, 186, 199
circumnuclear disk, 9, 53, 62
classical mechanics, 114, 115
CO emission, 16
co-rotating pattern, 155, 235
collimation, 208
collision radius, 87
collisionless Liouville's equation, 210
compressibility, 180

Compton limit, 44, 205
Compton scattering, 16, 43, 69, 179, 195, 197, 205, 231
Comptonization, 43, 197, 201, 232
contraction, 38, 123, 124, 128
contravariant tensor, 121
contravariant vector, 121, 126
convective instability, 177
coordinate transformation, 116, 137
coordinated observations, 76, 228
Copernican revolution, 25
cosmic distance scale, 26, 27
cosmic-ray electron, 16
Coulomb collision rate, 215
Coulomb forces, 64
counterclockwise rotation, 50, plate 9
covariant tensor, 122
Curtis, Heber D., 26
CXOGC J174538.0–290022, 61
CXOGC J174540.0–290027, 60
CXOGC J174540.9–290014, 61
cyclic conversion, 221
cyclosynchrotron, 163, 179, 206
cyclotron frequency, 166

dark matter, 1, 2, 7, 10, 24, 28, 31, 36, 111, 112, 114
dielectric, 223
diffuse X-ray emission, 13, 15, 16, 20, 62, 156, 158, 161–163
diffusion coefficients, 211, 212
diffusion equation, 210
diffusion-convection equation, 212
direct-orbit image, 242
discovery of Sagittarius A*, 25, 32, 204
disk equations, 186, 189
disk instability, 226, 228, 229, 231–234
disk magnetic field, 189–191, 194, 199, 200, 202, 204
disk structure, 164, 189, 197
disk torque, 186, 189
dispersion equation, 166, 181
distribution function, 92, 94, 113, 166, 195, 197, 209, 210
Doppler shift, 1, 118, 157, 193, 199, 201, 235, 240, 244
Downes, Dennis, 4, 27

dusty gas, 9, 10, 53, 56, plates, 1, 5, 6, and 9
dynamical center, 4, 8, 9, 58, 104

Earth orbit, 60, 109, 176
Earth's atmosphere, 8, 35, 55, 69
eccentric orbits, 107, 109, 110, 113
Eckart, Andreas, 1, 8, 29, 109, 112
Eddington luminosity, 65
Eddington, Sir Arthur, 264
EGRET, 22–24, 69
Einstein (satellite), 58
Einstein ring, 257
Einstein, Albert, 141
ejection of mass, 10, 64, 76, 156, 204, 217, 223, 226
electric field, 46–48, 130, 131, 187
electrodynamics, 114, 125, 230
electromagnetic acceleration, 208
electromagnetic wave, 46, 166, 218
elliptical polarization, 47, 221
enclosed mass, 1, 109, 111
energy cutoff, 216, 217
energy kinetic equation, 211
energy-at-infinity, 145
energy-momentum four vector, 128
epicyclic frequency, 181
episodic star formation, 108
equipartition, 167–169, 172, 174, 179, 183, 184, 187, 202, 206, 208, 218
equipartition magnetic field, 167, 169
equivalence principle, 137, 139, 251
ergosphere, 154
ESTEC, 12
Euler gamma function, 220
Eulerian perturbations, 180
European Southern Observatory, 8, 29, plate 5
European Space Agency, 68
event horizon, 3, 72, 78, 79, 144, 152, 154, 179, 199, 235, 236, 239
exponential integral function, 220
extragalactic nebulae, 26, 27
extragalactic sources, 32, 33
extraordinary and ordinary waves, 199
extraordinary component of intensity, 194, 195, 197, 199

Falcke, Heino, 43, 75, 207, 221, 227, 240
Faraday depolarization, 51, 165, 199, 223
Faraday rotation, 50, 51, 221
Faraday's law, 123, 172
Fatuzzo, Marco, 23, 196
Fe Kα line, 13, 15
Fermi shock acceleration, 208
field oscillation, 47
field-strength tensor, 130
field-tensor, 131
flare, 58, 71, 72, 74–76, 79, 80, 173, 197, 205, 208, 227, 231, 233–236, 241
flare duration, 234
flaring mechanism, 226
flat spacetime metric, 114
flattened isothermal sphere, 97
flux freezing, 167
flux modulation, 240
Fokker-Planck equation, 96, 210, 211, 213
formation of IRS 16, 101
Fort Davis, 40
four-dimensional spacetime, 115, 119, 125, 126, 129, 187
four-force, 127, 138
four-tensor, 130
four-vector, 119–123, 125–128, 129, 138
four-vector potential, 125
four-velocity, 138
Fourier transform, 49, 51, 55
frame dragging, 150, 153, 154, 245, 247, 259
free fall, 132
free-fall frame, 132, 133, 137, 139, 168
free-free absorption, 44
free-free emission, 9, 53
Fromerth, Michael, 16
full width at half maximum (FWHM), 41, 59, 62, 66, 68, 261

G359.54+0.18, 16, 18
galactic bulge, 9, 108
galactic center, 2–6, 8, 9, 11, 12, 16, 22, 24, 25, 27–29, 32, 34–36, 38, 49, 53, 59, 60, 62, 65, 68, 69, 74, 78, 86, 93, 104, 107–109, 111, plate 6, 156, 158, 165, 184, 193, 205, 206, 219, 220, 240
galactic coordinates, 5, 13, 27, 34, plate 7

galactic plane, 5, 6, 11, 13, 15, 19, 35, 39, plates 1, 2, and 7
galactic rotation, 104
Galaxy, 1–3, 7–9, 12, 14, 16, 17, 19, 20, 24, 29, 55, 57, 60, 62, 66, 79, 86, 105, 106, 108, plate 5, 157, 158, 160, 161, 185, 202, 234
Galilean relativity, 114, 115
Garching, 28
gas kinematics, 9
Gaussian noise, 81
GC441, 33
general relativity, 133, 137, 138, 141, 142, 150, 171, 240
generalized coordinates, 133
Genzel, Reinhard, 1, 8, 28, 29, 36, 55, 78, 80, 98, 99, 101, 102, 104–106, 109, 111, plate 5, 230, 236
geodesic, 251
Ghez, Andrea, 1, 36, 57, 101, 111
globular clusters, 25, 26, 28
Goldwurm, Andrea, 74, 235
Goss, Miller, 4, 5, 45, 75, 81, 82
GRANAT, 59
gravitational acceleration of S-stars, 29
gravitational confinement, 205
gravitational energy, 64, 142
gravitational friction, 108
gravitational lensing, 226, 255, 257
gravitational potential, 10, 32, 92, 94, 107, 162, 171, 201, 207
gravitational radius, 154, 227
gravitational redshift, 142, 143, 259–261
gravitational waves, 142
gravitomagnetic force, 248, 250
gravity, 10, 24, 32, 92, 109, 112, plate 6, 114, 132, 133, 140–142, 146, 150, 155, 158, 240
Greate Debate, 124
Green Bank Interferometer, 75
Green Bank, West Virginia, 32
GRS 1741.9–2853, 66, 67
gyro-resonant, 210

H₂ emission, 157
H₂O maser, 1, 20, 29
H-band flare, 78, 79

Haller, Joseph, 1
halo, 12, 26, 206, 208, 209, 217, 221, 223, 225
Hankel function, 197
harmonics, 231
Harvard College Observatory, 27, 28
Harvard-Smithsonian Center for Astrophysics, 28
Hawking radiation, 245
Hawley, John, 180, 183, 186, 201, 203, 204, 236
HCN emission, 9, 16, plate 6
Hegra ACT, 69
HeI stars, 101, 105
helicity of the field, 47
Herschel, John, 25
HESS, 69, 70, 220
high-energy emission, 65
High-Resolution Mirror Assembly (HRMA), 11
HII Region, 5
Hipparcos, 34
Hollywood, Jack, 78, 226, 240, 243
hot plasma, 157
Hoyle-Lyttleton theory, 162, 165, 178
hydrodynamic instabilities, 208
hydrodynamic simulations, 164, 165, 172, 178, 180, 186, 188, 192, 199
hydrostatic equilibrium, 188
hypervelocity stars, 99

IAU, 34
IBIS/ISGRI, 66
IC 443, 20
ideal MHD, 172
IGR J17456-2901, 67
Imaging Proportional Counter (IPC), 58
index of refraction, 50
inertia, 114, 117, 126–128, 132, 136, 138–140, 144, 166
inertial force, 114, 139
inertial frame, 137, 154
inertial mass, 132, 137
infall, 62, 64, 164, 166, 168
infalling clouds, 108
infrared counterpart, 28, 30
infrared flare, 78, 79, 226, 232

infrared flux, 56
infrared lightcurve, 226
infrared observations, 55, 58, 78
infrared sources, 9, 28
infrared variability, 71, 85
initial mass function, 7, 10, 104, 108, 109
instability, 80, 168, 173, 177, 180, 182, 183, 185, 201, 226, 228, 231–233, 239
instability growth rate, 174, 180, 185
intensity, 14, 16, 17, 19, 20, 22, 39, 41, 42, 44, 45, 48, 49, 59, 60, 65, 68, 79, plates 7 and 8, 168, 169, 180, 185, 193, 194, 196, 199, 200, 207, 241, 243
interferometer, 2, 32
International Gamma-Ray Astronomy Laboratory (INTEGRAL), 65, 68
interstellar medium, 10, 22, 39, 41, 54, 59, 158, 162
interstellar scattering, 35, 40
intrinsic size, 40–42, 62, 215, 236
invariance, 116, 119, 129, 132
inverse Compton scattering, 16, 44, 179, 192, 195, 197, 202, 230
ionized gas, 4, 8, 9, 20, plate 6
IRAS, 10
iron-line emission, 10, 13, 15
IRS 13, 75, 86, 101
IRS 16, 9, 86, 101, 157
IRS 16SW, 61
IRS 7, plate 4
island universe, 27
isosurface, 167
isothermal sphere, 93, 98

James Clerk Maxwell Telescope (JCMT), 53
jet, 65, 155, 202, 207, 227, 228, 245
Jokipii, Randy, 24, 84

Kα line, 21
K-band flare, 79
K-band flux, 78
K-band spectrum, 101
K-magnitude, 86
K-type giants, 101
Kassim, Namir, plate 1
Keck telescope, 55–57, 79

Keplerian period, 81
Keplerian region, 1, 80, 81, 164, 180, 181, 183–186, 190, 192, 195, 199, 202, 221, 223, 225, 228, 237, 239
Kerr black hole, 153, 154, 246, 261
Kerr metric, 84, 149–151, 247
Kerr, Roy, 149
kinematic viscosity, 187, 192
Kirchoff's law, 194
Klein-Nishina scattering, 196
Kowalenko, Victor, 169, 173
Kronecker delta, 123, 187
KS 1741–293, 65

lab-frame angle, 196
late type stars, 98, 104
late-B-type star, 100
LBVs, 103, 105
Leahy, Denis, 16
Lense-Thirring precession, 247, 248
light bending, 235, 240, 243, 251
light transit time, 214, 217
lightcurve, 4, 73, 77, 78, 80, 83, 84, 226, 238, 240
linear polarization, 46, 49, 50, 52, 54, 185, 221, 222, 225
linearized equations, 180, 182
Liouville's theorem, 209
Liu, Siming, 54, 78, plate 9, 155, 198, 205, 208, 218, 226–228, 235, 240
Lo, Fred, 4, 7, 37, 38, 44
local standard of rest (LSR), 33
long-term variability, 81, 251
Lorentz boost, 119
Lorentz factor, 44, 179, 196, 207, 208, 210, 214, 216, 218, 228, 231
Lorentz force, 171
Lorentz group, 116, 121
Lorentz invariant, 120
Lorentz transformation, 116, 117, 120, 122, 125, 171
Lorentz transformation of dimensionless quantities, 117
Lorenz gauge, 125
Los Alamos, 40, 158
Low Mass X-ray Binary (LMXB), 66
Lynden-Bell, Donald, 3, 34

M–0.02–0.07, 20, 22
M31, 4, 73
Mössbauer effect, 118
Maeda, Yoshitomo, 8, 19, 20, 22
magnetic bremsstrahlung, 163
magnetic dissipation, 168, 170
magnetic dynamo, 184
magnetic event, 233, 234
magnetic field, 20, 23, 24, 50, 104, 115,
    119, 131, 163, 166, 167, 169, 170,
    172–174, 176, 178–180, 183, 185–187,
    189–191, 194, 199, 202, 204, 206–208,
    210, 214, 215, 221, 230, 233, 234
magnetic field dissipation, 167
magnetic flare, 227
magnetic pressure, 168, 190, 203
magnetic reconnection, 7, 168, 173, 174,
    227
magnetized accretion, 169, 176, 237
magnetized disk, 176, 236
magnetohydrodynamic simulations, 170
magnetohydrodynamics (MHD), 170, 175,
    177, 180, 185, 200, 202, 209, 227
magnetorotational instability (MRI), 180,
    182, 189, 201, 202, 227, 228
Mapuche people, 8
marginally bound orbit, 149
marginally stable orbit (MSO), 202, 225,
    227, 234–239
Markoff, Sera, 23, 207, 226
masers, 1, 20, 28, 29
massive stars, 97, 108
Max-Planck-Institute, 28
Maxwell equations, 114, 129, 131, 132
Maxwell stress, 180, 185, 187
Maxwell stress tensor, 187
Maxwell-Boltzmann distribution, 44, 93,
    166, 194, 197, 208
McGinn, Martina, 1
Melia, Fulvio, 4, 16, 23, 44, 45, 54, 62,
    71, 78, 80, 84, plate 9, 114, 155, 157,
    163, 164, 168, 169, 174, 179, 187, 196,
    198, 200, 204–206, 209, 218, 227,
    234–236, 240
Menten, Karl, 28, 29
Mercury perihelion, 146
Mereghetti, Sandro, 12

metric tensor, 84, 124, 134, 135, 137,
    139–142, 144, 146, 149, 151, 154,
    193
MHD disk simulations, 201
microlensing, 251, 256
microquasar, 65
Milky Way, 2, 10, 25, 26
minispiral, 20, 53, plate 3
Minnett, Harry C., 27
Mira variables, 27, 28
mixed-morphology SNR, 8, 21, 156
Miyoshi, Makoto, 1
mm wave map, 45
mm wavelengths, 41, 43–45, 51, 53, 166,
    185, 216, 221, 259, 262
mm/sub-mm, 41, 44, 45, 51–53, 56, 78,
    plate 9, 166, 185, 202, 215, 217, 223,
    230, 232, 236
modified Bessel function, 195, 230
monotonic decrease in period, 235
Monte Carlo tests, 81
Moon, 60, 141
Morris, Mark, 7, 9, 19, 44, 49, 91, 104
Mount Wilson Observatory, 26
MPE SHARP camera, 29
multiwavelength, 9

N-body, 36, 87, 88, 100
Narayan, Ramesh, 39, 40, 84, 170, 175,
    178, 237
NASA, 11, 17, 18, plate 7
Nasmyth Adaptive Optics System
    (NACO), 55
National Radio Astronomy Observatory
    (NRAO), 2, 32, plates 2, 4, and 6
Naval Research Laboratory, plate 1
nebulae, 26
NeII, 4
neutral pion decay, 23, 70, 221
neutron stars, 13, 60, 64–66, 149
New Technology Telescope, 29
Newton's calculus, 141
Newton, Sir Isaac, 127, 132, 138, 140,
    141, 145, 147
Newtonian energy, 145
Newtonian force, 127, 140
Newtonian gravity, 146, 240

Newtonian mechanics, 147
NGC 4258, 1
no hair theorem, 150
Nobel Prize, 4, 60
nonthermal filaments, 16, plate 2
nonthermal radio source, 21
Northern Hemisphere, 38
Northwestern University, plates 3 and 6
nozzle, 207, 208
nuclear wind, 3
nucleon/antinucleon multiplicity, 219
numerical resistivity, 174

O-star, 107, 111
Of/LBV star, 105
OH maser, 20, 37, 50
Ohm's law, 171
Ohmic diffusion, 183
Oort's constants, 34
Ophiuchus, 25
orbital companion, 82
orbital decay, 109
orbital period, 80, 95
orbits, 1, 9, 32, 58, 95, 100, 104, 106,
    108–110, 113, plate 9, 146–148, 155,
    228, 236, 239, 244
Ordinary component of intensity, 194,
    195, 197, 199
ordinary waves, 199
outflows, 10, 12, 20, 156–158, 162, 201,
    202, 207, 208, 227

P-Cyg, 103
Paranal Observatory, 8, 97, plate 5
particle acceleration, 24, 209, 229, 231
particle diffusion, 209, 219, 221
particle distribution, 43, 44, 93, 166, 195,
    197, 207, 211, 216, 231
particle trajectories, 137, 140, 144, 148
PdV, 171, 191
periapse, 109, 113
period-luminosity zero point, 34
periodic modulation, 78, 81, 208, 225,
    226, 236, 240
periodicity, 58, 234, 240
permittivity, 166
permutation tensor, 130

Petrosian, Vahé, 209, 213, 218, 227
Petschek mechanism, 173
phase of a wave, 47, 117
phase-dependent amplitude, 47
phase-space density, 92, 96, 209
phenomenology of Sagittarius A*, 238
photodissociation, 157
photon geodesic, 251
photon index, 63, 72
physical laws in general relativity, 115,
    117, 125, 126
Piddington, Jack H., 27
Pie Town, 40
pion decay, 23, 69, 221
pion multiplicity, 219
Planck function, 42, 194
plane of polarization, 46
plasma instability, 226
plasma parameter, 213
plasma waves, 209, 210, 212
Plaza Hotel, 71
Poincaré sphere, 221, 222
point-spread function, 62, 261
Poisson's equation, 92, 140
polar funnels, 177
polarimetric imaging, 185, 195, 259, 262
polarimetry, 51, 56
polarization, 7, 46–54, 56, plate 9, 165,
    166, 177, 185, 195, 198, 199, 201, 211,
    221, 223
polarization filters, 49
polarized emission, 166, 197
poloidal, 84
position angle, 38, 50–52, 56, 194, 199,
    236
postulates of special relativity, 115
power spectral density (PSD), 81, 82
power-law distribution, 23, 64, 91, 97,
    212, 213, 216, 217, 227, 231
precession, 246
prograde, 79, 164, 236, 246
proper motion, 1, 10, 28, 30, 32, 34, 36,
    56, 82, 98, 99, 106, 221
proper time, 120, 127, 145
proton ejection, 217
proton-proton scattering, 220
protostars, 104

pseudo-Newtonian gravitational potential, 171, 201, 236
pulsars, 3, 5

quasars, 3, 4, 32, 33, 65, 72, 221
quasi-periodic modulation (QPO), 84
quiescent luminosity, 205

radiation pressure, 190
radiative efficiency, 207
radio filaments, 7, 16, plate 2
radio flare, 76
radio image, 28, plates 1, 2, and 6
radio modulation, 81, 85
radio morphology, 5
radio outburst, 75, 76
radio streamers, 7
radio structure, 3, 5, 20, 37, 41
radio telescope, 2, 3
radius of influence, 87
random-walk spatial diffusion, 214, 219
ray-tracing calculations, plate 9, 199, 260
Rayleigh-Jeans law, 42, 76, 190
Rees, Sir Martin, 3
refractive scintillation, 41, 54
Reid, Mark, 27–29, 32, 36
relative motion, 115
relativistic electrons, 7, 24, 202, 215, 217
relativistic force, 126, 128
relativistic kinetic energy, 129
relativity, 78, 114–118, 121, 125, 126, 128, 129, 131–133, 137, 138, 142, 143, 150, 171, 240
relaxation time, 87
resistive diffusion, 173, 180
resistive dissipation, 171
resistive MHD, 171, 172
resistivity, 171, 173, 174, 184, 185, 190
rest mass, 126, 145
rest-frame angle, 196, 197
retrograde orbit, 155, 164, 236
Reynolds stress, 174, 187, 202
Rieke, George, 1, 8, 44, 101
Rieke, Marcia, 1, 44, 101
rise time, 72, 229
rise/fall timescale, 72
ROSAT, 10–12, 59

rotation measure, 50, 51, 54, 165–167
RR Lyrae variables, 26, 27
Russell, Henry Norris, 27
RX J1745.6–2900, 59

S XV, 15
S-stars, 29, 110, 245
S2, 30, 31, 111
S2 star, 29, 109, 112, 113
Sagittarius A, 27, 49
Sagittarius A East, 19
Sagittarius A East (Sgr A East), 1, 7, 17, 19, 22–24, 60, 62, 69, 156, 162
Sagittarius A West, 2, 7–9, 19, 53, plate 4, 156
Sagittarius A*, 2, 4, 5, 7, 9, 10, 15, 17, 19, 25, 27, 29, 30, 32, 35, 36, 38–40, 42–45, 48–51, 53–56, 58–60, 62, 63, 65–68, 70–75, 77–81, 83, 84, 86, 97, 99, 101, 105, 107, 109, 112, plates 2, 4, and 8, 155, 156, 159, 161, 163, 164, 166, 167, 170, 172, 176, 180, 183, 190, 192, 196, 198, 204, 206, 208, 215–217, 221, 226, 228, 236, 238, 244
Sagittarius A*-IR (Sgr A*-IR), 57, 58
Sagittarius B1, 7
Sagittarius B2, 7
Sagittarius D, 7
Sagittarius, constellation of, 2
Salpeter, Ed, 3
Sanskrit, 60
scalar tensor, 121, 130
scaling form, 220
scatter-broadened image, 35, 37, 40, 261
scatter-broadened size, 40, 261
scattering optical depth, 179
scattering time, 214
Schwarzschild black hole, 236, 243
Schwarzschild metric, 142, 144, 146, 147, 151, 202
Schwarzschild radius, 72, 76, 78, 79, 84, 88, 112, 154, 171, 234, 241
Schwarzschild, Karl, 140, 144
Scorpius, 2, 25
SCUBA camera, 53
secondary leptons, 23
secular parallax, 34

SED, 56
self-absorbed region, 45, 216, 221, 224
Sellgren, Kris, 1, 91
Seyfert galaxy, 7, 45, 46, 72
shadow of the black hole, 240, 259
Shapley–Curtis debate, 26
Shapley, Harlow, 25–28
shock acceleration, 23, 24, 227
shocked gas, 160, 162
Sidoli, Lara, 12
single dish telescope, 41
SiO masers, 28, 29
SLX 1744–299/300, 66
Smooth Particle Hydrodynamics (SPH), 158, 248–250
SNR 0.9+0.1, 7
SO-16, 112
SO-stars, 111
solar motion, 34, 36
source geometry, 79
space shuttle, 60
spacetime four-vectors, 119
spacetime interval, 116, 119
spacetime scalar, 119
spacetime tensors, 121
special relativity, 78, 115, 116, 118, 121, 125, 126, 128, 129, 132, 138, 140, 142, 143
speckle imaging, 55
spectral index, 23, 40, 44, 54, 69, 75, 78, 231
spectral turnover, 44, 46, 208
speed of light, 115, 116
spherical coordinates, 116
spin parameter, 54, 84, 149–151, 154
spin-induced disk precession, 54, 85, 155, 246
spiral nebulae, 26
stable circular orbits, 148, 201
star formation, 7, 11, 104, 108
star-disk collisions, 226
star-star scattering, 94
starburst, 101, 104, 108
static limit, 152
stationary limit, 153
stellar collision rate, 95
stellar cusp, 87, 94, 108

stellar density, 92, 95, 99
stellar disks, 107
stellar distribution, 91, 94, 97, 98, 106
stellar dynamics, 1, 101, 105
stellar luminosity profile, 91
stellar number density, 28, 93, 95
stellar orbits, 109
stellar velocity, 1, 95
stellar winds, 10, 157
stochastic particle acceleration, 70, 194, 204, 207, 208, 213, 216, 217, 223, 234
Stokes parameter, 48, 49, 222, 223
stress, 38, 119, 174, 180, 185, 187, 202, 227, 237, 238, 240
strong lensing, 256
strong-field physics, 245
sun, 173
super-equipartition field, 169
super-magnetosonic, 237
supergiant, 86, 101
supermassive black hole, 1, 3, 4, 16, 32, 36, 45, 54, 57, 74, 87, 94, 104, 107, plates 2, 4, and 5, 185, 225
supernova, 21, 101, 156, 158, 162
supernova ejecta, 22, 24
supernova remnant (SNR), 1, 5, 7, 8, 15, 20, 21, 23, 24, 70, 74, 108, plate 1, 156
synchrotron, 7, 21, 23, 43–45, 53, 54, 78, plate 1, 163, 179, 192, 194, 195, 200–202, 204, 205, 210, 215, 216, 221, 223, 227, 228, 230, 232, 236, 259, 260, 262
synchrotron cooling time, 216
synchrotron self-Compton (SSC), 230, 232, 236

tearing mode instability, 168, 169, 173
tensor, 85, 121–125, 130, 134, 135, 138–142, 144, 146, 149, 151, 152, 154, 191, 192, 223
tensor algebra, 123
TeV, 69, 70, 217, 221, 233
thermal synchrotron, 194, 205
Thomson scattering, 64, 179, 196
three-dimensional hydrodynamics, 10, 65, 158, 160, 179, 190, 200, 201
three-space, 125

tidal disruption, 107, 109
tidal force, 88
tidal shear, 108
timelike interval, 152, 153
Townes, Charles, 4, 101
transformation of physical laws, 114, 117, 125, 126
turbulent cascade, 212
turbulent energy, 182, 213, 216
turbulent magnetic field, 173, 182, 185, 194, 204, 207, 209, 212
turbulent medium, 37
two-body collisions, 88
two-body relaxation, 87

UCLA, 56
ultraviolet continuum, 3, 9, 45, plate 6
University of Wisconsin-Madison, 11
unstable modes, 182, 183
uv coverage, 38

variability, 4, 38, 43, 53, 56, 58, 60, 71, 74, 75, 81, 85, 112, 164, 179, 225, 227, 236, 240
variable emission, 177
vector tensor, 121
velocity dispersion, 1, 87, 95, 111
Very Large Array (VLA), 4, 5, 20, 29, 32, 34–36, 43, 49, 51, 52, 54, 62, 75, 76, 81, plates 1, 2, and 8, 223
Very Large Telescope (VLT), 55, 56, 58, 79
Very Long Baseline Array (VLBA), 29, 32, 35, 38–40, 82, 215
very long baseline interferometry (VLBI), 1, 4, 35, 37, 38, plate 9, 262
viscosity, 171, 180, 186, 187, 189, 190, 193, 202, 227, 228, 235
viscous dissipation, 185, 189, 191
VLA monitoring program, 76
Vlasov equation, 166
VLT YEPUN telescope, 8, 78, 79, 97

W109, 33
W28, 20
W44, 20

W56, 33
Wang, Q. Daniel, 13, 15, 16, 18, 113
wave modes, 182, 214
wave phase, 117
wave spectrum, 212
waveband, 156, 241
wavevector, 46, 118, 125, 180
WC star, 101, 102
WCL star, 105
weak lensing, 256
Weinberg, Steven, 118
Westerbork Observatory, 4, 49
Western Arc, 62, 156
Whipple, 69
wind-wind collisions, 62, plate 8, 157, 158, 161, 164, 172
WN star, 101, 102
WNE star, 105
WNL star, 105
WO star, 101
Wolf-Rayet star, 101, 103, 105
World Trade Center, 71

X-ray background, 11
X-ray binary, 60, 66, 74
X-ray continuum, 20
X-ray counterpart, 16, 60, 61, 228
X-ray flare, 71, 72, 75, 76, 79, 208, 226, 228, 229, 232, 234, 236, 259
X-ray hardness ratio, 72
X-ray interferometry, 262
X-ray morphology, 10, 156
X-ray observations of Sagittarius A*, 58
X-ray plume, 10, 12
X-ray spectrum, 15, 22, 59, 63, 192, 197
X-ray variability, 71, 74
XMM-Newton, 68, 72, 74–77, 79, 235

Yohkoh, 173
Yusef-Zadeh, Farhad, 7, 10, 18, 19, 23, 37, 49, 51, plates 2–4, and 6, 157

Zeeman splitting, 20
Zhao, Jun-Hui, 45, 75, 77, 81, 82

Milton Keynes UK
Ingram Content Group UK Ltd.
UKHW020640290824
447545UK00007B/216